Chemistry in Context
Laboratory Manual and Study Guide

Graham C. Hill M.A. (Cantab)
Deputy Headmaster, Dr Challoner's Grammar School, Amersham,
formerly Head of Science, Bristol Grammar School

John S. Holman M.A. (Cantab)
Head of Science, Watford Grammar School

Nelson

To our mothers, Beth and Betty

Thomas Nelson and Sons Ltd
Nelson House   Mayfield Road
Walton-on-Thames   Surrey KT12 5PL

51 York Place
Edinburgh EH1 3JD

P.O. Box 18123   Nairobi   Kenya

Yi Xiu Factory Building
Unit 05-06, 5th Floor
65 Sims Avenue
Singapore 1438

Thomas Nelson (Hong Kong) Ltd
Toppan Building 10/F   22A Westlands Road
Quarry Bay   Hong Kong

Thomas Nelson (Nigeria) Ltd
8 Ilupeju Bypass   PMB 21303   Ikeja   Lagos

© G. C. Hill and J. S. Holman, 1982
First published by Thomas Nelson and Sons Ltd 1982

Reprinted 1983, 1984

ISBN 0-17-448059-8
NCN 210-3240-2

Typeset by Santype International Ltd,
Salisbury, Wilts.

Printed and bound in Hong Kong

Acknowledgement is due to the following
for permission to reproduce photographs.

D. J. Adam p. 165
Associated Press p. 124, 146
Beken of Cowes p. 153
British Petroleum p. 144
Daily Mail p. 124
Daily Telegraph p. 146
General Motors p. 149
John Holloway p. 128
Keystone Press Agency p. 126, 161
Mercedes-Benz p. 129
Ralph Morse p. 142
Norwegian State Tourist Board p. 123
C. Ridgers, V.R.U. p. 140, 152, 154,
    165, 166, 168, 171

Picture research by Sue Armley

Preface

This Laboratory Manual and Study Guide has been planned as a companion to *Chemistry in Context*, although it is self-contained and can be used quite independently of the main text.

The book has three major sections:

Section 1 Laboratory Investigations,
Section 2 Comprehension Questions,
Section 3 Objective Questions.

Section 1   Laboratory Investigations

In this section, we have tried to provide a course of practical work which is both central to the theoretical development of the subject and relevant to the social and industrial importance of chemistry. The experiments are appropriate for any modern chemistry course to A-level or its equivalent. Considerable thought has been given to the questions incorporated with each laboratory investigation. These should ensure that students think critically about their methods of working and their results, instead of performing operations mechanically.

The practical work is presented as separate, self-contained assignments which enable students to decide beforehand how to organize their time. Most of the investigations can be completed in one double period (i.e. 80 minutes). An approximate time allocation for each investigation is indicated in the margin below the list of requirements. Many of these assignments can be used as assessed practicals for those A-level courses which incorporate this method of assessment.

A list of the unknown substances for the inorganic and organic qualitative analysis exercises is available to teachers writing to the publishers from an educational establishment.

Section 2   Comprehension Questions

The ability to read, assimilate and comprehend written material is an important skill, not least for the scientist. Each of the twenty-five comprehension tests in this section consists of a passage concerning some area of chemistry or its applications, followed by questions. These questions mainly test the student's comprehension of the passage, but other intellectual skills such as recall and analysis are also required. The passages have been carefully chosen to consider areas of chemistry which are interesting and intelligible to A-level students, and also cover much of the material found in modern A-level syllabuses. Some of the passages are concerned with pure chemistry, but the majority deal with the applications of the subject. We have tried to use material illustrating the relevance of chemistry to everyday life in its literal sense, so the applications of chemistry in the home are stressed as well as its uses in industry. In this way, we hope we can bring the subject nearer to the experience of every student. We hope the passages are sufficiently interesting in their own right to make enjoyable background reading, whether or not the associated questions are answered.

Section 3   Objective Questions

Each of the thirty-two objective tests is based on a chapter in *Chemistry in Context*. Thus, Test 1, entitled 'Atoms, Atomic Masses and Moles' relates to the ideas and information presented in chapter 1 of *Chemistry in Context*.

Each test contains between 14 and 20 questions. These are mainly multiple choice items together with a few of the multiple completion type. These are the two forms of objective question which are used most frequently by examination boards. There are no true-false or assertion-reason items. The questions have been pretested thoroughly during the last few years at a number of schools. Keys to the questions and the facility coefficients for most of them are published separately and these are available to teachers writing to the publishers from an educational establishment. In general, the questions retained in these published tests have facility coefficients in the range 0·20 to 0·85 (i.e. the questions were answered correctly by 20 to 85% of the candidates tested).

Nomenclature, units and abbreviations

Throughout the text we have been guided by the Royal Society Publication (*Symbols, Signs and Abbreviations Recommended for British Scientific Publications*, London 1969) and we have used the International System of Units (SI) with very few exceptions. Modern nomenclature has been used throughout the book, following recommendations in the second edition of *Chemical Nomenclature, Symbols and Terminology* (1979) published by the Association for Science Education.

Concentration of solutions—use of the symbol M

The term 'concentration' is now taken by IUPAC to mean 'amount of substance per unit volume'. The units of concentration are therefore moles per cubic decimetre. The word 'molar' is restricted to the meaning 'divided by the amount of substance'. The units associated with the word 'molar' are therefore 'per mole'.

Consequently, it is incorrect to refer to a solution containing 0·1 moles of solute per cubic decimetre as 0·1 molar. Nevertheless, the abbreviation 0·1 M for 'having a concentration of 0·1 moles per cubic decimetre' continues to be used in schools even though the word 'molar' is no longer appropriate.

Accordingly, we have retained the symbol M to mean 'moles per cubic decimetre' in order to avoid lengthy and awkward phraseology.

Eye and face protection

The DES and many LEA's now recommend that eye protection is worn whenever chemicals are being handled.
Certainly, protective spectacles or face shields must be worn by teachers and pupils whenever there is a risk of danger to the eyes or face. For example, they should always be worn when heating chemicals, when handling acids or alkalis and when performing potentially exothermic reactions. Failure to wear adequate protection in such circumstances could be regarded as serious negligence.
Within the text generally, there is a reminder that eye protection must be worn whenever there is an obvious danger.
In planning and writing the laboratory investigations considerable thought has been given to recommended safety procedures. In this respect, we would like to acknowledge the help of Dr Peter Borrows (Chairman of the Laboratory Safeguards Sub-Committee of the Association for Science Education) for his careful scrutiny of section 1 (Laboratory Investigations) and his advice and suggestions concerning safety.

Once again we are greatly indebted to our publishers, particularly Elizabeth Johnston, Donna Evans and Anna Grayson, for their competent and helpful handling of the manuscript. Finally, we thank our wives, Elizabeth and Wendy for their continued support and encouragement.

Graham Hill
John Holman
January 1981

Contents

v

SECTION TWO · Comprehension questions

SECTION THREE · Objective questions

TO THE STUDENT
Safety in the Chemistry Laboratory

Your safety and the safety of those working with you in the laboratory are your responsibility. Safety is largely a matter of common sense, provided you know the hazards associated with the chemicals you are using. **To alert you to the more hazardous materials and operations we have printed safety warnings in the text in bold type.** It is sensible, however, to regard *all* chemicals as potentially hazardous.

Some important points to remember

Always wear eye or face protection when there is any danger to your eyes. For example, you **must** wear eye protection
1 when handling acids or alkalis,
2 when heating chemicals and
3 when performing potentially exothermic reactions.

Always handle flammable liquids, such as ethanol and propanone, with great care and keep them away from naked flames.

Always wash your hands after practical work.

Always report accidents, however small, to your teacher.

Always check the name on a bottle is exactly that of the chemical you require.

Always work steadily and without undue haste.

Never pipette liquids without a safety filter, unless instructed to do so.

Never get your clothes or head near a bunsen flame. Long hair should be tied back, and a laboratory coat should be worn whenever possible. Adjust the bunsen to give a luminous flame when you are not using it.

Never smell gases directly—only very cautiously by wafting fumes of the gas towards your nose.

Never point a test tube containing chemicals which you are heating towards yourself or anyone else. Never look directly down the tube.

Never try to force glass tubing when putting it into or removing it from corks or bungs. Always hold glass tubing in a cloth when performing these operations.

Never hold bottles by the neck. If a stopper is tight, get help. Do not try to force it off.

Never remove chemicals or equipment from the laboratory.

Never perform unauthorized experiments.

Never taste anything unless instructed to do so.

Never eat or drink in the laboratory.

SECTION ONE Laboratory investigations

Qualitative analysis of inorganic compounds

The qualitative analysis of inorganic compounds can be divided into four stages.
1 Preliminary tests
2 Making a solution of the substance under test
3 Testing for cations (some of the methods of identifying cations are introduced in practical 27).
4 Testing for anions

1 Preliminary tests

The purpose of these preliminary tests is to provide some general pointers as to the nature of the substance under observation. Some preliminary tests are described in practical 27 (Identifying cations).

a Appearance
The colour of a substance may give a clue to its identity. Table 1 lists the colours of some cations in their compounds. It is important to remember that the list is not exhaustive.

Colour	Inference
White or colourless	Probably does not contain a transition metal
Coloured	Probably contains a transition metal
Blue	Possibly Cu^{2+} or Ni^{2+}
Green	Possibly, Cu^{2+}, Ni^{2+}, Fe^{2+} or Cr^{3+}
Yellow	Possibly Fe^{3+}

Table 1 The colours of some cations in their compounds

b Flame tests
Table 2 shows the colourations imparted to a flame by certain cations during flame tests. The method of carrying out a flame test is described in practical 27.

Flame colour	Inference	Flame colour	Inference
Persistent yellow	Na^+	Blue Green	Cu^{2+}
Lilac	K^+		
Red/Orange	Ca^{2+}, Sr^{2+}, Li^+		
Pale green	Ba^{2+}		

Table 2 Colorations imparted to a flame by certain cations

c Action of heat
CARE Eye protection must be worn when solids are heated. Do not use more than 2 spatula measures (0·5–1·0 g) of a substance.
   When a substance is heated, gently at first and then strongly, in a hard glass test

tube, it is sometimes possible to make deductions about its constituents from simple tests and observations made during heating. Whilst heating look carefully for colour changes and try to identify any gases evolved. (Table 3)

Observation	Inference
Sublimation	NH_4^+ salts
Decrepitation (spitting and cracking)	Anhydrous salts that decompose with evolution of gas (e.g. $Pb(NO_3)_2$, $KMnO_4$)
Charring (carbon produced)	CH_3COO^- or $C_2O_4^{2-}$
Gas evolved which rekindles a glowing splint	O_2 produced by decomposition of oxide or nitrate
Gas evolved which turns lime water milky	CO_2 produced by decomposition of carbonate or hydrogencarbonate
Gas evolved which burns with a blue flame	CO from decomposition of oxalate (ethanedioate)
Brown gas evolved	NO_2 from a nitrate or Br_2 from a bromide
Colourless liquid condenses on cooler parts of tube	Probably water from the decomposition of a hydrate, a hydroxide or a hydrogencarbonate
Pungent gas evolved which is alkaline	NH_3 from decomposition of ammonium salt

Table 3 Observations and inferences from the action of heat on inorganic substances

2 Making a solution of the substance under test

If you are provided with a solid, you will need to make an aqueous solution of it before certain tests are carried out.

First, attempt to dissolve a little of the substance (about one flat spatula measure) in about 3 cm³ of water. If it does not dissolve, warm gently. If the solid is insoluble in water, try cold and then hot dilute nitric acid.

CARE Eye protection must be worn when a liquid is heated in a test tube. The test tube should not be more than one quarter filled.

Table 4 provides a summary of the solubilities of common inorganic substances.

All common salts of group I cations and NH_4^+ are soluble.

All nitrates are soluble.

All ethanoates are soluble.

All sulphates are soluble except those of Ba^{2+}, Ca^{2+}, Sr^{2+} and Pb^{2+}.

All $\begin{Bmatrix} \text{chlorides} \\ \text{bromides} \\ \text{iodides} \end{Bmatrix}$ are **soluble** except those of Ag^+, Cu^+, Pb^{2+}, Hg_2^{2+}.

All $\begin{Bmatrix} \text{carbonates} \\ \text{phosphates} \\ \text{sulphites} \end{Bmatrix}$ are **insoluble** except those of group I cations and NH_4^+.

All hydroxides are **insoluble** except those of group I cations, NH_4^+, Sr^{2+} and Ba^{2+}.

All sulphides are **insoluble** except those of group I and group II cations and NH_4^+.

Table 4 The solubilities of common inorganic substances. (A substance is described as soluble if its saturated solution exceeds a concentration of 0·1 M at 20°C.)

3 Testing for cations

a Reaction with sodium hydroxide solution

To 3 cm^3 of the solution obtained in part **2**, add 2 M sodium hydroxide slowly until the latter is in excess. Watch carefully for signs of a precipitate. (Remember that if the substance has been dissolved in acid, the first drops of alkali will be needed to neutralize excess acid.) Table 5 gives the results of tests with 2 M sodium hydroxide.

CARE Eye protection is essential when using NaOH. It is more dangerous to the eyes than acid of the same concentration.

Observation	Inference
No precipitate at any stage	Possibly a cation from group I or NH_4^+. (Distinguish Li^+, Na^+ and K^+ by flame tests. NH_4^+ salts give NH_3 if alkaline solution is warmed.)
A white precipitate **insoluble** in excess NaOH	Possibly Mg^{2+}, Ca^{2+}, Sr^{2+} or Ba^{2+} which can be distinguished by flame tests and the solubility of their sulphates and chromates.
A white precipitate soluble in excess NaOH	Possibly Al^{3+}, Pb^{2+}, Sn^{2+} or Zn^{2+} which can be distinguished by heating to dryness a portion of their hydroxides.
A coloured precipitate forms.	
Pale blue precipitate	Cu^{2+}
Green precipitate turning brown	Fe^{2+}
Rust-brown precipitate	Fe^{3+}
Grey-green precipitate	Cr^{3+}
Pale brown precipitate	Ag^+
Green precipitate	Ni^{2+}
Cream precipitate turning dark brown	Mn^{2+}

Table 5 Testing for cations with 2 M sodium hydroxide

b Reaction with sodium carbonate solution

Sodium carbonate is sometimes used to confirm the presence or absence of a group I cation or NH_4^+. If sodium carbonate solution is added to a solution containing only such cations, no precipitate forms because NH_4^+ and group I cations all have soluble carbonates. All other cations, however, will give a precipitate of their carbonate.

4 Testing for anions

Anions can be divided into three groups:
a Anions which react with dilute strong acids evolving gases or volatile liquids.
b Anions which do not react with dilute strong acids, but react with concentrated sulphuric acid evolving gases or volatile liquids.
c Anions which react with neither dilute strong acids nor with concentrated sulphuric acid.

a Reaction with dilute strong acids

Add 3 cm^3 dilute nitric acid to 2 spatula measures of the solid under investigation. If no reaction occurs in the cold, warm gently. If there is effervescence, smell cautiously and test the gases evolved. Table 6 gives the results of treating with strong dilute acid.

Observation	Inference
Gas evolved which turns lime water milky	CO_2 from CO_3^{2-} or HCO_3^-
Gas evolved has a pungent smell, is acidic and turns dilute $KMnO_4$ colourless	SO_2 from SO_3^{2-}
Gas evolved smells of bad eggs and gives a black precipitate with lead nitrate solution	H_2S from S^{2-}
Pungent acidic gas is evolved and a pale yellow precipitate forms	SO_2 and S from $S_2O_3^{2-}$
Dark brown acidic gas evolved	NO_2 from NO_2^-
Pale green gas evolved which bleaches indicator	Cl_2 from ClO^-
Acidic gas evolved which smells of vinegar	CH_3COOH from CH_3COO^- or $HCOOH$ from $HCOO^-$

Table 6 Testing for anions with dilute strong acids

b   Reaction with concentrated sulphuric acid

Warm a little of the solid under test with about $1\ cm^3$ of concentrated sulphuric acid.
CARE Eye protection must be worn.

Table 7 gives some typical results. Remember that if you have an anion which reacted with dilute strong acid in the last section, it will react again here but even more violently.

Observation	Inference
Pale yellow vapour	Possibly HNO_3 vapour from NO_3^-
Steamy fumes evolved which give dense white smoke with $NH_3(g)$	HCl from Cl^-
Brown-orange fumes	Br_2 from Br^-
Purple fumes	I_2 from I^-
Gases evolved which burn with a blue flame and turn lime water milky	CO and CO_2 from $C_2O_4^{2-}$
Gas evolved which burns with a blue flame	CO from $HCOO^-$

Table 7 Testing for anions with concentrated sulphuric acid

Confirmatory test for nitrate (NO_3^-)

To $3\ cm^3$ of a solution of the suspected nitrate add an equal volume of fresh iron(II) sulphate solution. Then, carefully pour $3\ cm^3$ of concentrated sulphuric acid down the inside of the test tube so that it forms a separate layer.
CARE Eye protection must be worn.

A brown ring forms slowly at the interface of the concentrated sulphuric acid and the aqueous layer if a nitrate is present.

Confirmatory test for halides (Cl^-, Br^-, I^-)

To $3\ cm^3$ of a solution of the suspected halide add an equal volume of dilute nitric acid and then silver nitrate solution.

A white precipitate turning grey-purple in sunlight and readily soluble in dilute ammonia solution confirms chloride.

A cream precipitate turning green in sunlight and soluble in excess dilute ammonia solution confirms bromide.

A yellow precipitate unaffected by sunlight and insoluble in dilute ammonia solution confirms iodide.

c   Testing for anions which do not react with acids

Sulphate, phosphate and chromate ions do not react with dilute strong acids or with concentrated sulphuric acid to give any gaseous products. If a substance is suspected of containing one of these ions, the following confirmatory tests can be used.

Sulphate (SO_4^{2-}) To 3 cm³ of a solution of the suspected sulphate, add an equal volume of dilute nitric acid. Then add barium nitrate solution.

A dense white precipitate forms if sulphate is present.

Phosphate (PO_4^{3-}) To 3 cm³ of a solution of the suspected phosphate, add ammonium molybdate solution followed by a few *drops* of concentrated nitric acid and warm.

A yellow precipitate forms slowly if phosphate is present.

Chromate (CrO_4^{2-}) To 3 cm³ of a solution of the suspected chromate, add dilute sulphuric acid.

If chromate is present, the yellow solution becomes orange due to the formation of dichromate. If 1 drop of dilute hydrogen peroxide is added to the dichromate, a short-lived blue colour appears which turns green with the evolution of oxygen.

Observation/deduction exercises with inorganic compounds

Introduction

The following exercises do not involve systematic analysis or a knowledge of the traditional methods of separation. They do, however, assume that you are familiar with the simple methods of qualitative analysis of inorganic compounds just discussed, and with the simple reactions of the following ions:

NH_4^+, Na^+, K^+, Mg^{2+}, Ca^{2+}, Sr^{2+}, Ba^{2+}, Al^{3+}, Pb^{2+}, Sn^{2+}, Zn^{2+}, Cu^{2+}, Fe^{2+}, Fe^{3+}, Cr^{3+}, Ni^{2+}, Mn^{2+}, Ag^+;

CO_3^{2-}, HCO_3^-, SO_3^{2-}, $S_2O_3^{2-}$, S^{2-}, NO_2^-, ClO^-, CH_3COO^-, $HCOO^-$, $C_2O_4^{2-}$, Cl^-, Br^-, I^-, NO_3^-, SO_4^{2-}, PO_4^{3-}, CrO_4^{2-}.

If the substance under investigation contains ions not included in the above list, you will not be expected to identify the ions but to draw conclusions of a general nature.

For each exercise, you should state the letter of the substance under investigation and then divide your page into three columns headed 'Test', 'Observation' and 'Inference'.

Remember to record those tests for which there are no positive results since important inferences can be made from such observations. For example; if you add dilute nitric acid followed by silver nitrate to an aqueous solution of the substance X and no noticeable change occurs, it is reasonable to conclude that X does *not* contain Cl^-, Br^- or I^-.

Substance A is a single compound

1 Heat a sample of **A** in a hard glass test tube.
   Identify any gases evolved. Keep the residue.

2 Carry out a flame test on **A**.

3 Make a solution in water of the residue left in part **1** and test separate portions of this solution with
   a dilute nitric acid.
   b dilute sodium carbonate.
   c silver nitrate solution.
   d barium nitrate solution.

4 Make a solution of **A** in water.
   Test separate portions of this solution with
   a potassium iodide solution.
   b manganese sulphate solution, then warm.
   c silver nitrate solution.

5 Devise and perform two more tests which provide further evidence for the cation and anion in **A**. Record the results of these tests.

Substance B is a sodium salt

1 Heat a sample of **B** in a hard glass test tube. Try to identify any gases evolved.

2 Treat the cold residue from part **1** with dilute hydrochloric acid. Identify the gas evolved.

3 Make a solution of **B** in water.
To separate portions of this solution add
a copper(II) sulphate solution and then warm.
b silver nitrate solution and then warm.
c potassium manganate(VII) solution.

4 What suggestions can you make about the anion in **B**?

Substance C is a single compound

1 Add dilute hydrochloric acid to **C** and then warm gently.

2 Heat a sample of **C** in a hard glass test tube. When there is no further change, leave the residue to cool.

3 Add dilute hydrochloric acid to the cold residue from part **2**.

4 Make a solution of **C** in water.
To separate portions of this solution add
a silver nitrate solution and then warm.
b barium chloride solution followed by dilute hydrochloric acid.
c sodium carbonate solution and then warm.

5 What conclusions can you make about the nature of **C**?

Substance D is a simple salt.

CARE: D is poisonous.

1 Heat a sample of **D** in a hard glass test tube and keep the residue.

2 Add dilute hydrochloric acid to the cold residue obtained in part **1**.

3 Warm a little **D** with dilute hydrochloric acid.

4 Make a solution of **D** in water.
To separate portions of this solution add
a dilute sulphuric acid and then warm.
b sodium carbonate solution.
c barium nitrate solution.
d potassium chromate(VI) followed by dilute ethanoic acid.

5 Carry out a flame test on **D**.

6 Heat a sample of **D** with a little ethanol and two or three drops of concentrated sulphuric acid.   **CARE Eye protection must be worn.**

7 What conclusions can you make about **D**?

Substance E

1 Make a solution of **E** in water.
To separate portions of this solution add
a sodium hydroxide solution slowly until it is present in excess.
b ammonia solution until it is present in excess.
c dilute nitric acid followed by silver nitrate solution.
d lead nitrate solution.
e dilute hydrochloric acid and then bubble hydrogen sulphide into the mixture. **CARE Hydrogen sulphide is poisonous. Work in a fume cupboard.** Keep the resulting product.

2 To the resulting product in part **e**, add ammonia solution.

3 What conclusions can you make about the nature of **E**?

Substance F

CARE Eye protection must be worn.

1 Add **F** to dilute hydrochloric acid and warm gently. Identify any gases evolved.

2 Cool the resulting mixture from part **1** and filter.

3 To separate portions of the filtrate from part **2** add
   a sodium hydroxide solution.
   b ammonia solution.

4 Add **F** to dilute sodium hydroxide and warm gently.
   CARE Eye protection must be worn.

5 What deductions can you make about **F**?

Substance G

1 Heat a small sample of **G** in a hard glass test tube.

2 Make a solution of **G** in water.
   To separate portions of this solution add
   a sodium hydroxide solution until alkaline.
   b dilute sulphuric acid.
   c dilute ethanoic acid followed by potassium chromate(VI) solution.
   d sodium bromide solution, warm the mixture and then cool.

3 What deductions can you make about **G**?

Substance H is a simple salt

CARE Eye protection must be worn

1 Working in a fume cupboard, add concentrated sulphuric acid to a small sample of **H**.
2 Mix a little of **H** with an equal bulk of manganese(IV) oxide and then, in a fume cupboard, add concentrated sulphuric acid.

3 Make a solution of **H** in water.
   To separate portions of this solution add
   a silver nitrate solution.
   b bromine water.
   c dilute sulphuric acid and then hydrogen peroxide.

4 What conclusions can you make about **H**?

Substance I is a mixture of two salts which contain different cations but the same anion.

1 Add dilute hydrochloric acid to a sample of **I**. Keep the resulting solution.

2 Filter the solution obtained in part **1**. To the filtrate add ammonia solution, a little at a time until it is present in excess. Boil the resulting mixture.

3 Heat a sample of **I** in a hard glass test tube. Keep the residue.

4 Transfer the residue from part **3** to a crucible lid or a piece of porcelain and heat it strongly in the air. Keep the residue.

5 When the residue obtained in part **4** is cold, treat a small portion of it with concentrated hydrochloric acid.

   CARE Eye protection must be worn.

6 What conclusions can you make about
   a the anion
   b the cations
   present in **I**?

Substance J is a mixture of two simple compounds containing the same cation but different anions.

1 Heat a sample of **J** in a hard glass test tube.

2 Test the solubility of **J** in water.

3 Add dilute sulphuric acid to **J** and heat to boiling. Filter the mixture and keep both the filtrate and the residue.

4 To separate portions of the filtrate obtained in part **3**, add
   a potassium iodide solution.
   b ammonia solution.
   c sodium hydroxide solution and then boil.   **CARE Eye protection must be worn.**

5 Warm the residue obtained in part **3** with concentrated sulphuric acid.
   CARE Eye protection must be worn.
   Add three times the volume of water to the solution and then filter.

6 To separate portions of the filtrate obtained in part **5**, add
   a silver nitrate solution.
   b barium nitrate solution.
   c potassium hexacyanoferrate(II) solution.

7 What conclusions can you make about the constituents of **J**?

Qualitative analysis of organic compounds

Organic compounds are often highly complex and their analysis can be a major operation involving sophisticated modern instrumental techniques. In this treatment, we are concerned with the identification of relatively simple organic compounds using test tube reactions.

The tests carried out on an organic compound can be summarized as follows. For a given compound, it is unlikely that every one of these tests will need to be done.

1 Appearance	7 Reaction with iron(III) chloride
2 Action of heat	8 Tests for unsaturation
3 Solubility in water	9 Tests for reducing agents
4 Action of dilute sodium hydroxide	10 Heating with soda-lime
5 Action of dilute hydrochloric acid	11 Reaction with copper sulphate
6 Action of concentrated sulphuric acid	

1   Appearance

a   Physical state

If the substance is a solid, it may be one of the following: aromatic acid, phenol, amide, amino acid, carbohydrate, carboxylic acid salt, amine salt, aliphatic acid with two or more functional groups.

If the substance is a liquid, it may be one of the following: alcohol, aliphatic carboxylic acid, ester, ether, aldehyde, ketone, halogenated hydrocarbon.

b   Smell

Smell the substance cautiously at first. If there is no obvious smell, warm it gently. Some commonly encountered smells:

Sharp, sour – carboxylic acids
Pleasant, spirity – alcohols
Pleasant, fruity – esters
Ammoniacal, fishy – amines
Antiseptic – phenol.

c Colour

Most of the organic compounds you are likely to encounter are colourless. Some exceptions:

Yellow – nitro compounds, triiodomethane

Blue–green – copper salts (other transition metals also give characteristic colours)

Certain phenols and aromatic amines are readily oxidized, so although they are colourless when pure, they may contain pink, yellow, brown, grey or black impurities.

2   Action of heat

Heat a little of the substance in the presence of air, on a crucible lid or on a piece of broken porcelain. Heat gently at first, then more strongly. Note whether the substance burns, and with what kind of flame, whether it melts or boils readily, whether it shows signs of decomposition and whether any gases are evolved. (Test with indicator paper.) Note the appearance of the residue.

a   Burning

Burns with a yellow flame but little smoke: aliphatic, with a low C : H ratio.

Burns with a yellow or orange smoky flame: aromatic or unsaturated, with a high C : H ratio.

Does not burn: carboxylic acid salt, or highly halogenated compound.

Note whether the substance imparts a characteristic colour to the bunsen flame: persistent yellow – Na^+, lilac – K^+, red – Ca^{2+}, Sr^{2+}, Li^+, apple green – Ba^{2+}, green–blue – Cu^{2+}

b   Residue

If there is a residue, heat it very strongly. If the residue is carbon it will burn away eventually. If the residual ash remains, it is probably a metal oxide or carbonate formed by the decomposition of a carboxylic acid salt.

c   Other observations

Substance chars readily: carbohydrate, carboxylic acid salt.

Violet vapour evolved: contains iodine.

3   Solubility in water

Shake a **small** quantity of the substance with cold water and see if it dissolves. (The mixture should not occupy more than one fifth of the test tube.) If the substance is a liquid, look to see whether two layers form. If it is insoluble in cold water, try heating. Test any solution with indicator paper.

a   Solids which are soluble in water include polar or hydrogen-bonded compounds (carbohydrates, amino acids, urea, phenols, carboxylic acids, low molecular mass amides) and ionic compounds (carboxylic acid salts, amine salts).

b   Liquids which are soluble in water include low molecular mass alcohols, aldehydes, ketones, carboxylic acids, acid chlorides and amines.

c   Substances which are much more soluble in hot water than cold include certain aromatic acids and substituted amides.

d   Substances forming acidic solutions include carboxylic acids, acid anhydrides, acid chlorides, readily hydrolysed esters, phenols and amine salts.

e   Substances forming alkaline solutions include aliphatic amines and salts of carboxylic acids.

Note   Some substances may be insoluble in water but soluble in acid or alkali, and vice-versa. Such behaviour provides a useful clue: see **4** and **5**.

4 Action of dilute sodium hydroxide solution

Add a small quantity of the substance to a few cm^3 of dilute sodium hydroxide, and warm if there is no obvious reaction in the cold.   **CARE Eye protection must be worn.**   See whether the substance dissolves and look for signs of hydrolysis. Test for evolved gases using indicator paper and by smelling cautiously.

a Solubility
If the substance dissolves in sodium hydroxide, but not in water, it is likely to be an acid, e.g. benzoic acid or phenol, though acids of low molecular mass dissolve in both. If it dissolves in water but not in sodium hydroxide it is probably an amine salt. Aromatic amine salts liberate an oil when treated with sodium hydroxide.

b Gases evolved
Ammonia is produced by ammonium salts and amides, the latter giving the gas only slowly on warming. Aliphatic amine salts liberate amines, which have a fishy, ammoniacal smell.

c Other effects
Aliphatic aldehydes may give a brown resin.

5 Action of dilute hydrochloric acid

Add a small quantity of the substance to a few cm^3 of dilute hydrochloric acid and warm if there is no obvious reaction in the cold. See whether the substance dissolves. Test for evolved gases using indicator paper and by smelling cautiously.

a Solubility
If the substance dissolves in acid, but not in water, it is likely to be an amine, e.g. phenylamine, though amines of low molecular mass are soluble in water as well as acid. If it is soluble in water but not in acid, it may be a salt of a carboxylic acid of fairly high molecular mass. If the substance is soluble in acid **and** alkali, but not in water, it may be an amino acid.

b Gases evolved
A sharp smelling acidic gas on warming with acid indicates a carboxylic acid salt.

6 Action of concentrated sulphuric acid

CARE Concentrated sulphuric acid is corrosive. Eye protection must be worn.

If the substance reacted vigorously with dilute acid, do not test it with concentrated sulphuric acid.

Put a small quantity of the substance in a test tube and add 1 cm^3 of concentrated sulphuric acid. Warm if no reaction occurs when cold. Test for the evolution of gases, particularly carbon dioxide (which turns lime-water milky) and carbon monoxide (which burns with a blue flame).
CARE carbon monoxide is poisonous

Concentrated sulphuric acid **dehydrates** some organic compounds:
   carbohydrates form carbon (charring).
   ethanedioates (oxalates) form CO and CO_2.
   methanoates (formates) form CO.

Sulphuric acid adds across double bonds – compounds containing carbon—carbon double bonds dissolve in the concentrated acid, often giving a brown coloured solution. Concentrated sulphuric acid will also displace weaker acids from their salts, such as ethanoates (acetates), just as dilute hydrochloric acid does.

7 Reaction with iron(III) chloride

The iron(III) chloride solution must be neutral. Add a few drops of ammonia solution to 3 cm^3 of $FeCl_3$(aq) until the first trace of a red precipitate appears. Filter or decant off the solution and add the neutral iron(III) chloride solution to an aqueous solution of the substance under test.

Red colour – methanoate or ethanoate ion
Violet colour – phenols
Buff-coloured precipitate – salt of aromatic acid, e.g. benzoate
Lime-green colour – ethanedioate ion

8 Tests for unsaturation

a Bromine water
Shake the substance with bromine water, which is decolourized by non-aromatic unsaturated compounds, e.g. alkenes, alkynes.

b Potassium manganate(VII)
Shake a little potassium manganate(VII) solution with the substance. The purple solution is decolourized by non-aromatic unsaturated hydrocarbons. (Manganate(VII) may turn brown rather than completely colourless.)

9 Tests for reducing agents

Test reagent	How to perform test	Positive result	Compounds giving positive result
acidified potassium manganate(VII)	Shake substance with manganate(VII) acidified with a little dil. H_2SO_4. Warm if necessary	purple manganate (VII) decolourized	• methanoate • ethanedioate • aldehydes • primary and secondary alcohols
acidified potassium dichromate(VI)	Shake substance with acidified dichromate(VI). Warm if necessary.	yellow dichromate(VI) turns green	• methanoate • ethanedioate • aldehydes • primary and secondary alcohols
Fehling's solution (Cu^{2+} complexed with tartrate ions in alkaline solution)	Mix 1 cm^3 of Fehling's solution *A* with 1 cm^3 of Fehling's solution *B*, add a little of the substance and boil cautiously **(CARE Eye protection must be worn.)**	orange-red precipitate	• aldehydes
Tollen's reagent (Ag^+ complexed with NH_3 in alkaline solution)	Add **one drop** of NaOH(aq) to 2 cm^3 silver nitrate solution, then add **just enough** dilute ammonia to redissolve the precipitate. Add a little of the substance and boil. After you have carried out the test, pour the contents of the tube down the sink and wash it away with plenty of cold water	silver formed, often as a mirror on the tube	• aldehydes

Table 1 Tests for reducing agents

10 Heating with soda-lime

Soda-lime is a solid alkali which decarboxylates, that is, it removes CO_2 from, carboxylic acids and their salts. Mix equal volumes of the dry substance and soda-lime and heat in a hard-glass test tube. Test for the evolution of decarboxylation products.
For example:

$$CH_3COO^-Na^+ + Na^+OH^- \longrightarrow CH_4 + Na_2CO_3$$

11 Reaction with copper sulphate solution

Most compounds containing the NH or NH_2 groups form intensely coloured complexes with Cu^{2+} ions. Add a few drops of the organic compound to copper sulphate solution. Most amines give deep blue colours, but phenylamine gives a green precipitate.

The following test, sometimes called the **biuret test**, is a sensitive test for peptide groups (whose structure is shown on the left) and is particularly useful for detecting proteins.

Add an equal volume of dilute sodium hydroxide to a few cm^3 of an aqueous solution of the substance under test. **CARE Eye protection must be worn.** Add **one drop** of very dilute (0·1 M) copper sulphate solution.

There are many other simple tests that can help in identifying an organic compound, but there is insufficient room to include all of them. Those listed above are the more important ones. See also the section on qualitative analysis of inorganic compounds.

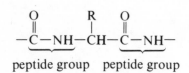

peptide group   peptide group

Observation/deduction exercises with organic compounds

In the following exercises you will be asked to perform certain tests on unknown organic compounds and from the results deduce as much as possible about the nature of the substance involved. The exercises assume that you are familiar with the simple reactions of the following types of compound:

alkenes, halogenoalkanes, alcohols, phenols, aldehydes, ketones, carboxylic acids, carboxylic acid salts, esters, amides, amines, amine salts.

Some of the substances may contain more than one functional group and the salts of carboxylic acids and amines will, of course, contain inorganic ions. You may need to refer to the section on qualitative analysis of inorganic compounds. In some of the exercises it may not be possible to identify the substance completely. In such cases, you should draw conclusions of a general nature.

For each exercise state the letter of the substance under test. Divide the page into three columns, headed 'Test', 'Observation', and 'Inference'. Remember to record negative as well as positive results.

Substance K

1 Put 3 cm^3 of water in a test tube. Add an equal volume of **K** and shake. Allow to settle, then decide whether **K** is **a** soluble, **b** insoluble, **c** partially soluble in water.

2 Put a few drops of **K** on a crucible lid and heat in a bunsen flame.

3 To 2 cm^3 of **K** in a test tube add 1 cm^3 of potassium dichromate(VI) solution and 1 cm^3 of dilute sulphuric acid. Boil gently for 3 minutes.

4 Add 1 cm^3 of glacial ethanoic acid to a similar volume of **K**. Add 1 or 2 drops of concentrated sulphuric acid and warm gently.
CARE Eye protection must be worn.
Allow to cool and then carefully pour the mixture into about 50 cm^3 of cold water in a beaker. Smell the vapour cautiously.

5 What can you deduce about the nature of **K**?

Substance L

1 Working in a fume cupboard, heat **L** on a porcelain crucible lid.

2 Add water drop by drop to **L**.

3 Make a solution of **L** in water and acidify with dilute nitric acid. Add silver nitrate solution followed by ammonia solution.

4 Dissolve **L** in the minimum possible amount of water. Add sodium hydroxide solution, then add excess dilute hydrochloric acid.

5 Make a solution of **L** in about 3 cm^3 of dilute hydrochloric acid. Add 1 cm^3 of sodium nitrite solution then boil carefully for a minute or two.

6 What can you conclude about **L**?

Substance M

1 Add **M** to water and shake.

2 Add a little **M** to about 20 cm^3 of dilute sodium hydroxide in a boiling tube. Shake. Divide into 3 portions.
   a To one portion add dilute hydrochloric acid in slight excess. Heat strongly to boiling point, then allow to cool.
   b To a second portion add a few drops of bromine water.
   c To a third portion add a little potassium manganate(VII) solution. Warm gently.

3 Working in a fume cupboard, heat a small quantity of **M** on a crucible lid, gently at first, then strongly.

4 What conclusions can you make about **M**?

Substance N

1 Heat a little **N** on a crucible lid, gently at first, then strongly.

2 Shake a little **N** with water.

3 To 2–3 cm^3 of an aqueous solution of **N** add an equal volume of neutral iron(III) chloride solution.

4 Put a little **N** in a boiling tube and add about 5 cm^3 of dilute sulphuric acid. Boil the solution, then pour it into a small beaker or evaporating basin and cautiously smell the vapour.

5 Put about 1 g of **N** in a test tube and add 2–3 cm^3 of ethanol. Carefully and gradually add about 1 cm^3 of concentrated sulphuric acid.
   CARE Eye protection must be worn.

   Gently warm for a minute or two, then pour the contents of the tube carefully into a beaker containing about 50 cm^3 of cold water. Cautiously smell the vapour.

6 What can you conclude about the nature of **N**?

Substance P

1 To a small amount of **P** in a test tube add 3 cm^3 of dilute sodium hydroxide and warm.   **CARE Eye protection must be worn.**

2 To a small amount of **P** in a test tube add 3 cm^3 of dilute sulphuric acid and warm. Keep the resulting mixture for test **3**.

3 Bring the solution resulting from **2** to the boil, then add dilute potassium manganate(VII) solution drop by drop.

4 Heat a small amount of **P** on a crucible lid.

5 Add 2 cm^3 of dilute sodium hydroxide solution to a small amount of **P**. To the resulting suspension add one drop of very dilute copper sulphate solution.

6 What can you conclude about substance **P**?

Substance Q

Substance **Q** is one of the carbonyl compounds listed in the table below. The following tests should enable you to decide which carbonyl compound it is.

1 Test the solubility of **Q** in water.

2 Add a few drops of **Q** to Fehling's Solution and boil.

3 Put a few drops of **Q** on a crucible lid and heat in a bunsen flame.

4 Add a few drops of **Q** to 5 cm^3 of a solution of 2,4-dinitrophenylhydrazine. (Details for the preparation of such a solution are given in practical 39.) Shake, and leave to stand for a minute or two. Filter off the precipitate using a Buchner funnel, wash it with a little methanol and dry it by sucking air through it. Recrystallize the precipitate by dissolving it in the minimum possible amount of a hot 1 : 1 ethanol-water mixture. **Heat the mixture in a water bath, with no naked flames anywhere near.** Allow the solution to cool, when crystals should appear. Filter and dry these crystals as before and then find their melting point using the method described in practical 38.

5 Identify the carbonyl compound **Q**.

Name	Formula	Boiling point/°C	Melting point of 2,4-dinitrophenyl-hydrazine derivative /°C
ethanal	CH_3CHO	21	164
propanal	CH_3CH_2CHO	48	156
butanal	$CH_3CH_2CH_2CHO$	75	123
pentanal	$CH_3CH_2CH_2CH_2CHO$	104	98
benzaldehyde	C_6H_5CHO	178	237
propanone	CH_3COCH_3	56	128
butanone	$CH_3COCH_2CH_3$	80	115
pentan-2-one	$CH_3CH_2CH_2COCH_3$	102	141
phenylethanone	$C_6H_5COCH_3$	202	250

Physical data for some carbonyl compounds and their 2,4-dinitrophenylhydrazine derivatives.

Substance R

1 Shake a small quantity of **R** with 3–4 cm^3 of water. Keep the resulting mixture.

2 To the mixture resulting from **1** add an equal volume of dilute sodium hydroxide solution and warm. **CARE Eye protection must be worn.**

3 Mix **R** with an equal amount of powdered soda lime and heat in a test tube.

4 To an aqueous solution of **R** add iron(III) chloride solution.

5 Put a little **R** in a test tube and add a few drops of methanol. Cautiously and gradually add 1 cm^3 of concentrated sulphuric acid. **CARE Eye protection must be worn.** Warm the mixture, then carefully pour into about 50 cm^3 of cold water in a beaker. Cautiously smell the vapour.

6 What are your conclusions about the nature of compound **R**?

Substance S

1 Add 2 cm^3 of **S** to an equal volume of water, shake and allow to settle.

2 Shake 2 cm^3 of **S** with an equal volume of bromine water.

3 Put 2 cm^3 of potassium manganate(VII) solution in a test tube and acidify with 1 cm^3 of dilute sulphuric acid. Add a few drops of **S** and shake.

4 Put 2 cm^3 of concentrated sulphuric acid in a test tube. **CARE Eye protection must be worn.** Slowly add about 10 drops of **S**.

5 Heat a little **S** on a crucible lid.

6 What are your conclusions concerning **S**?

PRACTICAL 1
An analysis of aspirin tablets

Introduction

Aspirin is an analgesic and antipyretic drug. Analgesics are drugs which relieve pain. Antipyretics are drugs which lower body temperature.

The main constituent of aspirin tablets is 2-ethanoylhydroxybenzoic acid (acetyl-salicylic acid, $CH_3COOC_6H_4COOH$, figure 1). Aspirin passes unchanged through the acidic conditions in the stomach but is hydrolysed to ethanoate (acetate) ions and 2-hydroxybenzoate (salicylate) ions by the alkaline juices in the intestines.

$$CH_3COOC_6H_4COOH + 2OH^- \longrightarrow CH_3COO^- + HOC_6H_4COO^- + H_2O$$

Salicylates lower body temperature rapidly and effectively in feverish patients (antipyretic action), but have little effect if the temperature is normal. They are also mild analgesics, relieving certain types of pain such as headaches and rheumatism.

Although the toxic dose from salicylates is relatively large, their uncontrolled use could be dangerous. Single doses of 5 to 10 grams of salicylate have caused death in adults, and 12 grams taken over a period of twenty-four hours produces symptoms of poisoning.

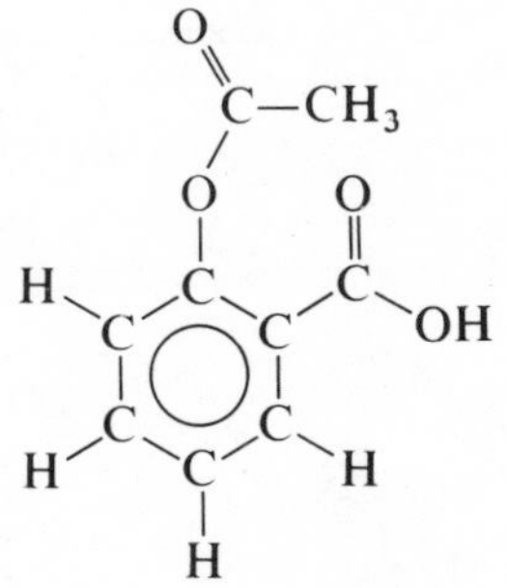

Figure 1 2-ethanoylhydroxybenzoic acid (acetyl-salicylic acid)

Principle

The object of this experiment is to determine the percentage of 2-ethanoylhydroxybenzoic acid (acetyl-salicylic acid) in aspirin tablets. A known amount of standard sodium hydroxide solution is used in excess to hydrolyse a known mass of aspirin tablets.

$$CH_3COOC_6H_4COOH + 2NaOH$$
$$\longrightarrow CH_3COONa + HOC_6H_4COONa + H_2O$$

The unused sodium hydroxide which remains is then titrated with standard acid. The amount of alkali required for the hydrolysis can now be calculated and from the above equation, the number of moles of acetylsalicylic acid which have been hydrolysed can be found.

Procedure

Work in pairs

Partner No. 1 should standardize the approximately 1·0 M NaOH used for the hydrolysis as follows:
Using a safety filler, pipette exactly 25 cm³ of the approximately 1·0 M NaOH solution into a 250 cm³ standard flask and make up to the mark. Now titrate 25 cm³ of this solution against 0·10 M hydrochloric acid using phenol red (or phenolphthalein) indicator.

Carry out one rough and two accurate titrations.

1 Record your results in a table similar to the one below.
2 Calculate the accurate molarity of the approximately 1·0 M NaOH.

Titration number	Rough	Accurate 1	Accurate 2
Final burette reading/cm³			
Initial burette reading/cm³			
Volume of 0·1 M HCl added/cm³			

Requirements

Each pair of students will need:
- Aspirin tablets (about 5)
- Pipette, 25 cm³, and safety filler
- Burette and stand
- Small beaker
- 2 standard flasks, 250 cm³ (one for each partner)
- Conical flask, 250 cm³
- Small funnel
- Bunsen burner
- Tripod and gauze

Approximately 1·0 M NaOH (30cm³)
0·10 M hydrochloric acid, (150 cm³)
Phenol red (or phenolphthalein) indicator

Time required   1 double period

15

Partner No. 2 should hydrolyse the aspirin as follows:
Weigh accurately between 1·3 g and 1·7 g of the aspirin tablets into a clean conical flask. (This will be about 5 tablets.) Using a safety filler, pipette 25 cm^3 of the approximately 1·0 M NaOH on to the tablets, followed by about the same volume of distilled water. Simmer the mixture gently for ten minutes to hydrolyse the acetyl-salicylic acid.   **CARE Eye protection must be worn.**

Now, cool the mixture and transfer with washings to a 250 cm^3 standard flask and make up to the mark with distilled water.

3   Record the weight of aspirin tablets taken.

4   Why is the mixture simmered gently and carefully during hydrolysis?
    Why is it unwise to boil it vigorously?

5   Why should the washings be transferred carefully to the 250 cm^3 standard flask?

Both partners should now estimate the quantity of unused NaOH after the hydrolysis as follows:
Pipette 25 cm^3 of the hydrolysed solution into a conical flask. Titrate this against 0·10 M hydrochloric acid using phenol red (or phenolphthalein) indicator.

6   Record your titration results in a table similar to that in question 1.

7   How many moles of NaOH
    a   are added to the flask before hydrolysis of the aspirin?
    b   remain after hydrolysis of the aspirin?
    c   are used in the hydrolysis of the aspirin?

8   How many moles of acetyl-salicylic acid have been hydrolysed?

9   What percentage of the aspirin tablets is acetyl-salicylic acid?

10   What might the remainder of the tablets be made of?

PRACTICAL 2
An analysis of iron tablets

Introduction

Iron is essential to the human body. Its principle role is as a constituent of haemoglobin, the oxygen-carrying agent in the blood. Iron is also present in a number of enzymes and co-enzymes involved in redox processes in the body.

Healthy adult males need little iron in their diet, but some groups in the population need substantial amounts in order to produce extra haemoglobin. Such people include growing children, pregnant and menstruating women, and individuals who have for various reasons lost considerable amounts of blood. A satisfactory intake of iron can normally be ensured by eating a suitable diet, because certain foods – liver, kidney, egg yolk and spinach for example – are rich in iron. Nevertheless, it is sometimes necessary to supplement the iron taken in the natural diet with 'iron tablets'.

Iron tablets bought at the chemist usually contain iron(II) sulphate (ferrous sulphate), a cheap, soluble form of iron. In this practical you will attempt to find the actual percentage of iron(II) sulphate in the tablets and then compare this result with the quantity stated on the bottle.

Assuming all the iron in the tablets is in the form of Fe^{2+}, it is possible to estimate the iron content by titration against potassium manganate(VII), $KMnO_4$.

1 Write an ionic equation for the reaction of Fe^{2+} with MnO_4^- in acid solution.

Requirements

Each pair of students will require:
- Iron tablets (5), sold by chemists as ferrous sulphate tablets
- 1·0 M sulphuric acid (200 cm³)
- 0·01 M potassium manganate(VII) (75 cm³). Dissolve 1·58 g in distilled water and make up to 1 dm³.
- 2 conical flasks (250 cm³)
- Standard flask (250 cm³)
- Burette and stand
- Pipette (25 cm³) and safety filler
- Filter funnel
- Filter paper
- Wash bottle and distilled water

Time required 1 double period

Procedure

Making a solution of the tablets
Weigh accurately five of the iron tablets, then dissolve them in about 100 cm³ of 1·0 M sulphuric acid in a conical flask. This will probably require heating, but do not heat more than necessary to dissolve the tablets.

2 Why should the tablets not be heated more than necessary?

3 Why are the tablets dissolved in sulphuric acid instead of water?

The outer coating of the tablets will probably not dissolve, so the solution will need filtering.

4 What do you think the outer coating might be?

Filter the mixture into a beaker, making sure you do not lose any of the solution, then wash out the conical flask with water and pour the washings through the filter. Finally pour distilled water over the residue and collect these washings as well. Pour the filtrate into a 250 cm³ standard flask, washing out the beaker and adding the washings to the standard flask. Make up to the mark with distilled water.

Titration with potassium manganate(VII)
Using a safety filler, pipette 25 cm³ of the iron(II) solution into a conical flask. Add about 25 cm³ of 1·0 M sulphuric acid and titrate with 0·01 M potassium manganate(VII) solution. Repeat until two consistent results are obtained.

5 How many moles of MnO_4^- were needed to react with 25 cm³ of your Fe^{2+} solution?

6 How many moles of Fe^{2+} were there in 25 cm³ of the solution?

7 How many moles of Fe^{2+} were there in all the tablets?

8 What mass of **a** iron **b** $FeSO_4$ **c** $FeSO_4.7H_2O$ is there in one tablet?

9 What mass of iron(II) sulphate is stated by the makers to be present in each tablet?

10 Compare your answer with the mass stated by the makers, and comment.

PRACTICAL 3

The principles of titrations involving iodine/thiosulphate

Determination of the percentage of copper in brass

Requirements

Each student or pair of students will need:

- Burette and stand
- Pipette (25 cm³) and safety filler
- Conical titration flask (250 cm³)
- Measuring cylinder (25 or 50 cm³)
- Small beaker
- Beaker (250 cm³)
- Standard flask (250 cm³)
- 2 test tubes
- Concentrated nitric acid
- Hydrogen peroxide solution
- Dilute sulphuric acid
- Distilled water
- Sodium carbonate solution
- Dilute acetic acid
- Approx. 1 M KI (30 cm³) (166 g KI in 1 dm³ of solution)
- 0·1 M Na₂S₂O₃ (100 cm³) (24·8 g Na₂S₂O₃ . 5H₂O in 1 dm³ of solution)
- Starch solution (6 cm³). (Make 2 g of soluble starch and 0·01 g of HgCl₂ into a thin paste with water. Add slowly to 1 dm³ of boiling water with stirring. Boil for a few minutes and cool.)
- Brass (about 3 g) (brass screws are ideal)
- Access to: Balance

Time required 1 double period

EXPERIMENT 1   The principles of titrations involving iodine/thiosulphate

Iodine/thiosulphate titrations are often used to determine the concentration of solutions of oxidizing agents.

A known volume of the oxidizing agent is first added to an excess of acidified potassium iodide solution, thus liberating iodine. The liberated iodine is then estimated by titration against standard sodium thiosulphate solution.

A   Add a few drops of hydrogen peroxide solution (an oxidizing agent) to a mixture of 3 cm³ KI(aq) and 3 cm³ of dilute H_2SO_4. Keep the final solution for part B.

1   Describe and explain what happens.

2 H_2O_2 acts according to the half-equation, $H_2O_2 + 2H^+ + 2e^- \longrightarrow 2H_2O$.

Write a half-equation for the oxidation of iodide ions to iodine.

B   Add sodium thiosulphate solution to the final mixture from part A until no further changes occur.

3   Describe and explain what happens.

4   Thiosulphate ions, $S_2O_3^{2-}$, are oxidized by iodine according to the half-equation,

$$2S_2O_3^{2-} \longrightarrow S_4O_6^{2-} + 2e^-.$$

Write an equation for the reduction of iodine by $S_2O_3^{2-}$.

5   How many moles of $S_2O_3^{2-}$ react with one mole of I_2?

In iodine/thiosulphate titrations, a standard solution of $Na_2S_2O_3$ is added to the iodine solution from a burette. In these circumstances, the colour would change from a pale yellow solution of I_2 to a clear solution of I^- ions at the end point. Generally, however, starch is added to improve the detection of the end-point. With iodine, starch forms a deep blue colour which disappears at the end point. The starch should not be added until the iodine has been reduced to a pale yellow colour, otherwise iodine becomes strongly adsorbed on to the starch and the titration is less accurate.

EXPERIMENT 2   Determination of the percentage of copper in brass

In the following experiment, a weighed sample of brass is first dissolved in nitric acid, forming a solution of copper(II) ions. When this solution is treated with aqueous potassium iodine, copper(I) iodide is precipitated as a white solid, and iodine is produced.

$$2Cu^{2+}(aq) + 4I^- \longrightarrow 2CuI(s) + I_2(aq)$$

The liberated iodine can be estimated using standard sodium thiosulphate solution. Knowing the amount of iodine formed, the mass of copper present in the original sample of brass can be determined.

CARE   Eye protection must be worn when working with concentrated HNO_3.

Weigh accurately somewhere between 2·5 g and 3·0 g of brass. Dissolve this in the minimum quantity of concentrated nitric acid in a 250 cm³ beaker.

CARE   Toxic nitrogen dioxide is evolved. This reaction must be carried out in a fume cupboard.

Transfer the solution, with washings from the beaker, to a 250 cm³ standard flask. Make up the solution to the mark and mix well.

Using a safety filler, pipette 25 cm³ of the brass solution into a conical flask and add sodium carbonate solution until a slight permanent precipitate is obtained. This neutralizes excess nitric acid in the solution. Dissolve the precipitate in the minimum volume of dilute acetic acid and then add 10 cm³ of approximately 1 M KI.

Finally titrate the liberated iodine against standard 0·1 M sodium thiosulphate adding about 2 cm³ of starch indicator when the iodine colour is pale yellow.

Repeat the experiment until two consistent results are obtained.

6   Describe what happens when
   a brass is treated with conc. HNO_3.
   b $Cu^{2+}(aq)$ is treated with $KI(aq)$.

7   Record your results in a table similar to the one below.

Titration number	Rough	Accurate 1	Accurate 2
Final burette reading/cm³			
Initial burette reading/cm³			
Vol. of 0·1 M $Na_2S_2O_3$ added/cm³			

Average accurate titration = _____ cm³

8   Why is it necessary to neutralize the excess nitric acid used to dissolve the brass? (Hint: Nitric acid is an oxidizing agent.)

9   Write an equation for
   a the reaction of copper in brass with concentrated nitric acid.
   b the neutralization of excess nitric acid with sodium carbonate solution.

10 How many moles of $S_2O_3^{2-}$ react during titration?

11 How many moles of I_2 react with this amount of $S_2O_3^{2-}$?

12 How many moles of Cu^{2+} ions liberate this amount of I_2?

13 How many moles of Cu^{2+} ions are there in the 250 cm³ of 'brass solution'?

14 How many grams of copper are there in 250 cm³ of 'brass solution'?

15 What is the percentage by mass of copper in the brass? ($Cu = 63·5$)

PRACTICAL 4

Determination of the concentration of chloride ions in sea water

Introduction

Every cubic decimetre (litre) of sea-water contains about 35 g of dissolved salts, though the amount varies according to locality.

Many different ions are present in sea-water, including gold ions. The table below shows some of the more abundant ions present. The commonest cation is Na^+ and the commonest anion is Cl^-. The object of this practical is to determine the concentration of chloride ions, Cl^-, in sea-water.

Some of the more abundant ions present in sea-water are:

Na^+	Cl^-
Mg^{2+}	SO_4^{2-}
Ca^{2+}	Br^-
K^+	HCO_3^-

The method used is the standard one for determining the concentration of chloride ions – titration with silver nitrate solution of known concentration. Silver ions form insoluble silver chloride when added to a solution containing chloride ions:

$$Ag^+(aq) + Cl^-(aq) \longrightarrow AgCl(s)$$

Each student or pair of students will require

- Burette and burette stand
- Small funnel
- Pipette (10 cm³)
- 2 beakers (100 cm³)
- 3 conical flasks (100 cm³ or 250 cm³)
- Graduated flask (100 cm³)
- White tile
- Sea-water (10 cm³)

  (If none is available, a suitable substitute can be made by weighing out the following salts and dissolving in water to give 1 dm³ of solution:

NaCl	27 g
$MgCl_2.6H_2O$	11 g
$MgSO_4.7H_2O$	13 g
KCl	0·75 g
KBr	0·10 g
$CaSO_4.2H_2O$	2 g
$NaHCO_3$	0·1 g)

- Potassium chromate(VI) solution (usual bench concentration)
- 0·05 M silver nitrate solution (50 cm³). Dissolve 8·49 g in distilled water and make up to 1 dm³.

Time required 1 double period

By adding silver ions until silver chloride is no longer precipitated, the amount of chloride in a solution can be found.

Potassium chromate(VI) can be used to indicate the end-point of the titration – the point at which all chloride ions have been precipitated. Silver ions combine with chromate(VI) ions to form a red precipitate of silver chromate(VI):

$$2Ag^+(aq) + CrO_4^{2-}(aq) \longrightarrow Ag_2CrO_4(s)$$

When both chloride ions and chromate(VI) ions are present, however, no silver chromate(VI) is precipitated until all the chloride ions have been removed. The sudden appearance of red silver chromate(VI) therefore indicates the end-point of the titration.

Procedure

Note It is particularly important in this practical to rinse all the glassware with distilled water before use. Silver nitrate is expensive, and is normally used in fairly low concentration. In this titration you will use 0·05 M $AgNO_3$(aq). To obtain sensible results, it is therefore first necessary to dilute the sea-water tenfold in order to give a concentration of chloride ions comparable to that of the silver nitrate.

Pipette 10 cm³ of sea-water into a 100 cm³ graduated flask. Make up to the mark with distilled water, stopper the flask and mix thoroughly.

Pipette 10 cm³ of the diluted sea-water into a conical flask and add about 10 drops of potassium chromate(VI) indicator. Rinse a burette with the silver nitrate solution, then fill it with the solution. Titrate the sea-water in the conical flask against the silver nitrate solution from the burette until a reddish tinge just begins to appear. You may find the end-point a little difficult to detect, so it is best to carry out a rough titration first and keep the result to remind you of the end-point colour when carrying out later, accurate titrations.

Repeat the titration until two consistent results are obtained.

1 What volume of 0·05 M silver nitrate was needed to react exactly with 10 cm³ of the diluted sea-water?

2 How many moles of silver ions are there in this volume of 0·05 M solution?

3 How many moles of chloride ions must there have been in the 10 cm³ of solution?

4 How many moles of chloride ions must have been in 1 dm³ of the original, undiluted sea-water?

5 What is the concentration, in g dm⁻³, of chloride ions in this sample of sea-water?

6 What are the major sources of error in this experiment?

7 Sea-water contains fluoride, bromide and iodide ions as well as chloride, although their concentrations are much lower. (Chloride is about 300 times more concentrated than bromide, the next commonest halide ion in sea-water.) Do you think the presence of these other halides will have affected the accuracy of your results? Explain.

8 It is essential in this experiment that glassware is washed with distilled water, not tap-water. Why is this?

9 How did chloride ions get into the oceans? Why do oceans contain high concentrations of dissolved salts, while inland lakes do not?

PRACTICAL 5
A test tube study of redox reactions

Introduction

Redox reactions involve electron transfer. When a substance is oxidized it loses electrons, and the substance which receives the electrons is reduced. If a substance has a strong tendency to lose electrons it behaves as a **strong reducing agent** because it will tend to reduce other substances by giving them electrons. If a substance has a strong tendency to gain electrons it behaves as a **strong oxidizing agent** because of its readiness to gain electrons from other substances and oxidize them. The intention of this practical is to investigate reactions between pairs of substances which can behave as oxidizing and reducing agents, and try to place these substances in order of strength as oxidizers or reducers.

Consider the reaction between chlorine water and a solution containing iodide ions. The mixture turns brown rapidly as iodide ions are oxidized to iodine molecules. At the same time chlorine molecules are reduced to chloride ions:

$$2I^-(aq) + Cl_2(aq) \longrightarrow I_2(aq) + 2Cl^-(aq)$$

Chlorine molecules and chloride ions can be regarded as a **redox pair** and this is summarized in the following half-equation:

$$Cl_2 + 2e^- \rightleftharpoons 2Cl^-$$

When chlorine acts as an oxidizer, it accepts electrons and the half-equation moves from left to right. When chloride ions act as a reducer they lose electrons and the half-equation moves from right to left. We can think of the I_2/I^- pair acting in the same way:

$$I_2 + 2e^- \rightleftharpoons 2I^-$$

In the reaction between chlorine and iodide, it is impossible for both the half-equations $Cl_2 + 2e^- \rightleftharpoons 2Cl^-$ and $I_2 + 2e^- \rightleftharpoons 2I^-$ to move in the same direction, because there would be nothing to provide the electrons needed by both half-equations. One of the half-equations must move from right to left instead of left to right.

The result of the reaction tells us that it is the iodine/iodide equation that moves from right to left. This is because chlorine molecules have a greater tendency to accept electrons than iodine molecules. In other words, chlorine is a stronger oxidizer than iodine. Of course, at the start of the reaction there was neither iodine nor chloride ions present, so the reaction could not have proceeded in the other direction in any case. In fact, though, if the reaction is carried out with chlorine, chloride ions, iodine and iodide ions all present, the result is still the same – chlorine oxidizes iodide to iodine. We can therefore place the two half-equations in order of the tendency to proceed to the right and this shows the order of strength of the oxidizing agents.

$$\uparrow \quad Cl_2 + 2e^- \rightleftharpoons 2Cl^-$$

increasing oxidizing strength $\quad | \quad I_2 + 2e^- \rightleftharpoons 2I^-$

In this practical you will attempt to place a number of redox pairs in order of oxidizing strength by carrying out suitable experiments. The redox pairs concerned are

A	$I_2 + 2e^-$	$\rightleftharpoons$	$2I^-$
B	$SO_4^{2-} + H^+ + e^-$	$\rightleftharpoons$	$H_2SO_3 + H_2O$
C	$ClO^- + H_2O + e^-$	$\rightleftharpoons$	$Cl^- + OH^-$
D	$Cl_2 + 2e^-$	$\rightleftharpoons$	$2Cl^-$
E	$Br_2 + 2e^-$	$\rightleftharpoons$	$2Br^-$
F	$Fe^{3+} + e^-$	$\rightleftharpoons$	Fe^{2+}

ClO^- is called chlorate(I) (hypochlorite) ion, and H_2SO_3 is sulphurous acid.

Requirements

Each student, or pair of students, will require:
- 6 test tubes
- Universal indicator paper
- Splints

The following solutions at usual bench concentration
- Iron(III) chloride
- Potassium iodide
- Bromine water
- Dilute sulphuric acid
- Sodium chlorate(I) (hypochlorite)
- Iron(II) sulphate
- Iodine in potassium iodide
- Starch
- Hydrogen peroxide
- Sulphurous acid (a solution of sulphur dioxide in water)
- Chlorine water
- Potassium bromide or sodium bromide solution

Time required 1 double period

1 Half-equations **B** and **C** are not balanced. Answer these questions with reference to each of **B** and **C**.
   a Which element is undergoing redox?
   b Work out the oxidation number of this element in both its oxidized and reduced forms.
   c Balance the half-equation.

Procedure

EXPERIMENTS 1 – 6   Placing the half equations in order of oxidizing strength

EXPERIMENT 1
Add a little of a solution of iron(III) ions to a solution of iodide ions. Describe what happens and test to see if iodine has been formed.
   The half-equations involved are:

$$\textbf{A} \quad I_2 + 2e^- \rightleftharpoons 2I^-$$
$$\textbf{F} \quad Fe^{3+} + e^- \rightleftharpoons Fe^{2+}$$

2 Use these half-equations to write a balanced ionic equation for the reaction that has occurred.

3 In this reaction, which of the half-equations moved from right to left instead of left to right?

4 Place **A** and **F** in order of oxidizing strength.

EXPERIMENT 2
Add a little sodium chlorate(I) solution to a solution containing iron(II) ions. Decide whether or not Fe^{2+} ions have been oxidized. (**Note:** when chlorate(I) acts as an oxidizing agent, hydroxide ions are produced. What effect will this have on iron(II) or iron(III) ions?)

5 Place half-equation **C** in its correct position relative to **A** and **F**.

6 Write a balanced ionic equation for the reaction.

EXPERIMENT 3
Add a little sodium chlorate(I) solution to a solution containing bromide ions. Decide whether or not bromide ions have been oxidized to bromine.

7 Can chlorate(I) ions oxidize bromide to bromine?

8 Place half-equation **E** in its correct position in the list.

9 Write a balanced ionic equation for the reaction.

EXPERIMENT 4
Add a little sulphurous acid to a solution containing iodine, I_2, and note the result.

10 Place half-equation **B** in its correct position in the list.

11 Write a balanced ionic equation for the reaction.

EXPERIMENT 5
Finally, decide the position of half-equation **D** by adding chlorine water to a solution containing bromide ions.

12 Place half-equation **D** in its correct position.

13 Write a balanced ionic equation for the reaction.

14 Use your final order of oxidizing power to predict whether chlorate(I) ions will oxidize iodide ions to iodine.

EXPERIMENT 6
Test your prediction experimentally.

EXPERIMENTS 7 and 8   Redox reactions of hydrogen peroxide
Hydrogen peroxide, H_2O_2, can behave as both an oxidizer and a reducer depending on the conditions. Half-equation **G** shows hydrogen peroxide behaving as an oxidizing agent.

$$\textbf{G} \quad H_2O_2 + 2H^+ + 2e^- \rightleftharpoons 2H_2O_2$$

Half-equation **H** shows hydrogen peroxide behaving as a reducing agent.

$$\textbf{H} \quad O_2 + 2H^+ + 2e^- \rightleftharpoons H_2O_2$$

When hydrogen peroxide acts as a reducing agent half-equation **H** will, of course, proceed in the **reverse** direction.

EXPERIMENT 7

Add a little hydrogen peroxide solution to a solution of sodium chlorate(I) and note what happens. Decide whether the hydrogen peroxide is behaving as an oxidizer or as a reducer.

15 Which half-equation describes the behaviour of hydrogen peroxide in this experiment, **G** or **H**?

16 Decide whether this half-equation should be placed above or below half-equation **C**. (**Note** It will be impossible to determine the exact position of the half-equation.)

17 Write a balanced ionic equation for the reaction of hydrogen peroxide with chlorate(I) ions.

EXPERIMENT 8

Acidify a little potassium iodide solution with dilute sulphuric acid. Add a little hydrogen peroxide solution. Describe what occurs.

18 Is hydrogen peroxide behaving as an oxidizer or as a reducer?

19 Which half-equation describes the behaviour of hydrogen peroxide in this reaction?

20 Decide whether this half-equation should be placed above or below half equation **A**.

21 Write a balanced ionic equation for the reaction of hydrogen peroxide with iodide ions in acid solution.

When hydrogen peroxide is left for some time it slowly decomposes into water and oxygen. This decomposition is speeded up by various catalysts, such as manganese(IV) oxide. The equation for the reaction is

$$2H_2O_2 \longrightarrow 2H_2O + O_2$$

22 Work out the oxidation number of oxygen in H_2O_2, H_2O and O_2.

23 Does the conversion of hydrogen peroxide to **oxygen** involve oxidation or reduction? What substance is acting as the oxidizer or reducer?

24 Does the conversion of hydrogen peroxide to **water** involve oxidation or reduction? What substance is acting as the oxidizer or reducer?

25 Show how the equation for the decomposition of hydrogen peroxide can be derived from half-equations **G** and **H**.

26 Explain what is meant by the term **disproportionation.**

Further work

If you have time, devise tests to establish more precisely the positions of **G** and **H** in the table.

Hydrogen bonding in the trichloromethane-ethyl ethanoate system

Requirements

Each student, or pair of students, will require:

- Reflux apparatus as shown in figure 3, or equivalent, fitted with 0–100°C thermometer
- Measuring cylinder (10 cm³)
- Measuring cylinder (25 cm³)
- Boiling tube
- Beaker (250 cm³)
- Cotton wool
- Trichloromethane (chloroform)
- Ethyl ethanoate (ethyl acetate)
- A central disposal point for solvent residues.
  (**Note** Mixtures of trichloromethane and propanone have been known to explode on standing. Avoid disposing of the solvent residues from this experiment in a container in which propanone is present.)

Time required 1–2 double periods

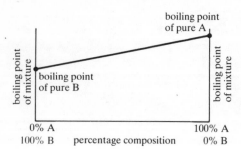

Figure 2 Boiling point–composition diagram for a mixture of liquids, A and B, which behaves ideally.

Introduction

Trichloromethane and ethyl ethanoate, whose structures are shown in figure 1, do not form hydrogen bonds with themselves but they do with each other, as the figure shows. In this practical, we shall investigate the formation of hydrogen bonds when the two liquids are mixed.

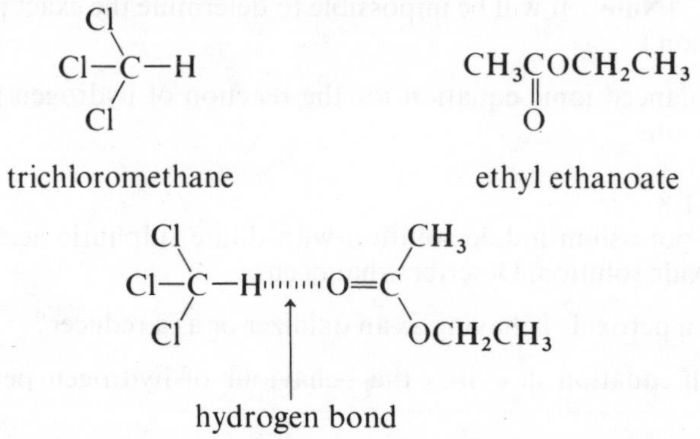

trichloromethane   ethyl ethanoate

hydrogen bond

Figure 1 Hydrogen bonding between trichloromethane and ethyl ethanoate.

Procedure

A The boiling point of mixtures of trichloromethane and ethyl ethanoate

In the first part of the practical, the boiling points of mixtures of trichloromethane and ethyl ethanoate are measured.

You might expect the boiling point of a mixture of liquids to be the average of the boiling points of the component liquids, weighted to take account of the proportions of the two liquids present. If this were so, a graph of boiling point against composition would be a straight line as shown in figure 2.

This sort of behaviour is indeed shown by some pairs of very similar liquids, for example hexane and heptane. They are said to behave as an **ideal mixture**. This experiment will enable you to see how mixtures of trichloromethane and ethyl ethanoate behave.

CARE Trichloromethane vapour is toxic and ethyl ethanoate is highly flammable. Always heat the liquids under reflux, and move your bunsen well clear before making additions of liquid to the apparatus.

Set up the reflux apparatus fitted with a thermometer as shown in figure 3. Put 10 cm³ of ethyl ethanoate in the flask and warm very gently until steady boiling and refluxing occurs. (It is particularly important that no vapour is lost from the apparatus.) Note the temperature. Move away the bunsen, allow the apparatus to cool a little, then add 2 cm³ of trichloromethane. Measure the boiling point of this mixture as before. Continue adding 2 cm³ portions of trichloromethane in this way, recording the boiling point of each mixture, until a total of 10 cm³ has been added. Then allow the mixture to cool and dispose of it as directed by your class teacher.

Repeat the experiment, starting with 10 cm³ of trichloromethane and making 2 cm³ additions of ethyl ethanoate, until a total of 10 cm³ has been added.

Tabulate your results, showing the boiling point of each mixture against the percentage composition of the mixture by volume.

Plot a graph of boiling point against percentage composition by volume.

1. Do trichloromethane and ethyl ethanoate form an ideal mixture?

2. Are the boiling points of the mixtures higher or lower than you would expect if the behaviour were ideal?

3. What does this suggest about the bonding between molecules of trichloromethane and ethyl ethanoate? Is it stronger or weaker than would be expected if the mixture were ideal?

4. At what point on your graph are the forces between ethyl ethanoate molecules and trichloromethane molecules at a maximum? Read off the percentage composition by volume of the mixture at this point. Convert this into a percentage composition by moles. To do this, you will need the following data.

Relative molecular masses
  ethyl ethanoate   88
  trichloromethane   119·5

Densities
  ethyl ethanoate   0·90 g cm^{-3}
  trichloromethane   1·48 g cm^{-3}

5. Do your answers to **3** and **4** agree with what you know about hydrogen bonding between trichloromethane and ethyl ethanoate?

B Finding an approximate value for the strength of the hydrogen bond between trichloromethane and ethyl ethanoate

When trichloromethane and ethyl ethanoate are mixed, hydrogen bonds are formed.

6. Would you expect a temperature rise or a temperature fall on mixing the liquids?

By measuring the temperature change when the liquids are mixed, you can estimate the energy given out when the hydrogen bond forms – in other words, the strength of the bond.

You will need to mix equimolar quantities of the two liquids. Using the data given in question **4**, calculate the volumes of liquids you need to mix in order to give an equimolar mixture whose total volume is between 20 cm^3 and 25 cm^3.

Set up an insulated boiling tube as shown in figure 4. Measure the calculated volume of the first liquid into the boiling tube and place the calculated volume of the second liquid in a measuring cylinder. Take the temperature of each liquid. Pour the contents of the measuring cylinder into the boiling tube, stir with the thermometer and note the maximum temperature reached.

7. Work out the temperature rise of each liquid.

8. Using the specific heat capacity data below, work out the quantity of heat needed to raise the temperature of each liquid by the observed amount.

Specific heat capacities

  ethyl ethanoate   1·92 J g^{-1} K^{-1}
  trichloromethane   0·96 J g^{-1} K^{-1}

9. Calculate the amount of heat that would have been given out if you had used one mole of each liquid.

10. What is the strength, in kJ mole^{-1}, of the hydrogen bond between trichloromethane and ethyl ethanoate?

11. How does this value compare with the strength of an ordinary covalent bond?

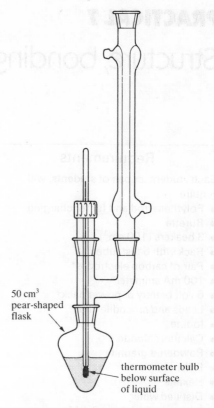

50 cm^3 pear-shaped flask

thermometer bulb below surface of liquid

Figure 3 Reflux equipment with thermometer.

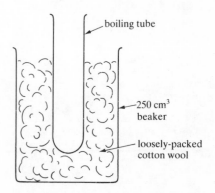

boiling tube

250 cm^3 beaker

loosely-packed cotton wool

Figure 4 Insulated boiling tube calorimeter.

PRACTICAL 7
Structure, bonding and properties

Requirements

Each student, or pair of students, will require:

- Polythene rod and fur for charging
- Burette
- 3 beakers (100 cm³)
- Rack with 6 test tubes
- Pair of carbon electrodes*
- 100 mA ammeter*
- 6 volt battery or power pack*
- Leads and crocodile clips*
- Iodine
- Calcium chloride
- Powdered graphite
- Ethanol
- Hexane
- Distilled water
- Silver nitrate (about 0·1 M)
- Ethyl ethanoate (ethyl acetate)
- Ethoxyethane (ether)
- Trichloromethane

* One set of these items could be shared by three or four pupils.

Time required 2 double periods

Introduction

Most of the physical and chemical properties of a substance can be related to the type of bonding present in that substance. In this practical you will investigate properties such as volatility and conductivity and try to explain these properties in terms of the bonding in the substance involved.

Substances that are ionically bonded contain positive and negative ions. Substances that are covalently bonded contain molecules; if these molecules contain atoms of different sorts, they may be polar due to the unequal sharing of electrons between the different atoms. The amount of polarity in a molecule, determined by the shape of the molecule and the relative electronegativities of the atoms in it, has a great effect on the properties of the substance.

Procedure

EXPERIMENT 1   The effect of a charged rod on liquid jets

Charge a polythene rod by rubbing it on a piece of fur. Hold the rod about 1 cm from a jet of water issuing from a burette and note what happens.

1   What sort of particles does water contain?

2   Are these particles polar?

3   Explain the effect of the charged rod on the jet of water.

4   What do you think would happen with a rod of opposite charge? Explain your answer.

Repeat the experiment using hexane, and then ethanol, instead of water.

5   Relate the behaviour of jets of these liquids towards a charged rod to the nature of the particles they contain.

6   Look up the boiling points and relative molecular masses of water, ethanol and hexane. Try to explain qualitatively the relative magnitudes of their boiling points in terms of polarity.

EXPERIMENT 2   Miscibility of liquids

Test the miscibility of **a** water and ethanol, **b** water and hexane, **c** hexane and ethanol.

7   Try to explain the results of experiment 2 in terms of the polarities of the molecules in the different liquids.

EXPERIMENT 3   Solubility of iodine in different liquids

Keep the solutions from this experiment and from experiments **4** and **5** for use in experiment 7.

Put a very small crystal of iodine in a test tube and shake with 5 cm³ of distilled water. Try to decide roughly how soluble iodine is in water. Repeat the experiment using ethanol as the solvent instead of water, and finally using hexane as the solvent.

8   What sort of particles does iodine contain?

9   What forces hold these particles together?

10   Explain the relative solubility of iodine in the three solvents.

11   Suggest a reason for the differing colours of solutions of iodine in different solvents. (If you have time, you could explore this question further in Experiment **8**.)

EXPERIMENT 4   Solubility of graphite in liquids

Repeat experiment 3, using powdered graphite instead of iodine. Answer questions 7 to 9 with reference to graphite instead of iodine.

EXPERIMENT 5   Solubility of calcium chloride in liquids

Repeat experiment 3, using calcium chloride instead of iodine. As solutions of calcium chloride are colourless, use the following procedure to decide how much has dissolved:

After shaking the crystal with the solvent, decant the liquid into a test tube, leaving the excess crystal behind. Add 2 cm^3 of silver nitrate solution to the decanted liquid and shake. From the amount of silver chloride precipitated, judge the relative solubility of calcium chloride in the three solvents.

Answer questions 7 to 9 with reference to calcium chloride instead of iodine.

EXPERIMENT 6   Volatility of iodine, graphite and calcium chloride

Heat a crystal of each of the solids in turn in a hard-glass test tube and judge their relative volatilities.

CARE   Iodine vapour is irritating and toxic. Use only a very small crystal of iodine and work in a fume cupboard.

12   Try to explain the relative volatility of these three solids in terms of the particles they contain and the forces between them.

EXPERIMENT 7   Conductivity

Collect each of the solutions produced in experiments 3, 4 and 5. Test the conductivity of each solution by pouring it into a 100 cm^3 beaker and immersing the carbon electrodes connected in series with a 6 volt battery and a 100 mA ammeter. (Test the same depth of solution each time.) Note the meter reading in each case. (If the solid showed no sign of dissolving at all, do not bother to perform the test in that particular case.) You should also test the conductivity of each of the pure solvents.

   NOTE   Keep the electrodes clean and make sure they are not contaminated with liquid from a previous test.

13   Try to give a qualitative explanation of the relative conductivities of the different solutions.

EXPERIMENT 8   The colour of iodine in solution

If you have time, try the following tests, then see if you can suggest an explanation for the different colours of iodine in different solvents.

A   Find the colour of iodine in solution in the following solvents: hexane, ethanol, ethoxyethane (ether), ethyl ethanoate, trichloromethane.

CARE Ethoxyethane and trichloromethane should be handled only in a fume cupboard. Ethoxyethane is highly flammable – keep it away from naked flames.

B   Add one drop of ethanol to a solution of iodine in hexane.

C   Add one drop of hexane to a solution of iodine in ethanol.

14   What is the colour of iodine vapour?

15   Try to suggest a theory to explain the variation of colour of iodine in different solvents.

PRACTICAL 8

Heats of neutralization

Requirements

Each student, or pair of students, will need:

- Plastic beaker (the sort used in drink dispensing machines is ideal)
- 0–100°C thermometer
- Measuring cylinder (50 cm^3)
- 2·0 M nitric acid (25 cm^3) (Make 128 cm^3 of concentrated acid up to 1 dm^3)
- 2·0 M hydrochloric acid (100 cm^3) (Make 172 cm^3 of concentrated acid up to 1 dm^3)
- 2·0 M sulphuric acid (50 cm^3) (Make 107 cm^3 of concentrated acid up to 1 dm^3)
- 2·0 M sodium hydroxide (75 cm^3) (Dissolve 80 g of pellets or flakes in distilled water and make up to 1 dm^3)
- 4·0 M sodium hydroxide (25 cm^3) (Dissolve 160 g of pellets or flakes in distilled water and make up to 1 dm^3)
- 2·0 M potassium hydroxide (25 cm^3) (Dissolve 112 g of pellets in distilled water and make up to 1 dm^3)

Time required 1 double period

Introduction

When an alkali neutralizes an acid, a salt and water are formed. Aqueous hydrogen ions (H^+(aq)) from the acid react with the hydroxide ions (OH^-(aq)) from the alkali, forming water. The identity of the salt will of course depend on the nature of the acid and alkali used. For hydrochloric acid and sodium hydroxide:

$$\underbrace{H^+(aq) + Cl^-(aq)}_{\text{hydrochloric acid}} + \underbrace{Na^+(aq) + OH^-(aq)}_{\text{sodium hydroxide}} \longrightarrow \underbrace{Na^+(aq) + Cl^-(aq)}_{\text{sodium chloride}} + H_2O(l)$$

Notice in this equation that the Na^+ and Cl^- ions are unchanged. The only chemical reaction occurring is between H^+ and OH^- ions:

$$H^+(aq) + OH^-(aq) \longrightarrow H_2O(l)$$

The combination of H^+ and OH^- ions in this way releases energy. In this practical, the enthalpy changes accompanying different neutralization reactions will be measured. Because the number of moles of water formed varies according to the acid and alkali used, it is the convention to measure enthalpies of neutralization in kJ per mole of water formed.

Procedure

EXPERIMENT 1   The reaction between hydrochloric acid and sodium hydroxide

Measure 25 cm^3 of 2·0 M hydrochloric acid into a plastic beaker. Record its temperature. Put 25 cm^3 of 2·0 M sodium hydroxide in a measuring cylinder and take its temperature. Now pour this into the acid, stir and take the final temperature.

1   What was the temperature rise? (If the acid and alkali were not at the same temperature initially, use the mean of their initial temperatures as the starting temperature.)

2   How much heat energy was delivered to the 50 cm^3 of final solution during the reaction? (Assume the specific heat capacity of the plastic cup is negligible, and that the specific heat capacity of the solution is 4·2 J K^{-1} g^{-1}.)

3   How many moles of water were formed by the reaction you have just carried out?

4   Work out the heat of neutralization for this reaction in kJ per mole of water formed.

EXPERIMENT 2   The reaction between hydrochloric acid and potassium hydroxide

Repeat the above experiment using 2·0 M potassium hydroxide instead of the sodium hydroxide.

5   Write an ionic equation like the one in the introduction to represent the reaction that has occurred.

6   Using the method outlined in questions **1** to **4** above, work out the heat of neutralization for this reaction.

EXPERIMENT 3   The reaction between nitric acid and sodium hydroxide

Repeat the first experiment using 2·0 M nitric acid instead of the hydrochloric acid.

7   Write an ionic equation to represent the reaction which has occurred.

8   Work out the heat of neutralization for this reaction.

9 Compare the values you have obtained for the heats of neutralization of the three reactions and explain the way the values compare.

Accepted values for the heats of neutralization of some acids and alkalis are shown in the following table.

Reaction				Heat of neutralization/ kJ mole^{-1}
HCl(aq) hydrochloric acid	+	NaOH(aq) sodium hydroxide	$\longrightarrow$ NaCl(aq) + H$_2$O(l)	-57.9
HBr(aq) hydrobromic acid	+	NaOH(aq) sodium hydroxide	$\longrightarrow$ NaBr(aq) + H$_2$O(l)	-57.6
HNO$_3$(aq) nitric acid	+	NaOH(aq) sodium hydroxide	$\longrightarrow$ NaNO$_3$(aq) + H$_2$O(l)	-57.6
CH$_3$COOH(aq) ethanoic acid	+	NaOH(aq) sodium hydroxide	$\longrightarrow$ CH$_3$COONa(aq) + H$_2$O(l)	-56.1
H$_2$S(aq) hydrogen sulphide	+	NaOH(aq) sodium hydroxide	$\longrightarrow$ NaHS(aq) + H$_2$O(l)	-32.2

10 How do the values for the first two reactions in this table compare with your own experimental results for the same reactions? Account for any errors in your results.

11 Do the values for the first three reactions in the table agree with the explanation you have given in question **9**?

12 Suggest why the heat of neutralization for the reaction involving ethanoic acid is slightly lower than the values for the first three reactions, and why the heat of neutralization for the reaction involving hydrogen sulphide is substantially lower. (The dissociation constants (K_a) for hydrochloric acid, hydrobromic acid and nitric acid are very large, that of ethanoic acid is 1.7×10^{-5} and that of hydrogen sulphide is 8.9×10^{-8}.)

You should now have some idea of how the relative magnitudes of heats of neutralization can be explained. You can use these ideas in the following experiment.

EXPERIMENT 4   The reaction between sulphuric acid and sodium hydroxide
Repeat the first experiment using 25 cm^3 of 2·0 M sulphuric acid and 25 cm^3 of 2·0 M sodium hydroxide. Record the temperature rise on mixing as before. Now repeat the experiment using 25 cm^3 of 4·0 M sodium hydroxide instead of 2·0 M.

13 Compare the temperature rises for the two experiments and explain their relative values.

Repeat the experiment using 25 cm^3 of 2·0 M hydrochloric acid and 25 cm^3 of 4·0 M sodium hydroxide.

14 Compare the temperature rise when 25 cm^3 2·0 M hydrochloric acid neutralizes 25 cm^3 2·0 M sodium hydroxide with the rise when 4·0 M sodium hydroxide is used instead. Explain their relative values. Compare your answer here with your answer to question **13**.

PRACTICAL 9

Determination of the enthalpy change for the thermal decomposition of potassium hydrogencarbonate to potassium carbonate

Requirements

Each student will need:

- Burette and stand
- Plastic beaker (approx. 100 cm^3) – the plastic cups used in drink dispensing machines are suitable.
- 2 clean, dry test tubes
- Thermometer (0–50°C by 0·2°C is ideal, but one covering 0–110°C by 1·0°C, can be used)
- 2 M hydrochloric acid (60 cm^3)
- Solid anhydrous potassium carbonate (3 g)
- Solid potassium hydrogencarbonate (3·75 g)
- Access to a balance

Time required 1 double period

Principle

The object of this experiment is to determine the change in heat content during the decomposition of potassium hydrogencarbonate. This enthalpy change is difficult to measure directly, so an indirect method is used.

1 Explain the following terms:
  a enthalpy.
  b enthalpy change.

2 Write an equation including state symbols for the thermal decomposition of potassium hydrogencarbonate to potassium carbonate, showing the products in their usual states under standard conditions.

3 What are the standard conditions for thermochemistry?

4 Why is it difficult to determine this enthalpy change directly?

You are provided with 2 M hydrochloric acid, solid potassium carbonate and solid potassium hydrogencarbonate. By determining the heat of reaction between potassium carbonate and hydrochloric acid and that between potassium hydrogencarbonate and hydrochloric acid, it is possible to obtain indirectly the enthalpy change for the decomposition of potassium hydrogencarbonate.

Procedure

EXPERIMENT 1   The reaction of potassium carbonate with hydrochloric acid

Using a burette, measure 30 cm^3 of approximately 2 M hydrochloric acid into a plastic beaker. Take the temperature of the acid and record this in a table similar to table 1 below.

Weigh accurately a test tube containing between 2·5 and 3·0 g of anhydrous potassium carbonate (K_2CO_3). Record your weighings in a table similar to table 1 below.

Now add the weighed portion of K_2CO_3 to the acid and stir the mixture carefully with the thermometer until all the solid has reacted. Rapid effervescence will occur. Be careful not to lose any of the reaction mixture by spilling. Record the maximum temperature of the solution after mixing and then re-weigh the empty test tube.

Mass of test tube + potassium carbonate	g
Mass of empty test tube	g
Mass of potassium carbonate used	g
Temperature of acid initially	°C
Temperature of solution after mixing	°C
Temperature change during reaction	°C

Table 1

5 Write an equation, including state symbols, for the reaction between potassium carbonate and hydrochloric acid.

6 From your results in table 1, calculate the heat evolved or absorbed during the reaction between the potassium carbonate and the acid.
(Assume that the specific heat capacities of all the solutions are the same as that of water (i.e. $4 \cdot 2$ J g^{-1} K^{-1}), and that the solutions have a density of $1 \cdot 0$ g cm^{-3}. Assume that the specific heat capacity of the calorimeter is negligible.)

7 Calculate the heat change for one mole of potassium carbonate.

8 Why is the exact concentration of the acid unimportant?

9 Why is it better to re-weigh the test tube after emptying out the solid than to record the mass of the clean test tube?

EXPERIMENT 2 The reaction of potassium hydrogencarbonate with hydrochloric acid

Repeat experiment 1 using an accurately weighed sample of potassium hydrogencarbonate between $3 \cdot 25$ g and $3 \cdot 75$ g in place of the potassium carbonate. Record all weighings and temperatures in a table similar to table 2 below.

Mass of test tube + potassium hydrogencarbonate	g
Mass of empty test tube	g
Mass of potassium hydrogencarbonate used	g
Temperature of acid initially	°C
Temperature of solution after mixing	°C
Temperature change during reaction	°C

Table 2

10 Write an equation, including state symbols for the reaction between potassium hydrogencarbonate and hydrochloric acid.

11 From your results in table 2, calculate the heat evolved or absorbed during the reaction between the potassium hydrogencarbonate and the acid. (You should make the same assumption as in question 5.)

12 Calculate the heat change for one mole of potassium hydrogencarbonate.

13 Mention three major sources of error in your experiments.

14 Draw an energy diagram linking the reaction of K_2CO_3(s) with HCl(aq), the reaction of $KHCO_3$(s) with HCl(aq) and the decomposition of $KHCO_3$(s).

15 What is the enthalpy change for the decomposition of potassium hydrogencarbonate?

16 What law have you used in answering question 14 and on what thermodynamic principle does this law depend?

PRACTICAL 10
Heats of solution

Requirements

Each student or pair of students will require:

- Sodium chloride
- Ammonium nitrate
- Sodium hydroxide (flakes or pellets)
- Sodium thiosulphate-5-water, $Na_2S_2O_3.5H_2O$
- Measuring cylinder, 100 cm^3
- Thermometer, 0–50°C
- Plastic beaker (the sort used in drink dispensing machines does very well)
- Pestle and mortar
- Weighing bottle
- Access to a balance

Time required 1 double period

Introduction

When an ionic solid dissolves in water to form an aqueous solution, a temperature change is always observed. In the first part of this practical we shall look in a simple, qualitative way at the magnitude of the temperature changes involved. In the second part, we will attempt to obtain an accurate value for the heat of solution of an ionic solid.

Procedure

EXPERIMENT 1 Temperature changes on dissolving different ionic solids in water

The following procedure should be carried out separately for each of the three solutes; sodium chloride, sodium hydroxide and ammonium nitrate.

Weigh out 0·1 mole of the solid. In the case of sodium hydroxide the pellets or flakes should be weighed in a sealed weighing bottle.

CARE Sodium hydroxide is very corrosive: avoid all contact with the skin and eyes. If any sodium hydroxide comes into contact with the skin, wash it immediately with plenty of cold water.

Measure 100 cm^3 distilled water into an insulated plastic beaker and measure its temperature. Quickly dissolve the 0·1 mole of solid in the water, stirring briskly with the thermometer, and record the final temperature.

Tabulate your results, showing the temperature rise for each solute.

1 Why, apart from the danger of skin contact, is it particularly important to weigh the sodium hydroxide in a sealed bottle?

2 Which solute, or solutes, gave out heat as they dissolved? Which absorbed heat?

When an ionic solid dissolves in water, two processes occur. First, the ions in the solid lattice must be separated from one another. As the ions carry opposite charges and attract one another, this process of bond-breaking requires the input of energy. The second process involves the interaction of the ions of the solute with polar water molecules – positive ions attract the negative ends of H_2O dipoles, and negative ions attract the positive ends. This process of bond-making releases energy. The sign and magnitude of the overall energy change depends on the relative sizes of the energy changes occurring in these two processes.

(See *Chemistry in Context*, section 11.14, for a more detailed consideration of the processes that occur when an ionic solid dissolves.)

3 Explain in terms of the two energy changes referred to above, the signs and relative magnitudes of the temperature rises that occurred when each of the three solutes dissolved.

   Note With some solutes, including sodium hydroxide, the situation is complicated by the formation of hydrates when the anhydrous solute dissolves. Leave this consideration aside when answering this question.

4 If you had used excess solute in your experiment, a saturated solution would have been formed, with undissolved solute in equilibrium with its aqueous solution, for example:

$$NH_4^+NO_3^-(s) + aq \rightleftharpoons NH_4^+(aq) + NO_3^-(aq)$$

By applying Le Chatelier's principle to this equilibrium, decide whether the solubility of ammonium nitrate will increase or decrease when the temperature is raised. How will the solubility of sodium chloride change when the temperature is raised?

EXPERIMENT 2 Measurement of the heat of solution of sodium thiosulphate

The heat of solution, ΔH_{soln}, of a solute is the heat change that occurs when one mole of the solute dissolves to form an infinitely dilute solution. In this part of the practical you are to devise and carry out a simple experiment to obtain as accurate a value as possible for ΔH_{soln} for hydrated sodium thiosulphate, $Na_2S_2O_3.5H_2O$. In principle this can be done by measuring the temperature change of the water in which the sodium thiosulphate dissolves, but you should bear the following points in mind when designing your experiment.

A Heat loss must be minimized. In particular, you will need good insulation and the sodium thiosulphate should be dissolved as quickly as possible. You may be able to think of other ways of reducing heat loss.

B It will not, of course, be possible to produce an infinitely dilute solution, and your solution will need to be concentrated enough to produce an accurately measurable temperature rise. You should think carefully about the best concentration to use.

C You can assume that the specific heat capacity of sodium thiosulphate solution is the same as that of water, i.e. $4.2 \text{ J g}^{-1} \text{ K}^{-1}$.

D You will need to bear in mind the apparatus available to you.

5 Write a brief description of your experiment and tabulate the results.

6 Work out a value for the heat of solution of sodium thiosulphate as follows.
   a Use the temperature change of the solution to calculate the heat change that occurred in the solution.
   b Knowing the amount of sodium thiosulphate used, calculate the heat change that would have occurred if one mole of sodium thiosulphate had been used.

7 What do you consider to be the major sources of error in your experiment?

8 How might these errors have been reduced if more sophisticated apparatus had been available?

9 How would you expect the value for ΔH_{soln} to change if you used anhydrous sodium thiosulphate instead of $Na_2S_2O_3.5H_2O$?

10 Strictly speaking, ΔH_{soln} applies to the formation of infinitely dilute solutions. What simple experiment could you perform with the solution you obtained to decide whether a significant error was introduced by not making it infinitely dilute?

PRACTICAL 11
Acids, bases and indicators

Requirements

Each student, or pair of students, will require:

- Rack and 6 test tubes

The following five solutions at usual bench concentration.

- Dilute hydrochloric acid
- Concentrated hydrochloric acid
- Dilute ethanoic (acetic) acid
- Dilute sodium hydroxide
- Ammonium chloride

- Sodium chloride
- Sodium ethanoate
- Phenol
- Sodium benzoate
- Ethanol
- Magnesium powder
- Distilled water
- Bromophenol blue indicator (if this is not available, methyl orange is equally suitable).

Time required 1 double period

Introduction

If we define acids as proton donors and bases as proton acceptors, it is clear that every acid must have a corresponding base. This is formed when the acid loses a proton:

$$\underset{\text{acid}}{HA} \rightleftharpoons H^+ + \underset{\text{base}}{A^-}$$

A^- is said to be the **conjugate base** of the acid HA, and vice-versa.

Acids differ markedly in their tendency to lose a proton. Acids which donate protons readily are said to be **strong acids**. Strong acids have weak conjugate bases, because the conjugate base will have little tendency to accept the proton back. Weak acids, on the other hand, have strong conjugate bases. For example, hydrochloric acid, HCl, is a strong acid whereas ethanoic acid, CH_3COOH, is weak. From this, it follows that the chloride ion, Cl^-, is a weak base whereas the ethanoate ion, CH_3COO^-, is a strong base. We can therefore arrange these two acid–base pairs in order of strength as acids and bases.

	Acid		Base	
increasing ↑ acid strength	HCl	$\rightleftharpoons H^+ + Cl^-$		increasing base strength ↓
	CH_3COOH	$\rightleftharpoons H^+ + CH_3COO^-$		

This idea has been extended in the table below to show a number of well-known acids and their conjugate bases in order of strength. The table includes several of the acids you will meet in this practical. Note that ethanol and water appear as acids in the table, even though both are so weak that they are not normally regarded as acids.

Name of acid	Acid	Base	Name of base
chloric(VII) acid	$HClO_4$	$\rightleftharpoons H^+ + ClO_4^-$	chlorate(VII) ion
hydrochloric acid	HCl	$\rightleftharpoons H^+ + Cl^-$	chloride ion
oxonium ion	H_3O^+	$\rightleftharpoons H^+ + H_2O$	water
methanoic acid	$HCOOH$	$\rightleftharpoons H^+ + HCOO^-$	methanoate ion
ethanoic acid	CH_3COOH	$\rightleftharpoons H^+ + CH_3COO^-$	ethanoate ion
ammonium ion	NH_4^+	$\rightleftharpoons H^+ + NH_3$	ammonia
phenol	C_6H_5OH	$\rightleftharpoons H^+ + C_6H_5O^-$	phenoxide ion
water	H_2O	$\rightleftharpoons H^+ + OH^-$	hydroxide ion
ethanol	CH_3CH_2OH	$\rightleftharpoons H^+ + CH_3CH_2O^-$	ethoxide ion

Conjugate acid–base pairs arranged in order of decreasing acid strength.

Procedure

EXPERIMENT 1   Proton-transfer reactions between acids and bases

1   Do **a** hydrochloric acid, **b** ethanoic acid, **c** ethanoate ions have characteristic smells?

   A   Add a little dilute hydrochloric acid to a little solid sodium ethanoate, and warm. Smell the reaction mixture cautiously.

2   Has a proton-transfer reaction occurred? If so, what substance has behaved as an acid, and what has behaved as a base?

B Add a little dilute ethanoic acid to a little solid sodium chloride and warm.

3 Is there any evidence that a proton-transfer reaction has occurred?

C Your observations in this experiment illustrate a simple rule: if an acid HA_1 is added to the conjugate base A_2^- of a weaker acid HA_2, a proton transfer reaction occurs, forming A_1^- and HA_2.

$$HA_1 + A_2^- \rightleftharpoons HA_2 + A_1^-$$
$$\text{stronger} \qquad\qquad \text{weaker}$$
$$\text{acid} \qquad\qquad\quad \text{acid}$$

This is because HA_1 has a greater tendency to donate protons than HA_2. This rule can be illustrated further using phenol.

CARE Phenol is corrosive. Eye protection is essential. Avoid all contact with the skin.

Put a few crystals of phenol in a test tube. Add an equal volume of water, cork the tube and shake.

4 Does phenol dissolve in water?

D Now add excess of a solution containing hydroxide ions.

5 What happens?

6 A proton-transfer reaction has occurred. What substance has acted as an acid, and what has acted as a base?

E Now add a few drops of concentrated hydrochloric acid.

7 What happens?

8 Another proton-transfer reaction has occurred. What substance has acted as an acid, and what has acted as a base this time?

9 Explain the reactions that have occurred in terms of the relative positions of hydrochloric acid, phenol and water in the table on page 34.

EXPERIMENT 2   Finding the position of benzoic acid in the table
Benzoic acid, C_6H_5COOH is a white solid which is nearly insoluble in water. Benzoate ions, however, are soluble.

Make a solution of sodium benzoate by dissolving 1·0 g in 10 cm³ water. Divide the solution between three test tubes. To the first tube add 2 cm³ dilute hydrochloric acid, to the second 2 cm³ dilute ethanoic acid, and to the third 2 cm³ of a solution containing ammonium ions. Record your results and write equations where appropriate.

10 Where should benzoic acid be positioned in the table?

EXPERIMENT 3   Acid strength and hydrogen ion concentration
In aqueous solution acids donate protons to water molecules, forming the oxonium ion, H_3O^+.

$$HA + H_2O \rightleftharpoons H_3O^+ + A^-$$

Oxonium ions are often written simply as $H^+(aq)$
The stronger the acid HA, the further this equilibrium will lie to the right, and the higher the concentration of $H^+(aq)$.

Prepare four test tubes as follows:
  Tube 1 – 4 cm³ dilute ethanoic acid
  Tube 2 – 4 cm³ ammonium chloride solution
  Tube 3 – 4 cm³ water
  Tube 4 – 4 cm³ ethanol

To each of these tubes add a **small pinch** of magnesium powder. Compare the reactions in the different tubes.

11 Write a general ionic equation for the reactions that occur.

12 Relate the vigour of each reaction to the acid strength of the liquid in the tube.

EXPERIMENT 4 Acid–base indicators

Indicators are substances which change colour according to the pH of the solution they are in. Indicators are themselves weak acids or bases which differ in colour from their conjugates. For example, phenolphthalein, an indicator which is colourless in acid but pink in alkali, is a weak acid which we might represent as H*Ph*. Un-ionized acid, H*Ph*, is colourless, but its conjugate base, *Ph*⁻, is pink.

$$\text{H}Ph\text{(aq)} \rightleftharpoons \text{H}^+\text{(aq)} + Ph^-\text{(aq)}$$
$$\quad\text{colourless} \qquad\qquad\qquad \text{pink}$$

When acid is added, the equilibrium moves to the left and the colourless form is in large excess. When alkali is added, H^+(aq) is removed, the equilibrium moves to the right and the pink form predominates. Hence phenolphthalein is colourless in acid but pink in alkali.

Put 2 cm³ dilute hydrochloric acid in a test tube and add 2 drops of the indicator bromophenol blue, which can be represented as H*B*.

13 In what form is the indicator present in the tube? What colour is this form? Add excess sodium hydroxide solution.

14 In what form is the indicator now present? What is the colour of this form?

Prepare five test tubes as follows:
Tube 1 – 4 cm³ dilute hydrochloric acid
Tube 2 – 4 cm³ dilute ethanoic acid
Tube 3 – 4 cm³ ammonium chloride solution
Tube 4 – 2 or 3 crystals of phenol (**CARE**) dissolved in 4 cm³ water
Tube 5 – 4 cm³ water

Add 2 drops of bromophenol blue indicator to each tube and record the colour.

15 Where would you place H*B* in the table of acids? Why is it not possible to position it exactly?

PRACTICAL 12

Electrochemical cells

Introduction

Many chemical reactions give out energy. This energy is usually in the form of heat, and when the heat is used, for example to drive an engine, the chemical is called a **fuel**. Occasionally a chemical reaction can be arranged so that the energy given out is in the form of electricity. Such an arrangement is called an **electric cell**.

Procedure

Introductory Experiment
Add a little zinc powder to a few cm³ of copper sulphate solution in a test tube. Note everything that happens.

In this reaction, copper(II) ions are being reduced by zinc:

$$\text{Zn(s)} + \text{Cu}^{2+}\text{(aq)} \longrightarrow \text{Zn}^{2+}\text{(aq)} + \text{Cu(s)}$$

Electrons are transferred from zinc atoms to copper ions. If the zinc atoms are kept apart from the copper ions, the electrons transferred from Zn to Cu^{2+} can be made to flow down a wire. In this way, the energy released as heat in the simple test tube reaction above can be made available as electricity.

EXPERIMENT 1 Constructing a zinc/copper cell
One way of doing this is shown in figure 1. A strip of zinc foil is placed in contact with zinc ions, and a strip of copper foil in contact with copper ions. When these two **half cells** are connected with a wire joining the metal electrodes and a salt bridge connecting the ionic solutions, the following reactions occur:

In the left-hand beaker $\quad \text{Zn(s)} \longrightarrow \text{Zn}^{2+}\text{(aq)} + 2\text{e}^-$

In the right-hand beaker $\quad \text{Cu}^{2+}\text{(aq)} + 2\text{e}^- \longrightarrow \text{Cu(s)}$

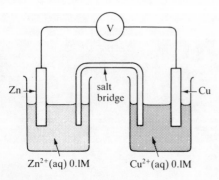

Figure 1 A simple cell.

The electrons flow down the wire from the left-hand to the right-hand beaker, giving an electric current. A potential difference is set up between the two half-cells. This potential difference is at a maximum when no current is flowing, and it is then called the **e.m.f.** of the cell, or E_{cell}. The value of E_{cell} depends on a number of factors, but when the concentrations of zinc and copper ions are both 1.0 M and the temperature is 298 K, E_{cell} is said to have its **standard** value, indicated by $E_{cell}^{\ominus}$.

Set up a zinc/copper cell (often called a Daniell cell) as shown in figure 1. Clean the zinc and copper foils with emery before use. Use 1.0 M zinc sulphate and 1.0 M copper sulphate for the electrolytes, and a strip of filter paper soaked in saturated potassium nitrate solution for the salt bridge. Measure the e.m.f. of the cell using a high-resistance voltmeter. You should be able to predict the polarity of the electrodes before connecting the meter.

1 What is $E_{cell}^{\ominus}$ for the zinc/copper cell?

2 Bearing in mind the relative tendencies of zinc, iron and copper to lose electrons, would you expect a zinc/iron cell to have a greater or smaller $E_{cell}^{\ominus}$ than a zinc/copper cell?

EXPERIMENT 2 Constructing a zinc/iron cell

Set up a zinc/iron cell using a similar arrangement to the previous one, with a Fe/Fe^{2+} half-cell instead of the copper one. (Keep the copper half-cell for a later experiment.) Use a clean iron nail for the iron electrode and an acidified 1.0 M iron(II) sulphate solution as the electrolyte. Use a fresh salt bridge. You should be able to predict the polarity of the cell before measuring its e.m.f.

3 Write an equation for the reaction occurring in this cell.

4 What is $E_{cell}^{\ominus}$ for the zinc/iron cell?

5 Assuming that each half-cell in a cell makes a fixed, independent contribution to E_{cell}, use your answers to questions **1** and **4** to predict $E_{cell}^{\ominus}$ for an iron/copper cell.

EXPERIMENT 3 Constructing an iron/copper cell

Set up an iron/copper cell using the Fe/Fe^{2+} and Cu/Cu^{2+} half-cells employed in the previous two experiments. Use a fresh salt bridge. Measure E_{cell}.

6 How does your measured value for $E_{cell}^{\ominus}$ compare with your predicted value?

EXPERIMENT 4 Effect of concentration on E_{cell}

When current is drawn from a Daniell cell, the reactions occurring at each electrode are:

$$Zn(s) \longrightarrow Zn^{2+}(aq) + 2e^- \quad \text{and} \quad Cu^{2+}(aq) + 2e^- \longrightarrow Cu(s)$$

However, when the e.m.f. of the cell is being measured, no current is being drawn and each electrode is at equilibrium:

$$Zn(s) \rightleftharpoons Zn^{2+}(aq) + 2e^- \quad \text{and} \quad Cu^{2+}(aq) + 2e^- \rightleftharpoons Cu(s)$$

7 Use Le Chatelier's principle to predict the changes which occur if the ion concentration in each equilibrium is reduced.

8 What will be the effect of decreasing the concentration of copper ions on the tendency of the copper electrode to accept electrons from the zinc?

9 What will be the effect on E_{cell} of decreasing the concentration of copper ions?

Test your prediction in question **9** as follows. Set up a Daniell cell as in the first experiment, and measure E_{cell} with the following concentrations of $Cu^{2+}(aq)$: 0.1 M, 0.01 M, 0.001 M. Use a fresh salt bridge each time. At the lower concentrations, it is very important to pay careful attention to cleanliness in the copper half-cells, as small amounts of impurities will give rise to large errors.

10 Does E_{cell} increase or decrease when Cu^{2+} concentration decreases?

11 Is the relation between E_{cell} and Cu^{2+} concentration a linear one? If not, what form does it take?

12 How would E_{cell} change when Zn^{2+} concentration decreases?

13 The cell shown in figure 2 is an example of a **concentration cell**. Use your experimental results to predict its e.m.f. Check your answer experimentally if you have time.

Requirements

Each student, or pair of students, will require:

- 3 beakers (100 cm³)
- 2 wire leads fitted with crocodile clips
- Access to a high resistance voltmeter such as a valve voltmeter. The millivolt scale of a pH meter is suitable. It is best to have one meter between three or four pairs.
- Strips of filter paper long enough to connect two 100 cm³ beakers
- Strip of zinc foil about 6 cm × 1 cm
- Strip of copper foil about 6 cm × 1 cm
- Iron nail
- Emery paper
- 1.0 M copper sulphate solution (50 cm³). Dissolve 250 g hydrated copper(II) sulphate in distilled water and make up to 1 dm³.
- 0.1 M copper sulphate solution (50 cm³)*
- 0.01 M copper sulphate solution (50 cm³)*
- 0.001 M copper sulphate solution (50 cm³)*
- 1.0 M zinc sulphate solution (50 cm³). (Dissolve 288 g hydrated zinc sulphate in distilled water and make up to 1 dm³)
- 1.0 M iron(II) sulphate solution (acidified). Dissolve 278 g hydrated iron(II) sulphate in 200 cm³ 1.0 M sulphuric acid and make up to 1 dm³ with distilled water.

* Made by diluting the 1.0 M solution as appropriate.

Time required 1–2 double periods

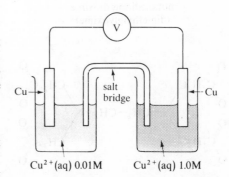

Figure 2 A concentration cell.

PRACTICAL 13
Complex formation and competition for cations

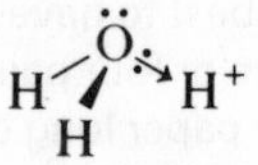

Figure 1 Co-ordinate bonding in an aqueous H^+ ion

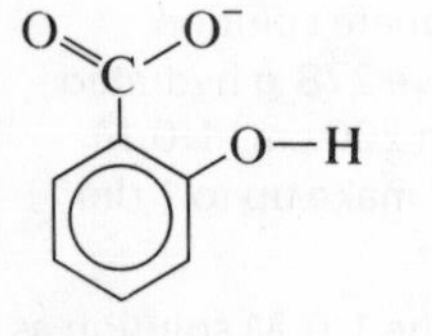

2-hydroxybenzoate ion (salicylate ion)

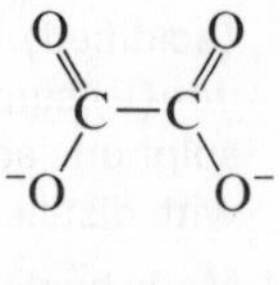

ethanedioate ion (oxalate ion)

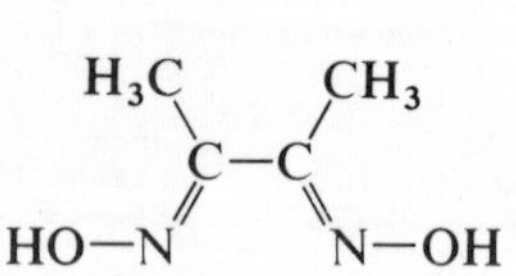

butanedione dioxime (dimethylglyoxime)

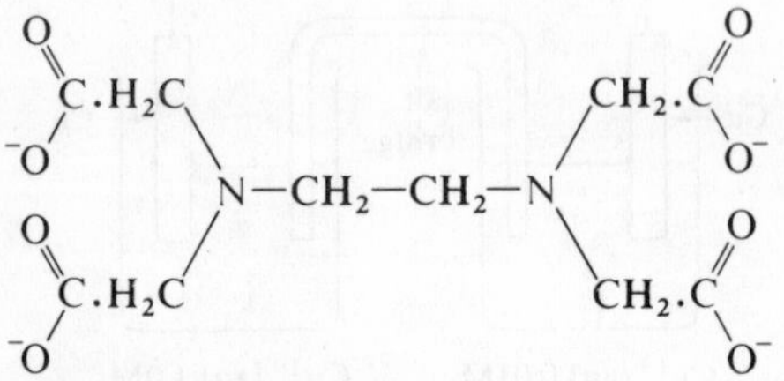

edta (ethylenediaminetetraacetate ion)

Figure 2 Some polydentate ligands

Introduction

In aqueous solution, H^+ ions are attached to polar water molecules by co-ordinate bonds forming H_3O^+ ions (figure 1). In the same way, other cations will also exist in aqueous solution as hydrated ions with formulae of the type, $[M(H_2O)_n]^{x+}$.

H^+ ions are very small and it is often assumed that each one associates with only one water molecule, but the larger size of other cations enables them to associate with two, four or even six water molecules. For example, Ag^+ ions exist in aqueous solution as $[Ag(H_2O)_2]^+$, Cu^{2+} ions exist as $[Cu(H_2O)_4]^{2+}$ and Fe^{3+} ions exist as $[Fe(H_2O)_6]^{3+}$.

Other polar molecules, besides water, can also co-ordinate with metal cations. Thus, in aqueous ammonia Cu^{2+} exists mainly as $[Cu(NH_3)_4]^{2+}$ and Ag^+ exists as $[Ag(NH_3)_2]^+$. Anions, such as Cl^-, OH^- and $S_2O_3^{2-}$, can also associate with cations in the same way as polar molecules like H_2O and NH_3. Thus, Ag^+ exists as $[Ag(S_2O_3)_2]^{3-}$ in sodium thiosulphate solution and Cu^{2+} ions form $[CuCl_4]^{2-}$ in concentrated hydrochloric acid.

Ions, such as $[Ag(S_2O_3)_2]^{3-}$ and $[Ag(NH_3)_2]^+$, in which a metal ion is associated with a number of anions or neutral molecules are known as **complex ions**. The anions and molecules co-ordinated to the central cation are called **ligands**. Each ligand contains at least one atom bearing a lone pair of electrons which can be donated to the central cation forming a co-ordinate (dative) bond.

Most ligands can form only one co-ordinate bond with a cation. These ligands, which include H_2O, NH_3 and Cl^-, are said to be unidentate because they have only 'one tooth' with which to attach themselves to the central cation in a complex. (The word 'dens' in Latin means tooth.) In some cases, however, a ligand can form two or more co-ordinate bonds to the central metal ion and such ligands are said to be polydentate, meaning 'many teeth' (figure 2).

Some ligands such as edta (ethylenediaminetetra-acetate) in $[Ag(edta)]^-$ can form as many as six co-ordinate bonds with the central ion. The complex ions formed between polydentate ligands and cations are known as **chelates** or **chelated complexes** from the Greek word 'chelos' meaning 'a crab's claw', because the polydentate ligands form a claw-like grip on the central metal ion.

The intention of this practical is to study the formation of complexes and to determine the relative strength with which different ligands form complex ions.

Procedure

EXPERIMENT 1   Complexes in which Fe^{3+} is the central ion

A   Put 10 drops of iron(III) chloride solution in a test tube.

1   What complex ion is present? What is its colour and what ligand does it contain?

B   Add potassium thiocyanate solution drop by drop until no further change occurs. (Keep half of the resulting solution for part **E**. Use the other half for part **C**.)

2   What is the colour of the solution now and what ligand is present in the complex ion?

C   If the solution from **B** is too deeply coloured, dilute it until the true colour of the complex is evident. Now, add sodium 2-hydroxybenzoate (salicylate) solution drop by drop until there is no further change in colour. Keep the solution for part **D**.

3   What is the colour of the solution now? What ligand is now present in the complex ion?

4 Which ligand, NCS^- or $HOC_6H_4COO^-$, competes more strongly for Fe^{3+} ions during complexing?

   D Take 10 drops of the solution from **C** and add a solution of edta drop by drop until there is no further colour change.

5 What colour is the edta – Fe(III) complex?

6 Write the ligands, NCS^-, $HOC_6H_4COO^-$, edta and water in order of increasing strength of co-ordination with Fe^{3+} ions.

7 Predict what would happen if edta solution were added to the solution obtained in **B**.

   E Add edta solution drop by drop to the solution retained in part **B** until there is no further colour change.

8 Was your prediction in question **7** correct?

EXPERIMENT 2   Investigating the relative strength of $C_2O_4^{2-}$ as a ligand with Fe^{3+}

Add ammonium ethanedioate (oxalate) solution drop by drop to 10 drops of iron(III) chloride solution until there is no further colour change.

9 What is the colour of the solution? What ligand is present in the complex ion?

Devise and perform test tube experiments to find the strength of the ethanedioate ion ($C_2O_4^{2-}$) as a ligand relative to NCS^-, $HOC_6H_4COO^-$, edta and H_2O.

10 Describe the experiments you performed and give the results.

11 Write the five ligands in order of increasing strength.

12 What structural features do the strongest co-ordinating ligands have in common?

EXPERIMENT 3   Complexes in which Ag^+ is the central ion

Some ligands react with cations to form precipitates of insoluble solids. Ag^+, for example, reacts with Cl^- to form insoluble AgCl. These precipitates can be regarded as neutral complexes. Being uncharged, they are less readily hydrated by polar water molecules than charged complexes and so they are less likely to dissolve in water.

Put ten drops of silver nitrate solution into each of six test tubes. To the first tube add ten drops of sodium chloride solution. To the second tube add ten drops of ammonia solution; to the third ten drops of potassium bromide; to the fourth sodium thiosulphate; to the fifth potassium iodide and to the sixth edta solution.

13 Make a table showing the ligand added to each tube and the formula, name, state and colour of each complex ion.

The relative strengths of the ligands used in the last experiment are:

$$edta > I^- > S_2O_3^{2-} > Br^- > NH_3 > Cl^- > H_2O$$

You are about to perform an experiment in which first a solution of sodium chloride, then one of ammonia, then one of potassium bromide, then sodium thiosulphate, then potassium iodide and finally edta are added in turn to a solution of Ag^+ ions.

14 Use the relative strengths of the ligands to predict what will happen.

Add sodium chloride solution drop by drop to ten drops of silver nitrate solution until no further change occurs. To the resulting solution add ammonia solution dropwise until no further change occurs. Now add potassium bromide dropwise until no further change occurs. Carry on in this fashion with first sodium thiosulphate, then potassium iodide and finally edta solution.

15 Were your predictions in question **14** correct?

Requirements

Each student or pair of students will need:
- 6 test tubes
- 2 teat pipettes
- Access to approximately 0·1 M solutions of the following:
- Iron(III) chloride (16·2 g $FeCl_3$ in 1 dm^3 solution)
- Potassium thiocyanate (9·7 g KSCN in 1 dm^3 solution)
- Sodium 2-hydroxybenzoate (salicylate) (16·0 g HOC_6H_4COONa in 1 dm^3 solution)
- Edta (disodium salt) (33·6 g $[CH_2N(CH_2COOH) . CH_2COONa]_2$ in 1 dm^3 solution)
- Ammonium ethanedioate (oxalate) (14·2 g $(COONH_4)_2 . H_2O$ in 1 dm^3 solution)
- Silver nitrate
- Sodium chloride
- Ammonia
- Potassium bromide
- Potassium iodide
- Sodium thiosulphate

Time required 1 double period

PRACTICAL 14
Determination of the formulae of complex ions

Requirements

For the colorimetric investigation each pair of students will need:

- 0·05 M CuSO$_4$ (60 cm^3) (12·5 g of CuSO$_4$. 5H$_2$O in 1 dm^3 of solution)
- 0·05 M 1,2-diaminoethane (60 cm^3) (3·0 g of H$_2$NCH$_2$CH$_2$NH$_2$ in 1 dm^3 of solution)

CARE The vapour from 1,2-diaminoethane is an irritant to the eyes and skin.

- 9 test tubes
- 2 × 10 cm^3 graduated pipettes or 2 burettes
- Colorimeter

(The suggested concentrations may require modification for the particular colorimeters used.)

For the complexometric investigation each pair of students will need:

- 0·1 M edta (disodium salt) – 80 cm^3 (33·6 g [CH$_2$N(CH$_2$COOH) . CH$_2$COONa]$_2$ in 1 dm^3 of solution)
- 0·1 M Ni^{2+} salt – 80 cm^3 (39·5 g NiSO$_4$. (NH$_4$)$_2$SO$_4$. 6H$_2$O in 1 dm^3 of solution)
- Burette
- Pipette (20 or 25 cm^3)
- Beaker (100 cm^3)
- 2 test tubes
- Murexide (shake 0·5 g of powdered murexide with 100 cm^3 of water. Allow the undissolved solid to settle and use the saturated supernatant. Prepare fresh solution daily.)
- Conical titration flask (250 cm^3)

Time required 1–2 double periods

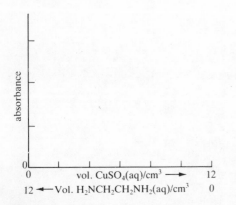

Figure 1

Introduction

Essentially, the determination of the formulae of a complex ion involves measurement of the number of ligands complexing with one metal ion. This can be investigated by several methods, the most important of which are:

a colorimetric methods requiring measurement of the colour intensity of the mixture as the proportion of metal ion to ligand is varied,

b titration methods involving competitive complexing.

The object of this experiment is first to investigate the stoichiometry of the copper(II)/1,2-diaminoethane complex using colorimetry, and then to investigate the stoichiometry of the nickel(II)/edta complex using a complexometric titration.

EXPERIMENT 1 Determination of the formula of the copper(II)/1,2-diaminoethane|complex by colorimetry

The stoichiometry of a complex ion can be investigated by colorimetry provided there is a significant difference in colour between the simple aqueous ions and the complex. Basic ideas regarding the principles and practice of colorimetry are discussed in the introduction to practical 18. When aqueous Cu^{2+} ions react with 1,2-diaminoethane (H$_2$NCH$_2$CH$_2$NH$_2$), a deeply-coloured complex ion is produced. We can write an equation for this reaction as

$$Cu^{2+}(aq) + xH_2NCH_2CH_2NH_2(aq) \longrightarrow [Cu(H_2NCH_2CH_2NH_2)_x]^{2+}(aq)$$

In order to determine the formula of the complex, we must find the value of x. This can be done using the method of continuous variation described below.

Procedure

Take nine test tubes and number them 1 to 9. Using these test tubes, make up mixtures of 0·05 M CuSO$_4$ and 0·05 M 1,2-diaminoethane with the compositions shown below. Shake each tube to ensure the solutions are thoroughly mixed. Notice that each tube contains a total volume of 12 cm^3.

Test tube number	1	2	3	4	5	6	7	8	9
Vol. of 0·05 M CuSO$_4$/cm^3	0·0	2·0	3·0	4·0	6·0	8·0	9·0	10·0	12·0
Vol. of 0·05 M 1,2-diaminoethane/cm^3	12·0	10·0	9·0	8·0	6·0	4·0	3·0	2·0	0·0

1 What is the colour of the copper(II)/1,2-diaminoethane complex ion?

2 What is the colour of the filter which would be most suitable for the experiment?

Follow the usual procedure to obtain zero absorbance (100% transmission) when tube 1, containing only 1,2-diaminoethane solution is placed in the colorimeter. Now measure the absorbance of the other mixtures relative to this zero reading.

3 Plot a graph of absorbance (vertical) against the volume of 0·05 M CuSO$_4$ reading from left to right, and, on the same horizontal scale, the volume of 0·05 M 1,2-diaminoethane solution reading from right to left (figure 1).

4 What are the volumes of the two solutions which, on mixing, give the maximum colour intensity?

5 In what molar proportions do Cu^{2+} and H$_2$NCH$_2$CH$_2$NH$_2$ react to form the complex?

40

6   Write an equation for the formation of the complex.

7   Draw the structural formula you would predict for the complex ion and relate this structure to that of $[Cu(NH_3)_4]^{2+}$.

EXPERIMENT 2   Determination of the formula of the nickel(II)/edta complex by competitive complexing

Add 5 drops of murexide solution (figure 2) to 5 cm^3 of an 0·1 M solution of a nickel(II) salt. Add edta solution to the mixture slowly until no further changes occur.

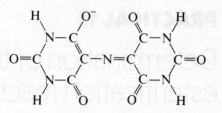

Figure 2 Murexide is ammonium purpurate. The anion ligand in murexide has the structure shown above.

8   What are the colours of Ni^{2+} ions complexed with **a** water, **b** murexide?

9   Explain the reactions that have occurred in terms of competitive complexing. (Remember that the final colour of the solution is due to the nickel(II)/edta complex and the non-complexed murexide.)

The colour of the nickel(II)/murexide remains as long as there are more than enough Ni^{2+} ions for the edta, because these excess Ni^{2+} ions will form a complex with murexide. As the nickel(II)/murexide complex is removed, the colour of the solution changes and the nickel(II)/edta complex forms. When all the nickel has complexed with edta, the non-complexed murexide appears in the solution.

Consequently, murexide can act as an indicator in nickel(II)/edta titrations.

10   Devise a titration to determine the volume of 0·1 M edta which will just react with 20 or 25 cm^3 of 0·1 M Ni^{2+} salt.

Discuss your suggested method with your teacher before performing a titration. (You will need to add the same proportion of murexide to nickel(II) solution as in the test tube reaction above. Mix the solutions very thoroughly during the titration. Near the end point, add the edta solution 0·5 cm^3 at a time. Titrate the solution to the same colour that you obtained in the test tube reaction.)

11   How many moles of Ni^{2+} were taken?

12   How many moles of edta reacted with this Ni^{2+}?

13   How many moles of edta react with one mole of Ni^{2+}?

14   What is the formula of the Ni^{2+}/edta complex?

15   Draw a diagram to represent the structure of the Ni^{2+}/edta complex.

Determination of the equilibrium constant for an esterification reaction

Requirements

Each student, or pair of students, will require:

- Reagent bottle with tight-fitting stopper
- 1·0 M sodium hydroxide (make 40 g of sodium hydroxide pellets up to 1 dm³ with distilled water)
- Burettes containing each of the following:
  Glacial ethanoic (acetic) acid
  Ethanol
  Ethyl ethanoate (ethyl acetate)
  3 M hydrochloric acid (make 258 cm³ of concentrated acid up to 1 dm³ with distilled water)
  Distilled water

(It is sufficient to have **one** burette containing each reagent.)

Time required Preliminary work – $\frac{1}{2}$ double period
Analysis (one week later) – 1 double period

Introduction

The practical determination of the equilibrium constant K for a reaction presents some difficulties. To find K for a reaction it is necessary to measure the concentration of reactants and products at equilibrium. Most methods for doing this involve removal of reactant or products, which changes their concentration and shifts the equilibrium. It is therefore ideal to use a method of measuring concentration that does not involve removing any component – colorimetry, for example.

The reaction under consideration here is the esterification of ethanoic acid (acetic acid) by ethanol:

$$CH_3COOH + CH_3CH_2OH \rightleftharpoons CH_3COOCH_2CH_3 + H_2O$$

ethanoic   ethanol   ethyl ethanoate
acid

The reaction is catalysed by H^+ ions. The advantage of studying this particular reaction is that it is kinetically slow, so it is possible to measure the concentration of a reactant by titration (which removes the reactant), without significantly disturbing the equilibrium in the relatively short time it takes to carry out the titration.

Known amounts of acid, alcohol and/or ester are mixed with dilute hydrochloric acid in a stoppered bottle and allowed to reach equilibrium at room temperature. After reaching equilibrium, which takes several days, the mixture is titrated with alkali in order to find the total amount of acid present. After allowing for the presence of the hydrochloric acid catalyst, the amount of ethanoic acid present at equilibrium can be calculated. From this the amounts of the other reagents can be calculated, and hence a value for the equilibrium constant can be found.

Procedure

Preliminary

Burettes are set up in the laboratory containing the following:

- Ethyl ethanoate (ethyl acetate)
- Concentrated ('glacial') ethanoic acid (**CARE**)
- Approximately 3 M HCl
- Ethanol
- Distilled water.

You will be asked to make up one of the mixtures in the table below. Run the liquids into the reagent bottle from the burette, and stopper immediately to prevent evaporation. Shake well, then allow the bottle to stand at room temperature for a week, to allow the mixture to reach equilibrium.
Note that the total volume in each experiment is 10 cm³.

After one week

Titrate the whole of the mixture in the bottle with 1·0 M sodium hydroxide, using phenolphthalein as indicator.

In order to find the exact concentration of the HCl catalyst, titrate 5 cm³ of the approximately 3 M HCl with 1·0 M NaOH.

Finally, weigh 5 cm³ of each of the liquids used to make up your mixture (including the HCl).

Experiment number	HCl/cm³	water/cm³	ethyl ethanoate/cm³	glacial ethanoic acid/cm³	ethanol/cm³
1	5	0	5	0	0
2	5	1	4	0	0
3	5	2	3	0	0
4	5	3	2	0	0
5	5	0	4	0	1
6	5	0	4	1	0
7	5	0	0	1	4
8	5	0	0	2	3

1   How many moles of ethyl ethanoate were there in your **original** mixture (i.e. before any reaction had occurred)?

2   How many moles of ethanoic acid were there in your original mixture?

3   How many moles of ethanol were there in your original mixture?

4   How many moles of water were there in your original mixture? (Remember to take into account the water present in the dilute HCl. You can find the weight of this by deducting the weight of pure HCl, found by titration, from the weight of 5 cm³ of 3 M HCl.)

5   How many moles of ethanoic acid were there in the mixture **at equilibrium**? (This can be found from the volume of 1·0 M NaOH needed to titrate the mixture, after subtracting the volume of NaOH needed to titrate 5 cm³ of approximately 3 M HCl.)

6   How many moles of ethanol were there in the mixture **at equilibrium**? (This can be found from your answer to question **5**, bearing in mind the number of moles of ethanol present in the initial mixture, and using the equation $CH_3COOH + CH_3CH_2OH \rightleftharpoons CH_3COOCH_2CH_3 + H_2O$.)

7   How many moles of ethyl ethanoate were there in the mixture at equilibrium?

8   How many moles of water were there in the mixture at equilibrium?

9   Work out the **concentrations** of water, ethyl ethanoate, ethanol and ethanoic acid in the mixture at equilibrium, using your answers to questions **5–8** and the fact that the total volume of the reaction mixture was 10 cm³.

10   What is the equilibrium constant for this reaction?

11   Compare your result with those obtained by other members of the group who started with different mixtures. Does the value of the equilibrium constant depend on the starting concentrations of the reactants?

12   ΔH for this reaction is $+ 17·5$ kJ mole⁻¹. Would you expect the value of K_c to be greater or smaller at 80°C than at room temperature?

13   The yield of ester formed in the reaction between ethanoic acid and ethanol is considerably increased if concentrated sulphuric acid is added to the reactants. Suggest a reason for this.

Determination of the dissociation constants of weak acids

Requirements

Each pair of students will require:

- 20 test tubes, 125 × 16 mm
- Test tube racks to hold the test tubes in nine pairs, one behind the other. (Alternatively, small sample bottles can be used. These have the advantage that they do not need a rack to stand in.)
- Teat pipette
- Measuring cylinder (10 cm³ or 25 cm³)
- Burette

Access to:

- Bromocresol green solution (0·1 g dissolved in 20 cm³ of ethanol and then diluted to 100 cm³ with water). (Bromocresol green is sometimes called bromocresol blue. Its pH range is 3·8 to 5·4.)
- Conc. hydrochloric acid
- 4 M sodium hydroxide
- pH meter

Communal burettes containing:

- 0·02 M ethanoic acid (1·20 g CH_3COOH in 1 dm³ of solution)
- 0·02 M sodium ethanoate (1·64 g CH_3COONa in 1 dm³ of solution)
- 0·02 M benzoic acid (2·44 g C_6H_5COOH in 1 dm₃ of solution). (Make up the benzoic acid solution in warm water and then allow to cool.)
- 0·02 M sodium benzoate (2·88 g C_6H_5COONa in 1 dm³ of solution)

Time required 1–2 double periods

Introduction

Although the pH of a solution provides some indication of the strength of the constituent acid or base, its value changes with concentration. Dissociation constants provide a much more reliable guide to the relative strengths of acids and bases because their values remain unaffected by concentration changes.

When the weak acid, HA, is dissolved in water the following equilibrium exists

$$HA(aq) \rightleftharpoons H^+(aq) + A^-(aq)$$

The dissociation constant of the acid, K_a, is given by

$$K_a = \frac{[H^+(aq)][A^-(aq)]}{[HA(aq)]}$$

The intention of this experiment is to determine an approximate value for the dissociation constant of a weak acid using an indicator, and then to obtain a more accurate value using a pH meter.

EXPERIMENT 1   Determination of the dissociation constant of ethanoic acid using the indicator bromocresol green

Principle

Many acid-base indicators are weak acids, and in aqueous solution the indicator, HIn, exists in equilibrium as

$$HIn(aq) \rightleftharpoons H^+(aq) + In^-(aq)$$

We can therefore write an expression for the dissociation constant of the indicator, $K_a(HIn)$ as

$$K_a(HIn) = \frac{[H^+(aq)][In^-(aq)]}{[HIn(aq)]} \qquad \ldots (1)$$

($K_a(HIn)$ for bromocresol green = 2×10^{-5} mole dm⁻³.)

Using an indicator with a known dissociation constant, it is possible to determine $[H^+(aq)]$ for a buffer solution of ethanoic acid and sodium ethanoate containing equimolar concentrations of CH_3COOH and CH_3COO^-.

Now
$$K_a = \frac{[H^+(aq)][CH_3COO^-(aq)]}{[CH_3COOH(aq)]}$$

but $[CH_3COO^-(aq)] = [CH_3COOH]$ in this case,

$$\Rightarrow K_a = [H^+(aq)]$$

Thus, for a solution containing equimolar concentrations of ethanoic acid and ethanoate ion, the hydrogen ion concentration is equal to K_a. K_a can therefore be obtained by using an indicator to measure the H⁺ ion concentration of such a solution.

Procedure

First, make up solutions A and B.

Solution A

1 drop of conc. HCl in 5 cm³ of bromocresol green solution. In solution A, virtually all the indicator is in the undissociated form as HIn.

Solution B

1 drop of 4 M NaOH in 5 cm³ of bromocresol green solution. In solution B, virtually all the indicator is in the dissociated form as In^-.

Arrange 18 test tubes in two parallel rows of nine tubes in a double test tube rack. It must be possible to look through each of the 9 pairs of tubes to see the combined colour from the solutions in both tubes.

Number the tubes as in the table below and use a burette to deliver 10 cm³ of water into each tube. Using the same pipette each time, add the required drops of A or B and mix thoroughly.

Tube No.	1	2	3	4	5	6	7	8	9
Vol. of water/cm³	10	10	10	10	10	10	10	10	10
Drops of solution A	1	2	3	4	5	6	7	8	9
Tube No.	10	11	12	13	14	15	16	17	18
Vol. of water/cm³	10	10	10	10	10	10	10	10	10
Drops of solution B	9	8	7	6	5	4	3	2	1

1 Why is it important to use the same pipette in adding drops to the eighteen tubes?

Measuring the K_a for ethanoic acid

Mix 5 cm³ of 0·02 M ethanoic acid with 5 cm³ of 0·02 M sodium ethanoate. Add 10 drops of bromocresol green solution and mix thoroughly. Compare the colour of this mixture with the colours seen through corresponding pairs of test tubes in the rack (i.e. tubes 1 and 10, 2 and 11, 3 and 12, etc.).

2 Which pair of test tubes matches the colour in the equimolar CH_3COOH/CH_3COONa mixture most closely?

3 What is the ratio of $\dfrac{[HIn]}{[In^-]}$ in the CH_3COOH/CH_3COONa mixture?

4 Use equation (1) to calculate the $[H^+(aq)]$ in the CH_3COOH/CH_3COONa mixture.

5 What is the value of K_a for ethanoic acid?

Now repeat the experiment using benzoic acid and sodium benzoate in place of ethanoic acid and sodium ethanoate.

EXPERIMENT 2   Determination of the dissociation constant of ethanoic acid using a pH meter

Principle

For an equimolar mixture of $CH_3COOH(aq)$ and $CH_3COONa(aq)$,

$$K_a = [H^+(aq)]$$

Thus, by measuring the pH of the equimolar CH_3COOH/CH_3COONa mixture, it is possible to calculate $[H^+(aq)]$ and hence K_a.

Procedure

Add 10 cm³ of 0·02 M ethanoic acid to 10 cm³ of 0·02 M sodium ethanoate. Mix thoroughly and measure the pH of the solution using a pH meter. (If necessary, calibrate the pH meter using a buffer solution of known pH.) Repeat the experiment using benzoic acid and sodium benzoate in place of ethanoic acid and sodium ethanoate.

6 What is the value of K_a for a ethanoic acid, b benzoic acid?

7 How do the values you obtained for K_a by this method compare with those obtained by the indicator method? Try to explain any discrepancies.

8 How do the values you obtained for K_a compare with the literature values?

9 What suggestions can you offer to explain the relative values of K_a for ethanoic acid and benzoic acid?

10 What advantages does the pH method of determining K_a have over the indicator method?

PRACTICAL 17

Determination of the order of a reaction

Requirements

Each pair of students will need:
- Burette and stand
- Beaker ($1\,dm^3$)
- Measuring cylinder ($500\,cm^3$)
- Measuring cylinder (25 or $50\,cm^3$)
- Measuring cylinder ($10\,cm^3$)
- Stirring rod
- Pipette ($10\,cm^3$) and safety filler
- Clock or watch with seconds hand
- Graph paper
- Small funnel
- 2 small beakers
- Approximately $0\cdot1$ M H_2O_2 ($30\,cm^3$)
- $0\cdot05$ M $Na_2S_2O_3$ ($100\,cm^3$)
- $1\cdot0$ M H_2SO_4 ($75\,cm^3$)
- $1\cdot0$ M KI ($45\,cm^3$)
- Starch solution ($2\,cm^3$)

Time required 1 double period

Introduction

Consider the reaction

$$xA + yB \longrightarrow \text{products.}$$

Experiments show that the reaction rate can be related to the concentrations of individual reactants by an equation of the form,

$$\text{Rate} = k[A]^m[B]^n$$

This expression is known as the **rate equation** or **rate law** for the reaction under consideration. m and n are constants whose values are usually 0, 1 or 2, and k is the **rate constant** for the reaction.

The index m is known as **the order of the reaction with respect to** A, and n is the order of the reaction with respect to B. The **overall order** of the reaction is $(m + n)$.

The reaction we shall be studying involves hydrogen peroxide and iodide ions in acid solution.

$$H_2O_2(aq) + 2I^-(aq) + 2H^+(aq) \longrightarrow 2H_2O(l) + I_2(aq)$$

The rate equation for this reaction can be written as

$$\text{Rate} = k[H_2O_2]^\alpha[I^-]^\beta[H^+]^\gamma$$

The order of the reaction with respect to one reactant may be investigated by having the other reactants present in large excess so that their concentrations remain effectively constant during the reaction. For example, in this experiment, the order of the reaction with respect to H_2O_2 is investigated by using very much larger concentrations of I^- and H^+.

Under these circumstances the reaction rate can be expressed as:

$$\text{Rate} = k'[H_2O_2]^\alpha,$$

where k' is a modified rate constant which includes the constant concentrations of I^- and H^+.

Principle

The object of this experiment is to determine the order with respect to H_2O_2 for the reaction discussed above.

Hydrogen peroxide is allowed to react with an acidified solution of potassium iodide in the presence of small amounts of starch and sodium thiosulphate:

$$H_2O_2(aq) + 2I^-(aq) + 2H^+(aq) \longrightarrow 2H_2O(l) + I_2(aq)$$

As soon as the iodine is produced, it reacts with thiosulphate and is converted back to iodide ions:

$$I_2(aq) + 2S_2O_3^{2-}(aq) \longrightarrow 2I^-(aq) + S_4O_6^{2-}(aq)$$

When the iodine produced in the reaction is in excess of the sodium thiosulphate, a blue colour suddenly appears as the iodine reacts with starch.

From the amount of $S_2O_3^{2-}$ added, we can calculate the amount of I_2 produced by the time the blue colour appears and hence the amount of H_2O_2 used up.

Knowing the original amount of H_2O_2, we can calculate $[H_2O_2]$ at the time the blue colour appears. The values of $[H_2O_2]$ at different times can then be used to find the order of reaction with respect to H_2O_2.

Procedure

Work in pairs.

Fill a burette with 0·05 M $Na_2S_2O_3$(aq).

Put 450 cm³ of distilled water in a 1 dm³ beaker and add 25 cm³ of 1·0 M H_2SO_4, 1 cm³ of starch solution and 15 cm³ of 1·0 M KI. Finally, add 2 cm³ of 0·05 M $S_2O_3^{2-}$(aq) from the burette and start the reaction by rapidly adding 10 cm³ of approximately 0·1 M H_2O_2 from a pipette, stirring during the addition. (Start timing when half of the H_2O_2 has been added.)

CARE Use a safety filler with the pipette.

When the I_2 produced in the reaction is in excess of the added $S_2O_3^{2-}$, a blue colour will suddenly appear. Quickly, note the time, add a further 2 cm³ of $S_2O_3^{2-}$(aq) and the blue colour will disappear.

Note the time again when the blue colour reappears and add another 2 cm³ of $S_2O_3^{2-}$(aq). Continue in this manner until 8–10 readings have been obtained.

Finally, determine the exact concentration of the H_2O_2 as follows. Pipette 10 cm³ of approximately 0·1 M H_2O_2 into a conical flask, **using a safety filler**, add 25 cm³ of 1·0 M H_2SO_4 and 15 cm³ 1·0 M KI and titrate this mixture against the 0·05 M $S_2O_3^{2-}$(aq) using starch as indicator.

1   Why must all other reactants (e.g. KI and H_2SO_4) be in large excess compared to the reactant for which the order is being determined?

2   Why does the concentration of I^- remain constant throughout the whole of the rate study experiment? (**Hint:** Look closely at the equations for the reactions involved.)

3   Why does the time interval between appearances of the blue colour become longer?

4   Why is it important to stir during the addition of the H_2O_2?

5   If the H_2O_2 solution is exactly 0·1 M and the $S_2O_3^{2-}$ solution is exactly 0·05 M then:
    10 cm³ of 0·1 M H_2O_2 will just react with 40 cm³ of 0·05 M $S_2O_3^{2-}$ during titration. Explain why this is so.

6   From the amount of $S_2O_3^{2-}$ added, we can calculate the amount of I_2 produced by the time the blue colour appears and hence the amount of H_2O_2 used up. Knowing the original amount of H_2O_2, we can now calculate the concentration of H_2O_2 at different times.

The concentration of H_2O_2 is most conveniently computed in terms of the volume of 0·05 M $Na_2S_2O_3$ added.

For example, suppose the H_2O_2 was exactly 0·1 M. Its initial concentration would then be equivalent to 40 cm³ of 0·05 M $S_2O_3^{2-}$. Suppose the blue colour first appeared after x secs, and appeared a second time after y secs. We can then say:

At $t = 0$ secs, 40 cm³ 0·05 M $S_2O_3^{2-}$(aq) $\propto [H_2O_2]$;
at $t = x$ secs, when blue colour first appears,

$$(40 - 2) = 38 \text{ cm}^3 \text{ } 0.05 \text{ M } S_2O_3^{2-}\text{(aq)} \propto [H_2O_2];$$

at $t = y$ secs, when blue colour appears a second time,

$$(40 - 4) = 36 \text{ cm}^3 \text{ } 0.05 \text{ M } S_2O_3^{2-}\text{(aq)} \propto [H_2O_2]; \text{ etc.}$$

$\Rightarrow$ At time, $t = 0$ secs, $[H_2O_2] \propto 40$;
at time, $t = x$ secs, $[H_2O_2] \propto 38$;
at time, $t = y$ secs, $[H_2O_2] \propto 36$; etc.

Record your results in a table similar to the one below. Remember to use your own value for the volume of $S_2O_3^{2-}$ needed to titrate 10 cm^3 of the approximately 0·1 M H_2O_2. This is unlikely to be exactly 40 cm^3.

vol. of 0·05 M $S_2O_3^{2-}$ added /cm^3	vol. of 0·05 M $S_2O_3^{2-} \propto$ [H_2O_2] /cm^3	Time at which blue colour appears /seconds
0	40	0
2	38	x
4	36	y

7   If the order of the reaction with respect to reagent X is zero, the corresponding rate equation is

$$\text{Rate} = k[X]^0$$

The **integrated** rate equation for this zero order reaction is

$$[X] = -kt + [X_0] \qquad \qquad \dots (1)$$

where $[X]$ is the concentration of substance X at time, t; $[X_0]$ is the original concentration of X at time 0 and k is the rate constant for the reaction.
Write similar integrated rate equations relating the concentration of reactant (say $[X]$) and time (t) for first and second order reactions.

8   Investigate whether the reaction is zero order, first order or second order with respect to H_2O_2, by plotting your results for this experiment on suitable graphs. (For example, in order to test whether the reaction is zero order, plot [H_2O_2] against t. (See equation (1).) If this gives a straight line, it suggests the reaction is zero order.

A colorimetric investigation of the rate of reaction between bromine and methanoic acid

Introduction

In this practical, the technique of colorimetry is introduced and then used to investigate the reaction between bromine and methanoic acid. Colorimetry can be employed to determine the concentration of any coloured substance. In comparison to gravimetric or titrimetric analysis, colorimetry has the advantage that it is relatively quick and easy to perform and does not require portions of a mixture to be removed for analysis. It is particularly useful in the study of reaction rates, where the concentration of a coloured species is constantly changing.

In a colorimeter, a narrow beam of light passes through the solution under test towards a sensitive photocell (figure 1). In most colorimeters, it is possible to select the most appropriate colour and wavelength of light by choosing a particular filter or by adjusting a diffraction grating. The current generated in the photocell is, of course, dependent on the amount of light transmitted by the solution which in turn depends upon the concentration of coloured solution under test. Thus, the current generated in the photocell will be greatest when the light transmitted by the coloured solution is greatest, i.e. when the coloured solution is least concentrated. Normally, however, the meter is not calibrated to show the light transmitted but the **absorbance**, because this is proportional to the concentration of the coloured solution.

i.e.   Absorbance (optical density) $\propto$ concentration of coloured species.

In the experiment which follows, the coloured species is bromine, so absorbance $\propto [Br_2]$.

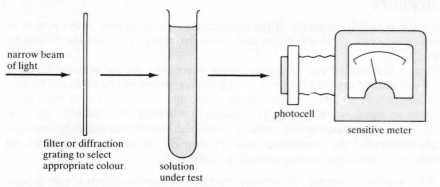

narrow beam of light

filter or diffraction grating to select appropriate colour

solution under test

photocell

sensitive meter

Figure 1 A simplified diagram of a colorimeter.

Principle

The reaction under investigation is

$$Br_2(aq) + HCOOH(aq) \longrightarrow 2Br^-(aq) + 2H^+(aq) + CO_2(g).$$

By using a concentration of methanoic acid which is ten times larger than that of bromine, it is possible to assume that the concentration of methanoic acid is constant throughout the experiment and to see how the concentration of bromine affects the reaction rate. If the reaction is zero order,

$$\frac{-d[Br_2]}{dt} = k_0,$$

where $-d[Br_2]/dt$ represents the reaction rate in terms of the rate of change of concentration of bromine and k_0 is the rate constant for a zero order reaction. Integrating this equation gives:

$$-[Br_2] = k_0 t + \text{constant} \qquad \qquad \ldots (1)$$

Requirements
Each pair of students will need:
- 0·1 M bromine, 16 g (5 cm³) Br_2 in 1 dm³ of solution (25 cm³)
- 1·0 M methanoic acid, 46g (37·7 cm³) HCOOH in 1 dm³ of solution (25 cm³)
  CARE Both methanoic acid and bromine are irritant, fuming liquids capable of causing serious burns. Their solutions should be prepared in a fume cupboard, and polythene gloves should be worn.
  (The concentrations of these two solutions may need modification with certain colorimeters.)
- Colorimeter with blue, green or turquoise filter
- Distilled water (10 cm³)
- 2 measuring cylinders (25 or 50 cm³)
- 1 boiling tube
- 2 matched thin-walled (soda) test tubes for use with the colorimeter
- Stop clock or watch reading in seconds

Time required 1 double period

Thus a graph of $[Br_2]$ (or absorbance) against t (time) should give a straight line.

If the reaction is first order with respect to bromine,

$$\frac{-d[Br_2]}{dt} = k_1[Br_2]$$

where k_1 is the rate constant for a first order reaction. Integration of this equation gives

$$-\ln[Br_2] = k_1 t + \text{constant}$$

or

$$-\lg[Br_2] = \frac{k_1 t}{2 \cdot 3} + \text{constant} \qquad \ldots (2)$$

Thus, a graph of $\lg[Br_2]$ (or lg absorbance) against t (time) should give a straight line.

If the reaction is second order with respect to bromine,

$$\frac{-d[Br_2]}{dt} = k_2[Br_2]^2$$

where k_2 is the rate constant for a second order reaction. Integration of this equation gives:

$$\frac{1}{[Br_2]} = k_2 t + \text{constant} \qquad \ldots (3)$$

Thus a graph of $1/[Br_2]$ (or the reciprocal of the absorbance) against t (time) should give a straight line.

By plotting three graphs, one for absorbance against time, one for lg(absorbance) against time and a third for the reciprocal of absorbance against time, we can determine the order of reaction with respect to bromine.

Procedure

Switch the sensitivity controls of the colorimeter to a minimum, insert a blue or green filter, switch on at the mains and allow five minutes for the instrument to warm up.

Insert a thin-walled test tube containing water into the colorimeter, cover it with a cap and adjust the sensitivity (using the higher sensitivity controls) to give zero absorbance (i.e. 100% transmission). Now, mix 10 cm^3 of 1·0 M methanoic acid with an equal volume of 0·1 M bromine in a boiling tube and start the stop watch. Shake the mixture gently, transfer some of it to a colorimeter tube and read the absorbance on the colorimeter after 15 seconds, 30 seconds, 45 seconds, 60 seconds, etc. Nine or ten readings should be sufficient.

1 What are the advantages of colorimetry over titrimetric analysis in rate experiments such as this?

2 Why is it necessary to use carefully matched tubes in the colorimeter?

3 Why is a blue or green filter chosen in this experiment? What general principles should be used in choosing a filter for a colorimetric investigation?

4 Tabulate your results. The columns of your table should show time, absorbance, lg(absorbance) and reciprocal of absorbance.

5 Using the same axes (but with different vertical scales), plot graphs of
   a absorbance against time.
   b lg(absorbance) against time.
   c reciprocal of absorbance against time.

6 What is the order of reaction with respect to bromine?

7 Why is the graph not absolutely straight even when there are no experimental errors? (**Hint:** What assumption has been made which is not wholly true?)

8 Using the appropriate rate equation, obtain a value for the rate constant for the reaction and state its units.

PRACTICAL 19

Determination of the activation energy for the reaction between bromide and bromate(V) ions

Introduction

During a reaction, bonds are first broken, and others are then formed. Energy is required to break certain bonds and start this process, whether the overall reaction is exothermic or endothermic. Particles will not always react when they collide because they may not possess sufficient energy for the appropriate bonds to break.

A reaction will only occur if the colliding particles possess more than a certain minimum amount of energy known as the activation energy, E_A.

During the 1930's, Zartmann, Ko and others showed that the distribution of kinetic energies amongst the particles of a liquid or gas was similar to that shown in figure 1.

Essentially, the curve is a histogram showing the number of particles within each small range of kinetic energy, and the area beneath the curve is proportional to the total number of particles in that energy range.

Consequently, the number of particles capable of reacting, with energy greater than the activation energy, E_A, is proportional to the shaded area beneath the curve at energies above E_A. Hence the **fraction** of particles with energy greater than E_A is given by the ratio:

$$\frac{\text{(shaded area beneath curve)}}{\text{(total area beneath curve)}}$$

Using the kinetic theory and probability theory, Maxwell and Boltzmann showed that the fraction of molecules with energy greater than E_A J mole^{-1} was given by $e^{-E_A/RT}$ where R is the gas constant, T is the absolute temperature and e is the exponential function.

This suggests that at a given temperature:

$$\text{Reaction rate} \propto e^{-E_A/RT}$$

But as k, the rate constant for a reaction, is a measure of the reaction rate, we can write

$$k \propto e^{-E_A/RT}$$
$$\Rightarrow k = Ae^{-E_A/RT}$$

This last expression is called the **Arrhenius equation**, because it was first predicted by the Swedish chemist, Svante Arrhenius in 1889.

Principle

The object of this experiment is to obtain E_A for the reaction between Br^- and BrO_3^- in acid solution.

$$5Br^-(aq) + BrO_3^-(aq) + 6H^+(aq) \longrightarrow 3Br_2(aq) + 3H_2O(l)$$

The time, t, required for the reaction to proceed to a given extent at different temperatures is found by adding a fixed amount of phenol and a small amount of methyl red indicator to the reaction mixture.

The bromine produced during the reaction reacts very rapidly with phenol (forming tribromophenol). Once all the phenol is consumed, any further bromine bleaches the indicator immediately.

Now, as $k \propto$ reaction rate $= \dfrac{\text{concentration change}}{\text{time}}$

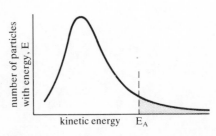

Figure 1

Requirements

Each pair of students will need:
- 0·01 M phenol, 0·94 g phenol in 1 dm^3 of solution (70 cm^3)
  CARE Avoid skin contact with phenol; it is preferable to wear polythene gloves.
- Br^-/BrO_3^- solution (0·1 M w.r.t. potassium bromide, KBr, and 0·02 M w.r.t. potassium bromate(V), KBrO$_3$) 11·90 g KBr and 3·34 g KBrO$_3$ in 1 dm^3 of solution (70 cm^3)
- 0·5 M H$_2$SO$_4$ (35 cm^3)
- Methyl red indicator (28 drops)
- Large beaker (1 dm^3 or 500 cm^3)
- Thermometer (0–110°C)
- 2 boiling tubes
- Measuring cylinder (10 cm^3 or 20 cm^3)
- Bunsen burner
- Tripod and gauze
- Clock or watch with seconds hand

Time required 1 double period

and as the concentration change in this experiment is constant,

$$k \propto \frac{1}{\text{time for methyl red to be bleached}}$$

$$\therefore \quad k = \frac{c}{\text{time for methyl red to be bleached}} = \frac{c}{t}$$

where c is a constant and t is the time taken to bleach the methyl red. Hence, $c/t = Ae^{-E_A/RT}$ and if we take logs to base e:

$$\ln\left(\frac{c}{t}\right) = \ln A + \ln e^{-E_A/RT}$$

$$\Rightarrow \ln c - \ln t = \ln A - \frac{E_A}{RT}$$

$$\Rightarrow \ln t = \ln c - \ln A + \frac{E_A}{RT}$$

$\ln c$ and $\ln A$ are constants, so a graph of $\ln t$ against $1/T$ has a gradient of E_A/R.

Procedure

Work in pairs.

Using a measuring cylinder, put 10 cm^3 of 0·01 M phenol and 10 cm^3 of the bromide/bromate(V) solution into a boiling tube, and then add 4 drops of methyl red indicator. In a second boiling tube, put 5 cm^3 of 0·5 M H_2SO_4.

Fill a 1 dm^3 beaker with water, warm to about 75°C and adjust the heating to keep the temperature of the water steady.

Immerse the two boiling tubes in the water and when their contents reach the water temperature (± 1°C), mix the contents of the tubes and time the reaction until the colour of the methyl red disappears.

Repeat the experiment at about 65°C, 55°C, 45°C, 35°C, 25°C and 15°C.

1 Arrange your results in a table, as below:

Temp./°C	Temp./K	$(1/T)/K^{-1}$	Time taken to bleach methyl red, t/seconds	$\ln t$

2 Why does the reaction not start until the contents of the boiling tubes are mixed?

3 Write an equation for the reaction between bromine and phenol.

4 What function does the methyl red play in the experiment? (**Hint:** it is not acting as an indicator in the accepted sense.)

5 Why is it unsatisfactory to measure the reaction rate at temperatures above 75°C?

6 Plot a graph of $\ln t$ (vertically) against $1/T$ (horizontally).

7 Measure the gradient of this graph and obtain the value of E_A. ($R = 8\cdot3$ J K^{-1} mole^{-1})

8 What thermodynamic sign does E_A have? What does this signify?

9 How is E_A related to the heat change for the forward reaction? Draw a simple energy diagram to make your answer clear.

10 What effect does E_A have on the equilibrium constant for the Br^-/BrO_3^- reaction?

PRACTICAL 20

Periodicity – a practical study of the third period

Introduction

The periodic table provides one of the most important unifying patterns to the study of chemistry. The intention of this practical is to study some of the trends in properties across the third period of the periodic table.

A   Periodicity in the properties of elements

The properties of the elements show a repeating pattern with increasing atomic number. The most obvious pattern is the steady change from metals on the left to non-metals on the right of each period and then back to metals again.

Procedure

Examine samples of the elements in period 3 (Na to Ar) and test the electrical conductivity of those elements for which the results are not given in table 1.

1   Copy and complete table 1. You will have to consult a data book, or section 3.8 in *Chemistry in Context*, for boiling point values.

	Na	Mg	Al	Si	P	S	Cl	Ar
Atomic number								
Physical state and appearance								
Boiling point/°C								
Conductivity at room temperature	good				poor		poor	poor
Structure (giant metallic, giant molecular or simple molecular)								
Type of element (metal, non-metal or metalloid)								

Table 1 The elements of period 3

2   How do the boiling points of the elements vary across the period from sodium to argon? How are the boiling points related to structure of the elements?

3   Describe the changes in structure and type of the elements across period 3.

B   Periodicity in the properties of oxides

You should know quite a lot about the properties of these compounds already.

4   Copy out table 2 and complete the first three rows.

Examine the solubility of each oxide in water. Add a **very small** measure of each solid to about 3 cm^3 of distilled water and shake thoroughly. (In the case of sulphur dioxide, bubble the gas through 3 cm^3 of water in a fume cupboard for a few seconds.)

CARE Eye protection must be worn when using Na$_2$O and P$_2$O$_5$.
Take care to avoid skin contact when using Na$_2$O and P$_2$O$_5$.

Requirements

- Samples of elements from period 3
- Simple apparatus to test electrical conductivity
- 1 rack + 4 test tubes
- Full-range indicator paper
- Samples of the following oxides: Na$_2$O (or NaOH), MgO, Al$_2$O$_3$, SiO$_2$, P$_2$O$_5$
- Access to SO$_2$ cylinder
- Samples of the following chlorides: NaCl, MgCl$_2$, AlCl$_3$, PCl$_3$, SCl$_2$ (or S$_2$Cl$_2$)
- Bunsen burner

Time required 2 double periods

Oxide formula	Na$_2$O	MgO	Al$_2$O$_3$	SiO$_2$	P$_2$O$_5$	SO$_2$	Cl$_2$O
State at room temp.							(g)
Appearance							yellow-red gas
Volatility							high
Conductivity of molten oxide	good	good	good	poor	poor	poor	poor
Solubility in water							dissolves readily
pH of solution in water							2
Classification of oxide (acidic, basic or amphoteric)							acidic
Structure of oxide (simple molecular, giant molecular, giant ionic)							simple molecular

Table 2 Properties of the oxides in period 3

If Na$_2$O is not available, use sodium hydroxide (NaOH). Some oxides will dissolve easily, but others will not dissolve at all. Test the pH of the solution obtained from each oxide using full-range indicator paper and record your results in table 2.

5 Which of the oxides of period 3 elements
   a form acidic solutions with water?
   b form alkaline solutions with water?
   c are insoluble in water?

6 Write equations for the reactions of the soluble oxides with water.

If an oxide is insoluble in water, its reactions with acids and alkalis can be used to decide its acid–base character. If it dissolves in acid, it must have reacted with it and can therefore be classified as basic. If the oxide dissolves in alkali, it can be described as acidic. If the insoluble oxide dissolves in both acids and alkalis, it is both basic and acidic. The adjective 'amphoteric' (from a Greek word meaning 'both') is used to describe these oxides.

Examine the solubility of the insoluble oxides in dilute hydrochloric acid and then in dilute sodium hydroxide. Remember to use **very small** measures of solid.
CARE Eye protection must be worn if the mixtures are heated.

7 Write equations for the reactions of the insoluble oxides with hydrochloric acid and sodium hydroxide where appropriate.

8 Complete the last two lines in table 2.

9 Describe the following trends in properties of the oxides across period 3.
   a their state.
   b their character (acidic/amphoteric/basic).
   c their structure.

C Periodicity in the properties of chlorides

Examine samples of the chlorides of period three elements.

10 Copy out table 3 (p. 55) and complete the first two rows. (You will **not** be expected to test silicon tetrachloride yourself; the results for this are already inserted.)
CARE Take great care with the chlorides of phosphorus and sulphur. Work in a fume cupboard and wear eye protection.
Warm separate **small** samples of the chlorides gently with a bunsen.
Investigate the relative volatility of the chlorides and put your results in table 3.
Investigate the effect of water on each chloride by adding a **small** quantity to about 3 cm^3 of distilled water in a test tube.
Test the pH of the mixture obtained using full-range indicator paper, then complete table 3.

Element	Na	Mg	Al	Si		P	S
Formula of chloride				$SiCl_4$			
Appearance and state of chloride				colourless liquid			
Volatility of chloride				High			
Action of water on chloride				very vigorous reaction. HCl fumes evolved.			
pH of solution of chloride in water				3			
Structure of chloride				simple molecular			

Table 3 Properties of the chlorides in period 3

11 Explain, with equations, the action of sodium chloride, aluminium chloride and phosphorus chloride with water.

12 Describe the trends across period three in
   a the formula of the chlorides.
   b the state of the chlorides.
   c the pH of the aqueous chloride solutions.
   d the structure of the chlorides.

PRACTICAL 21

A practical study of some group II elements

Introduction

The *s*-block of the periodic table contains the most reactive and, in chemical terms, the most typically metallic elements. All the elements in Group I are highly reactive, but those in Group II are slightly less so and show a rather more obvious trend in reactivity. In this practical you will study some of the properties of the elements of Group II and their compounds. The members of the group are shown in table 1.

We will concentrate on the three elements magnesium, calcium and barium. Beryllium will not be studied because its compounds are extremely toxic and very expensive.

Some physical data concerning the elements of Group II are given in table 2.

1 What type of ion is formed by Group II elements when they react?

2 Why do ionization energies decrease as you go down Group II?

3 Why do atomic radii increase as you go down Group II?

4 Which of the various sets of data in table 2 gives the most accurate indication of the likely reactivity trend within the group?

Try to use the data in the table when interpreting the results of your experimental work.

CARE Barium and its compounds are poisonous. Handle with care and wash your hands after the practical.

The elements of Group II	
Beryllium	Be
Magnesium	Mg
Calcium	Ca
Strontium	Sr
Barium	Ba
Radium	Ra

Table 1

Requirements

Each student, or pair of students, will require:

- Hard-glass test tubes (8)
- Test tube holders
- Angled glass bend with bung to fit test tube
- Beaker (400 cm³)
- Funnel
- Universal indicator solution
- Indicator paper
- Filter paper
- Splints
- Magnesium ribbon
- Magnesium powder
- Calcium, granules or turnings
- Barium metal, cut into small pieces (The teacher may prefer to demonstrate the reaction of barium with water.)
- Magnesium oxide
- Calcium hydroxide
- Barium hydroxide
- Hydrated magnesium chloride
- Hydrated calcium chloride
- Hydrated barium chloride
- Magnesium carbonate
- Calcium carbonate
- Barium carbonate
- Lime-water

Approximately 0·1 M solutions of:

- Mg^{2+} (25 g $Mg(NO_3)_2 . 6H_2O$ or 20 g $MgCl_2 . 6H_2O$ per dm³)
- Ca^{2+} (25 g $Ca(NO_3)_2 . 4H_2O$ or 10 g anhydrous $CaCl_2$ per dm³)
- Ba^{2+} (25 g $Ba(NO_3)_2$ or 25 g $BaCl_2 . 2H_2O$ per dm³)

Approximately 1·0 M solutions of:

- OH^- (40 g NaOH per dm³)
- CO_3^{2-} (100 g anhydrous Na_2CO_3 per dm³)
- SO_4^{2-} (300 g $Na_2SO_4 . 10H_2O$ per dm³)

250 cm³ of each of these six solutions is enough.

Time required 1–2 double periods

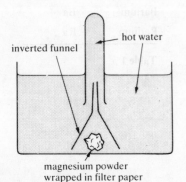

Figure 1 Investigation of the reaction of magnesium with water.

inverted funnel
hot water
magnesium powder wrapped in filter paper

Element	Be	Mg	Ca	Sr	Ba
Electron structure	$(He)2s^2$	$(Ne)3s^2$	$(Ar)4s^2$	$(Kr)5s^2$	$(Xe)6s^2$
First ionization energy/kJ mole⁻¹	900	740	590	550	500
Second ionization energy/kJ mole⁻¹	1800	1450	1150	1060	970
Third ionization energy/kJ mole⁻¹	14800	7700	4900	4200	—
Atomic radius (metallic radius)/nm	0·11	0·16	0·20	0·21	0·22
Ionic radius (M^{2+})/nm	0·030	0·065	0·094	0·110	0·134
Hydration energy (M^{2+})/kJ mole⁻¹	—	−1891	−1561	−1414	−1273
Standard electrode potential, $E^{\ominus}(M^{2+}(aq)\|M(s))$/volts	−1·85	−2·37	−2·87	−2·89	−2·91

Table 2 Some properties of the Group II elements.

Procedure

EXPERIMENT 1   Reaction of the elements with water

Put a very small piece of calcium metal into a large beaker of cold water. Observe the reaction and identify the products. Repeat, using a small piece of clean magnesium ribbon and then a small piece of barium metal.

CARE Eye protection must be worn. Your teacher may prefer to demonstrate the reaction of barium with water.

5   Write equations for any reactions which occur.

6   How do the metals differ in the vigour with which they react with water?

7   In all their reactions the Group II metals behave as reducing agents. What is being reduced in this reaction?

8   Explain the reactivity trend among the three metals in this experiment using some of the data in table 2.

9   What other factor, not listed in the table, may affect the relative reactivity of the metals with water?

The reaction of magnesium with water was probably very slow. You can investigate this reaction further by setting up the experiment shown in figure 1. Leave the experiment for half an hour or so, then test the products.

EXPERIMENT 2   Acid–base character

Place a very small quantity (about 0·01 g) of magnesium oxide, calcium hydroxide and barium hydroxide in three separate test tubes. Add 10 cm³ distilled water to each tube and shake. Add 2 drops of universal indicator solution to each tube and mix. Record the pH values indicated for the three tubes.

10   How does the acid–base character of the hydroxides vary within the group?

11   Write a general ionic equation to represent the equilibrium between the undissolved solid hydroxide and its aqueous ions.

12   Why is it valid to use magnesium oxide instead of magnesium hydroxide in this experiment?

13   What is milk of magnesia and what is it used for?

14   Look at the values for ionic radius in table 2. Which Group II ion will have the strongest attraction for OH^- ions? What effect will this have on the basic strength of its hydroxide?

EXPERIMENT 3   Hydrolysis of the chlorides

Ionic chlorides dissolve in water forming simple hydrated ions. Many covalent and partly covalent chlorides, however, are hydrolysed, giving hydrogen chloride and the oxide or hydroxide. For example, aluminium chloride reacts vigorously with water as follows:

$$AlCl_3 + 3H_2O \longrightarrow Al(OH)_3 + 3HCl$$

The extent of hydrolysis of the Group II chlorides can be estimated by heating the hydrated chloride and testing for hydrogen chloride gas.

Working in a fume cupboard, strongly heat about 1 cm depth of each of the hydrated chlorides of magnesium, calcium and barium in separate, dry hard-glass test tubes. Test for the evolution of hydrogen chloride.

15 Are any of the chlorides hydrolysed? Is there a trend in the tendency towards hydrolysis?

16 Which of the chlorides shows the greatest covalent character? Explain this tendency towards covalency in terms of some of the data on table 2.

EXPERIMENT 4   Thermal stability of the carbonates
Strongly heat about 1 cm depth of each of the dry carbonates of magnesium, calcium and barium separately in the apparatus shown in figure 2. Continue heating strongly for several minutes. Note how rapidly gas is evolved, and the extent to which the lime-water becomes milky. Remember to remove the tube from the lime-water as soon as heating is stopped.

17 Write equations for any reactions which occur.

18 What trend do you detect in the thermal stability of the carbonates of the elements of Group II?

EXPERIMENT 5   Solubility of some compounds of Group II elements
To investigate the solubility of Group II compounds, solutions containing the appropriate anions and cations are mixed. If the compound is insoluble, a precipitate will form.

Put 2 cm^3 of a 0·1 M solution of each of the Group II cations under investigation (Mg^{2+}, Ca^{2+}, Ba^{2+}) in separate test tubes. Add an equal volume of a 1·0 M solution of hydroxide ions, and mix. Note whether a precipitate is formed, and if so, how dense it is. Repeat the experiment twice, using first a 1·0 M solution of sulphate ions and then a 1·0 M solution of carbonate ions, instead of the hydroxide ions. Tabulate your results.

19 What trends do you notice in the solubility of
a hydroxides,
b sulphates,
c carbonates?

If you have time, you could extend this experiment by investigating the solubility of some other compounds -- say chromates and oxalates. You might then be able to formulate a more general rule concerning solubility trends within the group.

20 Using the results and conclusions you have derived from this practical, predict the following properties of   a beryllium,   b strontium, and their compounds:

    i reaction with water
    ii acid–base character of the hydroxide
    iii tendency of the chloride to hydrolyse
    iv thermal stability of the carbonate
    v solubility of the hydroxide
    vi solubility of the carbonate
    vii solubility of the sulphate

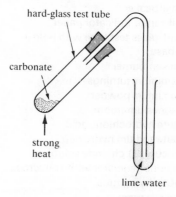

Figure 2 Investigation of the thermal stability of carbonates.

PRACTICAL 22
Aluminium and its compounds

Requirements

- Rack with four test tubes
- Small beaker
- Full-range indicator paper
- Hard-glass test tube with hole in its base
- Bunsen burner
- Aluminium turnings
- Aluminium powder
- Magnesium ribbon
- Dilute hydrochloric acid
- Dilute sodium hydroxide
- Mercury(II) chloride solution
- Chlorine generator with anhydrous CaCl₂ drying tube
- Ceramic wool
- Anhydrous aluminium chloride
- Anhydrous magnesium chloride
- Magnesium oxide
- Aluminium oxide
- 0·1 M Al³⁺
- 0·1 M Mg²⁺

Time required 1–2 double periods

Introduction

Aluminium has been variously described as a metalloid, a 'poor metal' and an 'amphoteric metal'. The intention of this experiment is to explore these descriptions of aluminium and to compare the chemistry of aluminium and its compounds with that of magnesium and its compounds.

Much of the chemistry of aluminium compounds is dictated by the charge density of the Al^{3+} ion which is so highly polarising that many aluminium compounds are covalent. (See *Chemistry in Context*, section 15.2.) Furthermore, the energy required to form Al^{3+} from Al is so large that Al^{3+} only forms in combination with anions of high charge density such as O^{2-} and F^-, or when it is hydrated as $[Al(H_2O)_6]^{3+}$.

EXPERIMENT 1   Reaction with dilute hydrochloric acid

Add about 5 cm³ of dilute hydrochloric acid to a few aluminium turnings in a test tube. If the aluminium has not started to react after five minutes, warm the mixture gently.

Repeat the test using magnesium ribbon in place of aluminium turnings. (CARE)

1 Explain your observations giving appropriate equations.

2 Look up the standard electrode potentials of hydrogen, aluminium and magnesium. These suggest that both aluminium and magnesium should react readily with dilute hydrochloric acid. What explanation can you offer for the slow reaction of aluminium?

EXPERIMENT 2   Reaction with dilute sodium hydroxide solution

Add about 5 cm³ of dilute sodium hydroxide solution to a few aluminium turnings (or a spatula measure of aluminium powder if this is available). If there is no action after five minutes, warm carefully.
CARE Eye protection must be worn.

3 Record your observations and identify the gas evolved.

4 Write an equation for the reaction which occurs.

5 Repeat the test using magnesium in place of aluminium and compare the two reactions.

6 Why should aluminium pans never be cleaned with washing soda (sodium carbonate)? (Hint: test the pH of sodium carbonate solution.)

EXPERIMENT 3   Reaction with oxygen

Place a piece of aluminium foil in a small beaker and cover it with mercury(II) chloride solution.   **CARE Mercury compounds are very poisonous.**

Leave it for a few minutes and then wash the foil with distilled water. Leave the wet foil in the air for a few minutes.

7 Describe and explain your observations with appropriate equations.

8 Why does mercury(II) chloride solution clean the surface of the aluminium foil so effectively? (How does $HgCl_2(aq)$ react with Al_2O_3 and Al? Hint: test the pH of $HgCl_2(aq)$ with full-range indicator paper.)

9 The processes, $Al \longrightarrow Al^{3+}$ and $O_2 \longrightarrow O^{2-}$ are both very endothermic. Why then does Al_2O_3 form and why is it so stable?

10 Why does aluminium not corrode away in air like iron?

11 What are the principal uses of aluminium? On what properties of aluminium (besides its resistance to corrosion) do these uses depend?

EXPERIMENT 4   Reaction with chlorine
(CARE This experiment must be performed in a fume cupboard.)
Arrange the apparatus in figure 1 in a fume cupboard. Clamp the tube horizontally, heat the aluminium and then pass chlorine over the heated metal. Once the reaction has started, remove the bunsen so that the vigour of the reaction can be observed. Try to collect the product at the end of the tube.

When the test tube has cooled, remove the product (Al_2Cl_6) and keep it for the next experiment.

12   Describe and explain your observations.

13   Why is the plug of ceramic wool used?

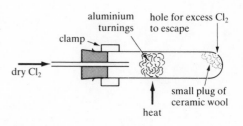

Figure 1 Reaction with chlorine.
(To make dry Cl_2, add conc. HCl to
$KMnO_{4(s)}$ covered with a layer of water.
Dry by passing through anhy. $CaCl_2$.)

EXPERIMENT 5   Comparing aluminium chloride and magnesium chloride

A   Action of heat on the anhydrous chlorides
Heat a spatula measure of anhydrous aluminium chloride in a test tube, **gently** at first, and then more strongly. Repeat the experiment using anhydrous magnesium chloride.

14   Describe the action of heat on Al_2Cl_6 and $MgCl_2$. What does this suggest about the bonding and structure in each chloride?

15   Draw dot and cross diagrams to represent the electronic structure of Al_2Cl_6 and $MgCl_2$.

B   Action of water on the anhydrous chlorides
Put one spatula measure of anhydrous aluminium chloride in a test tube and add water drop by drop.
CARE Eye protection must be worn.

16   Describe what happens. Does the tube become warm? What is the final pH of the solution?

17   Write an equation for the reaction of Al_2Cl_6 with water. Repeat the experiment with anhydrous magnesium chloride.

18   Compare the action of water on Al_2Cl_6 and $MgCl_2$.

EXPERIMENT 6   Comparing the acid/base character of aluminium oxide and magnesium oxide
Examine the reaction of small quantities of aluminium oxide and magnesium oxide with water and test the pH of the solutions obtained.

19   Describe the results of your tests. Can either of these oxides be described as alkaline?

Examine the reactions of these oxides first with dilute hydrochloric acid and then with dilute sodium hydroxide. Use only a small quantity (about 0·1 g) of the oxide and see if it dissolves on warming with 3 cm^3 of dilute hydrochloric acid and then with 3 cm^3 of dilute sodium hydroxide.
CARE Eye protection must be worn when warming with dilute acid or alkali.

20   Which of these oxides can be described as
    a basic,   **b** acidic,   **c** amphoteric?

21   Write equations for the reactions which occur when aluminium oxide and magnesium oxide are warmed with dilute hydrochloric acid and dilute sodium hydroxide.

EXPERIMENT 7   Comparing the acid/base properties of hydrated Al^{3+} and Mg^{2+} ions
In solution, Al^{3+} and Mg^{2+} ions are thought to be hydrated by six water molecules as $[Al(H_2O)_6]^{3+}$ and $[Mg(H_2O)_6]^{3+}$ respectively. These hydrated ions can act as Brønsted-Lowry acids by donating protons as follows:

$$[Al(H_2O)_6]^{3+} + H_2O \rightleftharpoons [Al(H_2O)_5OH]^{2+} + H_3O^+$$

$$[Mg(H_2O)_6]^{2+} + H_2O \rightleftharpoons [Mg(H_2O)_5OH]^{+} + H_3O^+$$

Pour 3 cm^3 of 0·1 M Al^{3+}(aq) into one test tube and 3 cm^3 of 0·1 M Mg^{2+} in another. Test the pH of each solution using full range indicator paper.

22 Which is the stronger Brønsted-Lowry acid – hydrated Al^{3+} or hydrated Mg^{2+}? Explain your results.

Add dilute sodium hydroxide to the 3 cm³ of 0·1 M Al^{3+}(aq) until the precipitate which forms has dissolved.

23 Describe and explain what happens. Write equations where appropriate. (The full formula of the precipitate is $Al(OH)_3(H_2O)_3$ and the final state of Al^{3+} is as aqueous $[Al(OH)_4(H_2O)_2]^-$.)

24 Why is $[Al(OH)_4(H_2O)_2]^-$ soluble in water, whilst $Al(OH)_3(H_2O)_3$ is insoluble?

Repeat the last test using 0·1 M Mg^{2+}(aq) in place of 0·1 M Al^{3+}(aq).

25 Describe and explain what happens. Why does the precipitate remain insoluble in excess sodium hydroxide in this case?

26 What are the main differences between the chemistry of aluminium and its compounds and that of magnesium and its compounds?

27 Is aluminium best described as a metalloid, a 'poor metal' or an 'amphoteric metal'? Explain your answer.

PRACTICAL 23
The halogens

The intention of this practical is to compare the properties and reactions of the halogens, a group of reactive non-metals in the periodic table.

A   Preparation of the halogens

All the methods of preparing halogens involve oxidizing halide ions.

$$2Hal^- \longrightarrow Hal_2 + 2e^-$$

CARE Particular care is needed in the following experiments. Wear eye protection at all times, use a fume cupboard if possible, and do not use greater quantities than those given.

Action of manganese(IV) oxide or potassium manganate(VII) on concentrated hydrohalic acid (HCl, HBr, HI).

Warm 1 cm^3 of concentrated hydrochloric acid with a spatula measure of manganese(IV) oxide.

1   Describe what happens and write an equation (or half equations) for the reaction occurring.

Add a few drops of concentrated hydrochloric acid to half a spatula measure of potassium manganate(VII).

2   Write an equation (or half equations) for the reaction occurring.

3   Compare the action of manganese(IV) oxide and potassium manganate(VII) as oxidizing agents in these two simple experiments.

4   Write the halide ions in order of their ease of oxidation.

5   Explain the relative ease of oxidation of the halide ions.

B   Physical properties of the halogens

6   Make a table comparing the colour, state at room temperature, melting point and boiling point of the halogens fluorine, chlorine, bromine and iodine.

7   What trends are evident in the physical properties of the halogens as their relative atomic mass increases?

C   Chemical properties of the halogens

EXPERIMENT 1   Reactions of halogens with water
Working in a fume cupboard, pass chlorine through 5 cm^3 of water for a few seconds. Test the pH of the resulting solution with full-range indicator paper.

8   What conclusion can you draw concerning chlorine water from this test?

9   Chlorine water contains dissolved chlorine, together with a mixture of two acids which are formed when chlorine reacts with water. What are these two acids called? Write an equation for the reaction of chlorine with water.

Shake one drop of bromine with 5 cm^3 of water in a test tube until the bromine dissolves. Test the resulting solution with full-range indicator paper.
CARE Keep the bromine bottle in a fume cupboard and avoid skin contact by wearing polythene gloves.
Repeat the experiment using a small crystal of iodine in place of the bromine.

Requirements

Each student will require:
- Rack with 4 test tubes
- Bunsen burner
- Manganese(IV)oxide
- Potassium manganate(VII)
- Supply of chlorine
- Bromine
- Iodine
- Conc. hydrochloric acid
- Full-range indicator paper
- Chlorine water
- Bromine water
- 1,1,1-trichloroethane
- Supply of hydrogen sulphide from a Kipp's apparatus

Approximately 1·0 M solutions of the following substances:
- Iron(II) sulphate
- Sodium hydroxide
- Potassium chloride
- Potassium bromide
- Potassium iodide

Time required 1 double period

10 Compare the ease with which the three halogens dissolve in water, the pH of their resulting solutions and their action as bleaching agents.

11 What specific uses does chlorine have as a bleach?

EXPERIMENT 2   Solubility of the halogens in 1,1,1-trichloroethane

Shake 3 cm^3 of chlorine water, $Cl_2(aq)$, with 1 cm^3 of 1,1,1-trichloroethane in a test tube.

Repeat the experiment using first 3 cm^3 of bromine water and then a small crystal of iodine in place of the 3 cm^3 of chlorine water.

12 Tabulate and explain your observations in the three simple experiments.

EXPERIMENT 3   Halogens as oxidizing agents

a   Reactions with hydrogen sulphide
(CARE Do these experiments in a fume cupboard.)

Pass hydrogen sulphide from a Kipp's apparatus through 3 cm^3 of chlorine water. Compare the final solution with clear chlorine water.

13 Describe what happens and write an equation for the reaction which occurs.

Repeat the experiment using bromine water in place of chlorine water.

14 Compare the reactions of bromine and chlorine with hydrogen sulphide.

b   Reactions with iron(II) sulphate solution

Put 3 cm^3 of iron(II) sulphate solution into each of two test tubes. To one test tube add 3 cm^3 of $Cl_2(aq)$ and to the other 3 cm^3 of water. This second test tube acts as a control.

15 What is a 'control'?

16 What do you observe in
  a tube one,
  b the control tube?

Investigate whether Fe^{2+} ions have been oxidized to Fe^{3+} by adding 1 cm^3 of sodium hydroxide solution to each tube.

17 What evidence does the addition of sodium hydroxide provide for the oxidation of Fe^{2+} ions by chlorine?

18 Write an equation for the reaction of chlorine with iron(II) sulphate. Repeat the experiment using $Br_2(aq)$ in place of $Cl_2(aq)$.

19 Describe and explain the reactions of $Br_2(aq)$.

EXPERIMENT 4   Relative reactivity of the halogens as oxidizing agents

Investigate the action of $Cl_2(aq)$ on
  a KBr(aq)
  b KI(aq)
and then the action of Br(aq) on
  a KCl(aq)
  b KI(aq).

Use 3 cm^3 of each reacting solution and then shake each of the four resulting solutions with 1 cm^3 of 1,1,1-trichloroethane.

20 Tabulate the results of the four tests showing the solutions mixed, the changes which occur before 1,1,1-trichloroethane is added, and the final colour of the 1,1,1-trichloroethane layer.

21 Write equations for the reactions where appropriate.

22 What is the order of the halogens as oxidizing agents?

23 Look up the electrode potentials of the halogens for the Hal_2/Hal^- systems. Do these confirm your deductions about their relative oxidizing power?

PRACTICAL 24

Some properties of group IV elements and their compounds

Introduction

More than any other group of the Periodic Table, Group IV (table 1) shows the trend from non-metallic to metallic elements as the group is descended. The elements change from a non-metal (carbon) to metalloids (silicon and germanium) and then to metals (tin and lead).

In this practical the trend in properties is illustrated by the change in physical properties of the elements and by the trend in acid–base nature of their oxides.

Another feature of Group IV is the occurrence of two oxidation states, $+2$ and $+4$. You will investigate the relative stability of these two oxidation states, and its variation within the group. In this respect we shall concentrate on the elements at the bottom of the group, tin and lead.

The elements in Group IV
C
Si
Ge
Sn
Pb

Table 1

Procedure

EXPERIMENT 1   Physical properties of the elements

A   Appearance
Examine samples of carbon, silicon, tin and lead. Germanium may also be available, but it is expensive and much rarer than the other four elements.

1   Copy out table 2 and record in it the appearance of each element. Indicate the form and nature of each sample, (e.g. graphite or charcoal, foil or granules, etc.).

	C	Si	Ge	Sn	Pb
Appearance					
Volatility					
Melting point					
Malleability					
Structure					

Table 2 Physical properties of Group IV elements

B   Volatility
Heat a small sample of each element in turn in a hard-glass test tube. See if you can melt the element.

2   Record in the table the ease with which each melts, then look up and record the melting points of the elements.

C   Malleability
If suitable samples of the elements are available, test their malleability using a hammer.

3   Record your results in the table.

4   What type of structure does each element have? Record your answer in the table.

Requirements

Each student or pair of students, will require:

- Rack with 6 hard-glass test tubes
- Boiling tube
- Beaker (250 cm^3)
- Dropping pipette
- Filter funnel and paper
- Universal indicator
- Splints
- Hammer
- Samples of carbon (graphite), silicon, tin (foil or granules), lead (foil or shot), germanium if available.
- Lead(II) oxide (litharge)
- Lead(IV) oxide (lead dioxide)
- Tin(IV) oxide (stannic oxide)
- Tin(IV) chloride (stannic chloride)
- Concentrated methanoic (formic) acid
- Concentrated sulphuric acid
- Concentrated hydrochloric acid
- Concentrated sodium hydroxide (about 8 M – make 80 g of pellets or flakes up to 250 cm^3)
- Ice

The following solutions at usual bench concentration:

- Tin(II) chloride (stannous chloride)
- Lead(II) nitrate
- Potassium iodide
- Iodine in potassium iodide
- Dilute hydrochloric acid
- Dilute nitric acid
- Dilute sodium hydroxide

Time required 1–2 double periods

EXPERIMENT 2 Relative stability of 2+ and 4+ oxidation states

A Oxides of carbon

Working in a fume cupboard, generate a small quantity of carbon monoxide by adding 1 cm^3 of concentrated sulphuric acid to 1 cm^3 of concentrated methanoic (formic) acid in a test tube.

CARE Both acids are corrosive. Eye protection must be worn. Carbon monoxide is poisonous.

As soon as steady evolution of carbon monoxide has started, hold a lighted splint to the mouth of the tube.

5   What happens? To what is the carbon monoxide being changed? Write an equation.

6   Is carbon dioxide readily decomposed to carbon monoxide on heating?

7   Which is the more stable oxide of carbon?

B Oxides of tin and lead

Heat small samples of lead(IV) oxide and tin(IV) oxide in separate hard-glass test tubes. Test to see if oxygen is evolved and note the appearance of the residue.

8   What happens, if anything, in each case? Identify the residues and write equations for any reactions which have occurred.

9   Which is the more stable oxide of tin, SnO or SnO$_2$?

10   Which is the more stable oxide of lead, PbO or PbO$_2$?

11   Why is PbO named lead(II) oxide, not lead monoxide, whereas CO is named carbon monoxide, not carbon(II) oxide?

C Oxidizing power of tin(IV) and lead(IV) compounds

Prepare a solution of tin(IV) chloride by shaking a small quantity (about 0·1 g) of solid SnCl$_4$ with about 5 cm^3 of water. (Work in a fume cupboard – hydrogen chloride is evolved.) Acidify a 1 cm^3 portion of this solution with an equal quantity of dilute hydrochloric acid, then add 1 cm^3 of potassium iodide solution. Observe carefully.

Repeat the experiment using lead(IV) chloride in place of tin(IV) chloride. Lead(IV) chloride is rather unstable (see experiment 3) and must be specially prepared. Put 5 cm^3 of concentrated hydrochloric acid in a test tube, **(CARE Eye protection must be worn)** and cool for 10 minutes in a beaker of iced water. Add a very small quantity of lead(IV) oxide and shake. The reaction produces lead(IV) chloride. Take 1 cm^3 of this solution, add an equal volume of potassium iodide solution and observe carefully.

12   Do
   a tin(IV) chloride,
   b lead(IV) chloride,
   oxidize iodide ions to iodine?

13   Which is the more powerful oxidizer, lead(IV) chloride or tin(IV) chloride?

14   Write equations for any reactions which have occurred.

15   Would you expect carbon tetrachloride, CCl$_4$, to oxidize iodide ions? Explain.

D Reducing power of tin(II) and lead(II) compounds

Acidify 1 cm^3 of tin(II) chloride solution with an equal volume of dilute nitric acid. Add 1 cm^3 of iodine solution drop by drop. Observe to see if the iodine is decolorized. Repeat the experiment using lead(II) nitrate solution instead of tin(II) chloride.

16   Do
   a Sn^{2+}
   b Pb^{2+}
   reduce iodine to iodide ions?

17   Write equations for any reactions which have occurred.

18   Do your results here agree with your answer to question **13**? Explain.

19 Why does carbon dichloride, CCl_2, not exist?

20 What generalizations can you make about the relative stability of the $+2$ and $+4$ oxidation states of the Group IV elements?

EXPERIMENT 3 Chlorides of lead

A Lead(II) chloride

Put a small quantity (about $0 \cdot 1$ g) of lead(II) oxide in a test tube and add about 5 cm^3 of dilute hydrochloric acid. Warm until most of the lead(II) oxide dissolves. Now cool the tube under the tap. What happens? Warm the tube again. What happens?

21 What can you say about the solubility of lead(II) chloride in water?

B Lead(IV) chloride

Put about 5 cm^3 of concentrated hydrochloric acid in a boiling tube **(CARE Eye protection must be worn)** and cool it in a beaker of iced water for about 10 minutes. Add about $0 \cdot 5$ g of lead(IV) oxide and shake the tube to dissolve the oxide. If some oxide is left undissolved, decant or filter the solution of lead(IV) chloride into a tube cooled in iced water.

22 Write an equation for the reaction of lead(IV) oxide with cold hydrochloric acid.

Gently warm a 1 cm^3 portion of the solution and cautiously identify the gas which is evolved. Cool the tube under the tap and note what happens.

23 Identify the two products of the thermal decomposition of lead(IV) chloride and write an equation.

24 Compare the stability of lead(II) chloride and lead(IV) chloride.

Add excess dilute sodium hydroxide solution to 1 cm^3 of your lead(IV) chloride solution. (You will need to add enough sodium hydroxide to neutralize the excess acid as well.)

25 Identify the precipitate that forms.

26 Lead(II) chloride is a white solid, melting point $498°C$. Lead(IV) chloride is a yellow oil, melting point $-15°C$. What do these facts suggest about the structure and bonding of these two chlorides?

EXPERIMENT 4 Acid–base nature of oxides

27 Copy out table 3 and using the results of experiment **3B** fill in the reaction of lead(IV) oxide with acid.

28 Using your knowledge of the chemistry of carbon dioxide, fill in the entries for this compound.
A Investigate the reaction of lead(IV) oxide with alkali by cautiously warming a **very small** portion of the oxide with concentrated sodium hydroxide solution. **(CARE Eye protection must be worn.)**
B Investigate the reaction of tin(IV) oxide with alkali and acid by warming separate small portions with concentrated sodium hydroxide and concentrated hydrochloric acid. **(CARE)**

29 Complete the table. Classify CO_2, SnO_2 and PbO_2 as acidic, basic or amphoteric oxides.

30 What trend can you identify in acid–base behaviour of the oxides as you move down Group IV?

31 How does this trend relate to the change in metal/non-metal character as you move down the group?

Oxide	Reaction with acid	Reaction with alkali
CO_2		
SnO_2		
PbO_2		

Table 3 Reaction of Group IV oxides with acid and alkali

PRACTICAL 25
The oxidation states of vanadium and manganese

Requirements

Each student, or pair of students, will require:

- 2 boiling tubes with bung
- Rack with 4 test tubes
- Filter paper and funnel
- Small beaker
- Ammonium vanadate(V) (ammonium metavanadate) (NH_4VO_3)
- 2 M NaOH (dilute NaOH)
- 1·0 M H_2SO_4 (dilute H_2SO_4)
- Granulated zinc or zinc amalgam
- Concentrated HNO_3
- Concentrated H_2SO_4
- 0·01 M $KMnO_4$
- 0·1 M $KMnO_4$
- Manganese(IV) oxide (MnO_2)
- Hydrated manganese(II) sulphate

Time required 1–2 double periods

Introduction

The atoms of transition metals have electrons of similar energy in both the $3d$ and $4s$ levels. This means that one particular element can form a number of different stable ions by losing or sharing different numbers of electrons. Thus, all the transition metals in the first series (titanium to copper) can exhibit two or more oxidation states in their compounds. The intention of this experiment is

A to investigate the chemistry of vanadium and manganese.

B to apply a knowledge of electrode potentials to the reactions studied.

C to prepare some vanadium and manganese compounds in their less common oxidation states.

1 The electronic structure of vanadium is (Ar) $3d^3$ $4s^2$. Write the electronic structures of V^+, V^{2+} and V^{3+}.

2 Which of these ions has the most stable electronic structure? Explain your answer.

3 Vanadium(III) compounds are much more common than vanadium(II) and vanadium(I) compounds. What factors are responsible for this?

Oxidation states of vanadium

Step-by-step conversion of vanadium(V) to vanadium(II)

Dissolve 1 g of ammonium vanadate(V) (NH_4VO_3) in 10 cm^3 of 2 M sodium hydroxide in a boiling tube and then add 20 cm^3 of 1·0 M sulphuric acid. This solution contains dioxovanadium(V) ions, VO_2^+, in acid solution.

$$VO_3^-(aq) + 2H^+(aq) \longrightarrow VO_2^+(aq) + H_2O(l)$$

Now add a few pieces of granulated zinc or a few cm^3 of zinc amalgam to the mixture. Close the boiling tube with a rubber bung and shake until no further changes occur. Remove the bung to release pressure from time to time.

CARE Zinc amalgam contains mercury and is poisonous.

4 Describe what happens as the tube is shaken.

5 Copy and complete table 1, recording the colours of the various oxidation states to which vanadium is successively reduced. (**Note** The first green colour which you see is a mixture of the original vanadium(V) solution and vanadium(IV).)

Ion	VO_2^+	VO^{2+}	V^{3+}	V^{2+}
Name	dioxovanadium (V) ion	oxovanadium (IV) ion	vanadium (III) ion	vanadium (II) ion
Oxidation state				
Colour				

Table 1 Oxidation states of vanadium

The standard electrode potentials for the reactions involved are:

$$VO_2^+ + 2H^+ + e^- \longrightarrow VO^{2+} + H_2O \qquad E^\ominus = +1.00 \text{ V}$$

$$VO^{2+} + 2H^+ + e^- \longrightarrow V^{3+} + H_2O \qquad E^\ominus = +0.32 \text{ V}$$

$$V^{3+} + e^- \longrightarrow V^{2+} \qquad E^\ominus = -0.26 \text{ V}$$

$$Zn^{2+} + 2e^- \longrightarrow Zn \qquad E^\ominus = -0.76 \text{ V}$$

6 Explain why zinc is capable of reducing VO_2^+ to V^{2+}.

7 Which of the stages in the reduction of vanadium(V) to vanadium(II) would you expect to be achieved least readily? Explain your answer.

When no further colour changes occur, filter the mixture to obtain 5 cm^3 of the final solution. Now add concentrated nitric acid drop by drop to the 5 cm^3 of filtrate, shaking carefully.

8 Describe what happens. To which oxidation state does nitric acid oxidize vanadium(II)?

9 Are your results consistent with standard electrode potentials? Explain.

$$NO_3^- + 2H^+ + e^- \longrightarrow NO_2 + H_2O \qquad E^\ominus = +0.80 \text{ V}$$

Oxidation states of manganese

You have possibly met and used the following compounds of manganese already: potassium manganate(VII) $KMnO_4$, manganese(IV) oxide, MnO_2 and manganese(II) sulphate, $MnSO_4$.

10 Prepare an oxidation number chart for manganese and write $KMnO_4$, MnO_2, $MnSO_4$ and Mn on the appropriate line of the chart.

The commonest oxidation states of manganese are $+7$, $+4$ and $+2$. In the next experiment, we shall investigate the preparation of manganese(VI) and manganese(III) compounds.

The preparation of manganese(VI) compounds

11 Use the following electrode potentials to predict whether Mn(VI) might be prepared by reacting Mn(VII) with Mn(IV) in *acid* solution. Explain your answer.

In acid:

$$\underset{\text{Mn(VII)}}{MnO_4^-} + e^- \longrightarrow \underset{\text{Mn(VI)}}{MnO_4^{2-}} \qquad E^\ominus = +0.56 \text{ V}$$

$$4H^+ + \underset{\text{Mn(VI)}}{MnO_4^{2-}} + 2e^- \longrightarrow \underset{\text{Mn(IV)}}{MnO_2} + 2H_2O \qquad E^\ominus = +2.26 \text{ V}$$

12 Would increasing the concentration of either MnO_4^- or H^+ improve the chances of preparing Mn(VI)? Explain.

13 Explain why the reaction is more likely to yield Mn(VI) in *alkaline* solution.

14 Do the following electrode potentials suggest that Mn(VI) might be prepared from Mn(VII) and Mn(IV) in *alkaline* solution? Explain.

In alkali:

$$MnO_4^- + e^- \longrightarrow MnO_4^{2-} \qquad E^\ominus = +0.56 \text{ V}$$

$$2H_2O + MnO_4^{2-} + 2e^- \longrightarrow MnO_2 + 4OH^- \qquad E^\ominus = +0.59 \text{ V}$$

15 Would increasing the concentration of either MnO_4^- or OH^- improve the chance of preparing Mn(VI)?

Test your predictions as follows: Put 10 cm^3 portions of 0.01 M $KMnO_4$ into each of two boiling tubes. Add 5 cm^3 of dilute sulphuric acid to one tube and 5 cm^3

of dilute sodium hydroxide to the other. Now add a little manganese(IV) oxide to each tube and shake for two minutes. Filter each mixture into a clean test tube.

16 In which tube has a reaction occurred? (The colour of Mn(VI) is green.)

To the green solution add 5 cm^3 of dilute sulphuric acid.

17 Describe and explain what happens.

The preparation of manganese(III) compounds

18 Use the following electrode potentials to predict whether Mn(III) could be prepared by reacting Mn(II) and Mn(IV) in *acid* solution. Explain your answer.

In acid:

$$4H^+ + MnO_2 + e^- \longrightarrow Mn^{3+} + 2H_2O \qquad E^\ominus = +0.95 \text{ V}$$

$$Mn^{3+} + e^- \longrightarrow Mn^{2+} \qquad E^\ominus = +1.51 \text{ V}$$

19 Would increasing the concentration of H^+ or Mn^{2+} inprove the chance of preparing Mn(III)? Explain.

20 Use the following electrode potentials to predict whether Mn(III) could be prepared by reacting Mn(II) and Mn(IV) in *alkaline* solution.

In alkali:

$$2H_2O + MnO_2 + e^- \longrightarrow Mn(OH)_3 + OH^- \qquad E^\ominus = +0.20 \text{ V}$$

$$Mn(OH)_3 + e^- \longrightarrow Mn(OH)_2 + OH^- \qquad E^\ominus = -0.10 \text{ V}$$

21 Would increasing the concentration of OH^- inprove the chance of preparing Mn(III)? Explain.

22 Write an overall equation for the reaction including state symbols. Why is the rate of this reaction very slow?

Another possible way of obtaining Mn(III) is by reacting Mn(II) with Mn(VII).

In acid:

$$8H^+ + MnO_4^- + 5e^- \longrightarrow Mn^{2+} + 4H_2O \qquad E^\ominus = +1.51 \text{ V}$$

$$Mn^{3+} + e^- \longrightarrow Mn^{2+} \qquad E^\ominus = +1.51 \text{ V}$$

23 Do these electrode potentials suggest that Mn^{3+} could be obtained by reacting MnO_4^- with Mn^{2+}? Explain.

24 Would increasing the concentration of acid improve the chance of preparing Mn(III)?

Procedure

Dissolve 0.5 g of hydrated manganese(II) sulphate in about 2 cm^3 of dilute sulphuric acid and add 10 drops of concentrated sulphuric acid. **CARE Eye protection must be worn.** Cool the test tube in cold water and then add 5 drops of 0.1 M potassium manganate(VII).

25 Describe what happens. What is the colour of aqueous Mn^{3+} ions?

Pour the final solution into 50 cm^3 of water.

26 Describe and explain what happens.

PRACTICAL 26

Copper

Introduction

Before looking at the two oxidation states of copper, carry out some simple test tube experiments to remind yourself of some of the basic reactions of this metal.

A Introductory experiments

a Hold a piece of clean copper foil with tongs and heat it in a bunsen flame.

b Drop one copper turning into $2\,cm^3$ of dilute nitric acid and warm. Identify the gas evolved.

c Add dilute sodium hydroxide solution drop by drop to $2\,cm^3$ of copper sulphate solution, until the sodium hydroxide is in excess.

d Add ammonia solution drop by drop to $2\,cm^3$ of copper sulphate solution until the ammonia is in excess.

e Add concentrated hydrochloric acid drop by drop to $2\,cm^3$ of copper sulphate solution. **CARE Eye protection must be worn.** Continue adding concentrated hydrochloric acid until no further changes occur, then add water gradually, again until no further changes occur.

1 For each of the reactions **a** to **c**, decide what has been formed and write an equation.

2 Reaction **d** finally results in the formation of a complex ion containing Cu^{2+} and ammonia.

$$Cu(H_2O)_4^{2+}(aq) + 4NH_3(aq) \rightleftharpoons Cu(NH_3)_4^{2+}(aq) + 4H_2O(l)$$
pale blue   deep blue

Explain what you observed in reaction **d** in terms of this equilibrium.

3 Reaction **e** also involves the formation of a complex ion – in this case the ligand is Cl^-:

$$Cu(H_2O)_4^{2+}(aq) + 4Cl^-(aq) \rightleftharpoons CuCl_4^{2-}(aq) + 4H_2O(l)$$
pale blue   green

Intermediate ions such as $[CuCl_3(H_2O)]^-$ may also be formed. Explain your observations in reaction **e** in terms of this equilibrium.

4 What is the oxidation state of copper in all the compounds formed in these introductory experiments?

5 What other oxidation states does copper have besides this one?

The atomic number of copper is 29, and its 29 electrons are arranged in orbitals thus:

$$1s^2\ 2s^2\ 2p^6\ 3s^2\ 3p^6\ 3d^{10}\ 4s^1$$

6 How many electrons would you expect a copper atom to lose in forming its compounds, in view of what you know about the tendency of atoms to achieve stable electron structures?

7 Does it surprise you that the commonly encountered oxidation state of copper is $2+$? Explain.

8 Given the electron structure of the Cu^+ ion, what colour would you expect copper(I) compounds to be?

The electronic structure of copper suggests that its stable oxidation state should be $1+$, yet experience shows that the $2+$ state is the more stable under normal conditions. Nevertheless, Cu(I) compounds do exist, and in this practical you will prepare some of them and examine their reactions. Your results may help you understand why, under normal conditions, Cu(II) is more common than Cu(I).

Requirements

Each student or pair of students will require:
- Rack and 6 test tubes
- 2 boiling tubes
- 2 hard-glass ignition tubes
- Beaker ($100\,cm^3$)
- Beaker ($250\,cm^3$)
- Filter funnel
- Filter paper
- Dropping pipette
- Copper foil
- Copper turnings
- Potassium sodium 2,3-dihydroxy-butanedioate (tartrate) (Rochelle Salt)
- Glucose
- Copper(II) oxide
- Copper(II) chloride (anhydrous)
- Copper(II) bromide (anhydrous)

The following solutions at usual bench concentration:
- Dilute hydrochloric acid
- Dilute sulphuric acid
- Dilute nitric acid
- Concentrated hydrochloric acid
- Sodium hydroxide
- Ammonia
- Copper(II) sulphate, about 0·25 M
- Potassium iodide
- Sodium thiosulphate

Time required 2 double periods

69

B   Copper(I) and copper(II)

EXPERIMENT 1 The preparation of copper(I) oxide

Copper(I) oxide is normally made by reducing Cu^{2+} ions in alkaline solution. This reaction forms the basis of Fehling's (and Benedict's) test for reducing sugars.

Put about 5 cm^3 of copper(II) sulphate solution in a boiling tube. In a separate boiling tube mix 5 cm^3 of sodium hydroxide solution and about 1 g of potassium sodium tartrate (Rochelle salt). Add this solution to the copper(II) sulphate until the precipitate formed at first just dissolves. You now have Fehling's solution. Now add about 1 g of glucose to the mixture and boil gently until an orange–red precipitate forms.

CARE Eye protection must be worn.

Allow the precipitate to settle (centrifuge if necessary), then decant the solution and wash the precipitate with distilled water. Keep the precipitate for use in experiment **2**.

The orange–red precipitate is copper(I) oxide.

9   What substance acted as the reducing agent in converting Cu(II) to Cu(I)?

10   What is the purpose of the tartrate ions in this experiment?

11   Would you describe copper(I) oxide as stable or unstable relative to copper(II) oxide under normal conditions? Give your evidence.

EXPERIMENT 2   The reactions of copper(I) oxide and copper(II) oxide with acids

Put a small amount (about 0·1 g) of your copper(I) oxide in each of three test tubes. Put a similar amount of copper(II) oxide in each of three further tubes. Use the tubes to investigate the reaction of each oxide with dilute hydrochloric acid, dilute sulphuric acid and dilute nitric acid in turn. Add the acid slowly to the oxide until acid is present in excess, then warm gently and observe **closely** to see what is formed. (In some cases there is more than one product.)

12   Tabulate your observations.

13   What generalization can you make about the reactions of copper(II) oxide with acids?

14   What evidence is there that   **a** Cu^+ ions,   **b** Cu^{2+} ions are formed when copper(I) oxide reacts with dilute sulphuric acid?

15   What other product is formed in the reaction of copper(I) oxide with dilute sulphuric acid?

16   What happens to the Cu^+ ion when copper(I) oxide reacts with dilute sulphuric acid?

17   Is the reaction with dilute nitric acid similar? If not, why not?

18   What is the oxidation state of copper in the product from the reaction of copper(I) oxide with dilute hydrochloric acid?

Your results in experiment **2** should show that copper(I) can exist under certain conditions but under other conditions the Cu^+ ion disproportionates.

19   What is meant by disproportionation?

20   To what does Cu^+ disproportionate in the reaction of copper(I) oxide with dilute sulphuric acid?

21   Under what circumstances do copper(I) compounds remain stable without disproportionating?

The relative stability of copper(I) and copper(II) is discussed further in *Chemistry in Context*, section 19.10.

EXPERIMENT 3 The preparation of copper(I) chloride

The Cu^+ ion normally disproportionates in aqueous solution, though it is quite stable under non-aqueous conditions. Copper(I) chloride is insoluble, so it is not subject to the disproportionation that Cu^+ undergoes in aqueous solution. Furthermore, copper(I) forms a complex ion, $CuCl_2^-$, in the presence of excess chloride ions, and this increases the stability of copper(I) chloride relative to copper(II) chloride. All this means that copper(I) chloride is quite stable and easily prepared. In fact, it can be prepared by reversing the normal disproportionation of copper(I) and reacting copper(II) with copper metal.

Put about 0·5 g of copper(II) oxide in a boiling tube and add 5–10 cm³ of concentrated hydrochloric acid. **CARE Eye protection must be worn.** Warm to obtain a clear green solution of copper(II) chloride. Now add about 1 g of copper turnings and boil gently for five minutes or so. Filter the resulting solution into about 200 cm³ of distilled water in a beaker. Allow the precipitate of copper(I) chloride to settle, then decant off the water.

22 Write an equation for the formation of copper(I) chloride from copper metal and copper(II) chloride.

23 Does copper(I) chloride appear to be a typical transition metal compound? Compare your answer with the answer you gave to question **8**.

24 What similarities are there between copper(I) chloride and silver chloride, AgCl? Why is it reasonable to expect copper(I) chloride to resemble silver chloride?

25 Bearing in mind the comparison between copper(I) chloride and silver chloride, predict the effect of ammonia solution on solid copper(I) chloride.

Test your prediction by adding a few cm³ of ammonia solution to a little solid copper(I) chloride in a test tube.

26 Record your results and try to explain what has happened.

Although copper(I) chloride is insoluble in water, it forms a soluble complex ion in the presence of excess chloride ions:

$$CuCl(s) + Cl^-(aq) \rightleftharpoons CuCl_2^-(aq)$$

Add 2 cm³ of concentrated hydrochloric acid to a small amount of your copper(I) chloride in a test tube. Note what happens, then pour the contents of the tube into about 50 cm³ of cold water in a small beaker.

27 Record your observations and explain them in terms of the above equilibrium.

28 Re-examine the copper(I) chloride. Are there any signs of reversion to the copper(II) state?

EXPERIMENT 4 The thermal decomposition of copper(II) halides

The previous experiments show that copper(I) chloride is reasonably stable with respect to copper(II) chloride. Perhaps copper(II) chloride will decompose to copper(I) chloride when it is heated.

Working in a fume cupboard, heat very small samples of anhydrous copper(II) chloride and copper(II) bromide in separate hard-glass ignition tubes. Heat gently at first, then more strongly. Observe closely to see if any gases are evolved, and note the appearances of the residues.

29 Do the halides decompose on heating? Write equations for any reactions that occur.

30 Which halide decomposed more readily?

31 How stable would you expect copper(II) iodide to be relative to copper(I) iodide?

EXPERIMENT 5   The preparation of copper(I) iodide

Add 3 cm^3 of potassium iodide solution to 3 cm^3 of copper(II) sulphate solution in a test tube. Note carefully what happens, and allow the contents of the tube to settle. Add sodium thiosulphate solution to the tube until the solution is clear and you can tell the colour of the solid that has been formed.

32   What substances are produced when KI reacts with $CuSO_4$? Write an equation for the reaction.

33   Does the result of this reaction agree with your answer to question **31**?

34   In this practical you have encountered copper in both oxidation states, $+1$ and $+2$. Summarize the conditions which favour the existence of each of the two oxidation states.

35   Is copper a transition metal?

PRACTICAL 27
Identifying cations

Introduction

Many common cations can be identified from the colour of their hydroxides and/or the colour they produce during flame tests.

The intention of this practical is to develop a simple, systematic method of identifying the following cations, provided there is only one cation present in the substance under test:

Al^{3+}, Ag^+, Ba^{2+}, Ca^{2+}, Cr^{3+}, Cu^{2+}, Fe^{2+}, Fe^{3+}, K^+, Mg^{2+}, Mn^{2+}, Na^+, NH_4^+, Ni^{2+}, Pb^{2+}, Zn^{2+}.

Appearance

The colour of a substance may give a clue to its identity and the cation it contains as shown in table 1 below.

Colour	Inference
white or colourless	probably does not contain a transition metal cation
not white or transparent	probably contains a transition metal cation
blue	possibly Cu^{2+}, Ni^{2+}
green	possibly Cu^{2+}, Cr^{3+}, Fe^{2+}, Ni^{2+}
yellow	possibly Fe^{3+}

Table 1 The colours of substances and their constituent cations

1   Why are the solutions of most transition metal cations coloured?

Preparing a solution of the substance

If you are provided with a solid, it is most convenient to carry out a flame test on this, but you will need to make an aqueous solution of the substance before other tests are carried out.

Put one spatula-full of the substance (i.e. 0.2–0.4 g) in a test tube and shake this with about 5 cm^3 of distilled water. If the solid will not dissolve in the cold, warm gently.

If the solid is insoluble in water, try cold and then hot dilute nitric acid. **CARE Eye protection must be worn.** If this fails, consult your teacher. It may be necessary to use concentrated hydrochloric acid.

Flame Tests

Dip a nichrome wire in concentrated hydrochloric acid (**CARE Eye protection must be worn**) and then heat it in a roaring bunsen flame until it no longer imparts colour to the flame. The wire is now clean. Mix a little barium chloride with concentrated hydrochloric acid on a watch glass, moisten the wire with this solution and hold the wire in the bunsen flame. Clean the wire thoroughly again and repeat the test with the chlorides of Ca^{2+}, Cu^{2+}, Mg^{2+}, K^+ and Na^+ in turn.

2   What are the flame colours (if any) of the following metal ions?
    a Ba^{2+} **b** Ca^{2+} **c** Cu^{2+} **d** Mg^{2+} **e** K^+ **f** Na^+

3   Why do certain metal ions give flame colours in this way?

4   One of the six metal ions tested gives no flame colour. Why is this?

Requirements

Each student will require:
- Conc. hydrochloric acid
- The following solids for flame tests: Barium chloride, calcium chloride, copper chloride, magnesium chloride, potassium chloride, sodium chloride, ammonium chloride.
- Aqueous solutions (approx. 0.1 M) of the following cations: Al^{3+}, Ag^+, Ba^{2+}, Ca^{2+}, Cr^{3+}, Cu^{2+}, Fe^{2+}, Fe^{3+}, K^+, Mg^{2+}, Mn^{2+}, Na^+, NH_4^+, Ni^{2+}, Pb^{2+}, Sn^{2+}, Zn^{2+}
- Sodium hydroxide solution (approximately 2.0 M)
- Unknown soluble solids containing a single cation and labelled X, Y, etc.
- Nichrome wire
- Bunsen burner
- Watch glass
- Rack with four test tubes
- Filter funnel
- Filter paper

Time required 2 double periods

Testing with sodium hydroxide solution

Add a few drops of sodium hydroxide solution to 3 cm^3 of Al^{3+}(aq).
CARE Eye protection must be worn.

5   What happens? Write an equation for the reaction.

6   Does the precipitate remain or dissolve with excess NaOH(aq)?

Repeat the experiment with solutions containing the following cations in place of Al^{3+}:

Ag$^+$, Ba^{2+}, Ca^{2+}, Cr^{3+}, Cu^{2+}, Fe^{2+}, Fe^{3+}, K$^+$, Mg^{2+}, Mn^{2+}, Na$^+$, NH$_4^+$, Ni^{2+}, Pb^{2+}, Sn^{2+}, Zn^{2+}.

Record your results in a table similar to table 2. (The table has been completed for Al^{3+} to show you what is required.) Write the formulae and colours of precipitates, write equations for reactions and say whether precipitates remain or dissolve when excess sodium hydroxide is added.

Cation tested	What happens when a few drops of NaOH(aq) are added to 3 cm^3 of the cation solution?	What happens when excess NaOH(aq) is added?
Al^{3+}	white precipitate of Al(OH)$_3$ forms. Al^{3+}(aq) + 3OH$^-$(aq) $\longrightarrow$ Al(OH)$_3$(s)	white ppte. dissolves forming a clear solution Al(OH)$_3$(s) + OH$^-$(aq) $\longrightarrow$ Al(OH)$_4^-$(aq)

Table 2 Reactions of various cations with sodium hydroxide solution

7   Which cations give no precipitate with NaOH(aq)?

8   How can these three cations be distinguished using flame tests?

9   Which cations give a white precipitate with NaOH(aq) which remains with excess NaOH(aq)?

10   How can these three cations be distinguished using flame tests?

11   Which cations give distinctive coloured precipitates with NaOH(aq)?
Al^{3+}, Pb^{2+}, Sn^{2+} and Zn^{2+} give white precipitates with NaOH(aq) which dissolve in excess. These four ions cannot be distinguished by flame tests.
Precipitate further samples of Al(OH)$_3$, Pb(OH)$_2$, Sn(OH)$_2$ and Zn(OH)$_2$. Filter each precipitate and heat a sample of it to dryness. Copy out and complete table 3.

Precipitate	Colour before heating	What happens on heating?	Colour after heating
Al(OH)$_3$			
Pb(OH)$_2$			
Sn(OH)$_2$			
Zn(OH)$_2$			

Table 3

12   How can Al^{3+}, Pb^{2+}, Sn^{2+} and Zn^{2+} be distinguished?

Identifying the cation in an unknown substance

Use the information you have gathered in the previous tests to identify the cation in each of the unknown substances provided by your teacher.

Further work
If time permits, develop a table similar to the one that you have obtained for the reactions of cations with NaOH, showing their reactions with ammonia solution.

PRACTICAL 28
The oxidation states of nitrogen

Introduction

As the element, N_2, nitrogen is remarkably unreactive. Yet many compounds of nitrogen exist, and in these compounds nitrogen shows several different oxidation states. In this practical you will meet a number of these states. Some of the compounds and ions you will encounter appear in table 1.

Name	Formula	Appearance
nitrogen gas	—	—
ammonia	—	—
ammonium ion	—	—
nitrogen dioxide	NO_2	pungent, toxic red-brown gas
dinitrogen oxide	N_2O	colourless gas
nitrogen monoxide	NO	colourless gas which turns brown in air
nitric acid	—	—
nitrate ion	—	—
nitrous acid	HNO_2	blue in aqueous solution
nitrite ion	NO_2^-	colourless in aqueous solution

Table 1 The formula and appearance of some important compounds and ions containing nitrogen

1 Copy out table 1, filling in the missing data and giving the oxidation state of nitrogen in each compound.

A   Redox reactions of nitric acid and nitrates

CARE Concentrated nitric acid is very corrosive. Eye protection must be worn when using it. Any acid splashed on your skin or clothes should be washed off immediately with plenty of water.

Pure nitric acid is colourless. The acid slowly decomposes in sunlight:

$$4HNO_3 \rightleftharpoons 2H_2O + 4NO_2 + O_2$$

The nitrogen dioxide formed in this way causes the nitric acid to turn yellow. This reaction is reversible. This can be shown by adding a little water to some concentrated nitric acid and observing what happens to the yellow colour.

The reduction of nitrate ions produces a number of different products, depending on the reducing agent, the pH and the concentration of the nitrate ions.

EXPERIMENT 1   Reaction of nitric acid with copper
Working in a fume cupboard, add **a few drops** of concentrated nitric acid (about 14 M) to **one** copper turning in a test tube.

2 What nitrogen compound is formed? What is the oxidation state of nitrogen in this compound?

3 Write a half equation for the reduction of nitrate ions in acid solution in this reaction.

4 Write a half equation for the oxidation of copper in this reaction.

5 Combine the two half equations to give a full ionic equation for the reaction.

Carefully dilute 2 cm³ of concentrated nitric acid with an equal volume of water to produce a 7 M solution. Add two or three copper turnings to this acid and observe the gas that is evolved very closely.

Requirements

Each student or pair of students will require:

- Rack + 6 hard-glass test tubes
- Beaker (250 cm³)
- Conical flask (100 cm³)
- Glass rod
- Splints
- Universal indicator paper
- Copper turnings
- Potassium nitrate
- Copper nitrate
- Sodium nitrite
- Ammonium dichromate(VI) (to be used in teacher demonstration only)
- Devarda's alloy
- Copper wire, about 24 swg
- Ice

The following solutions at usual bench concentration:

- Dilute sulphuric acid
- Dilute sodium hydroxide
- Dilute nitric acid
- Concentrated nitric acid
- Potassium iodide
- Potassium manganate(VII)

- Access to a fume cupboard

Time required 1 double period

6 What nitrogen compound is produced as the first product of this reaction? What is the oxidation state of nitrogen in this compound?

7 Write a half equation for the reduction of nitrate ions in acid solution in this reaction.

8 Combine this half equation with a half equation for the oxidation of copper to give an overall equation for the reaction.

9 Suggest why different products are formed when copper reacts with nitric acid at different concentrations.

10 What was the colour of the solution produced when the copper dissolved in the acid in each reaction? Account for any differences.

EXPERIMENT 2   The action of heat on metal nitrates

Working in a fume cupboard, heat small amounts of first potassium nitrate and then copper(II) nitrate in separate hard-glass test tubes. Warm the solids gently at first and then more strongly if there is no noticeable reaction. Try to identify the gases evolved and the solid residue in each case. If you cannot immediately identify the residue, add a little dilute sulphuric acid and observe the result. You may need to wait until you have completed experiment 4 before you can interpret the reaction of the residue with acid.

11 Name the products of the thermal decomposition of
    a potassium nitrate
    b copper(II) nitrate.

Write equations for each reaction and give the oxidation numbers of nitrogen in each of the products.

EXPERIMENT 3   The reduction of nitrate ions in alkaline solution

Put $2 \, cm^3$ of dilute (2 M) nitric acid in a test tube and add $5 \, cm^3$ of dilute sodium hydroxide.   **CARE Eye protection must be worn.**   Add a spatula measure of Devarda's alloy (an alloy of copper, aluminium and zinc) to the solution and warm. Identify the gas evolved.

12 To what have the nitrate ions been reduced? What is the oxidation number change in this reaction?

13 Which of the components of Devarda's alloy is most likely to have acted as the reducing agent in this reaction?

B Redox reactions of nitrous acid

EXPERIMENT 4   The preparation and reactions of nitrous acid

Cool about $10 \, cm^3$ of dilute sulphuric acid in a test tube by standing the tube in an ice bath for five minutes. Add the cooled acid to about 1 g of sodium nitrite in a second tube. The resultant solution contains nitrous acid, HNO_2.

14 Record the appearance of the solution of nitrous acid.

15 What is the oxidation number of nitrogen in
    a sodium nitrite
    b nitrous acid?
    Is the formation of nitrous acid from sodium nitrite a redox reaction? Explain your answer.

16 Write an ionic equation for the reaction.

Transfer about one third of the nitrous acid solution to another tube and warm gently. Identify the gas evolved.

17 What has been formed? What is the oxidation number of nitrogen in this compound?

18 The thermal decomposition of nitrous acid is a disproportionation reaction. What is meant by disproportionation?

19 Another nitrogen-containing product must also have been formed in this disproportionation reaction. Will it contain nitrogen in a higher or lower oxidation state than in nitrous acid? Explain your answer. Suggest what this other product may have been.

20 Would you expect nitrous acid to behave as an oxidizing agent, a reducing agent, or both? Explain your answer.

Add half of the remaining nitrous acid solution to a little potassium iodide solution.

21 Identify the product.

22 Has nitrous acid behaved as an oxidizing agent or as a reducing agent in this reaction?

Add the remaining portion of nitrous acid solution to a little potassium manganate(VII) solution.

23 Describe what happens.

24 Has the nitrous acid behaved as an oxidizing agent or as a reducing agent in this reaction?

25 Why does nitric acid not undergo disproportionation reactions?

C Redox reactions of ammonia and ammonium ions

26 Does ammonia burn in air under normal conditions? Under what conditions can it be made to burn?

EXPERIMENT 5 The catalytic oxidation of ammonia

Working in a fume cupboard, put about $10 \, cm^3$ of concentrated ('0·880') ammonia solution into a $100 \, cm^3$ conical flask.

CARE Concentrated ammonia solution is corrosive and smells powerfully of ammonia. Ammonia gas is irritant and poisonous. Avoid spilling the solution or breathing its fumes.

Make a spiral of medium gauge copper wire and suspend it from a glass rod so that it hangs just above the surface of the ammonia solution as shown in figure 1. Warm the conical flask slightly to get a good supply of ammonia gas, then heat the copper spiral to red heat (do not heat it too strongly or it will melt). Replace the spiral and observe closely.

27 What evidence is there that an exothermic reaction is occurring?

28 To what is the ammonia being oxidized?

29 What is acting as the oxidizing agent in this reaction?

30 Write an equation for the reaction.

31 What is the industrial significance of this reaction?

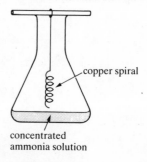

Figure 1 The catalytic oxidation of ammonia.

EXPERIMENT 6 Oxidation of ammonium ions by dichromate(VI) ions

Note: Because of the possible danger of inhaling chromate dust which may become airborne during this reaction, the experiment should be demonstrated by your teacher.

Working in a fume cupboard, gently heat about 1 g of ammonium dichromate(VI) crystals, $(NH_4)_2Cr_2O_7$, in a hard-glass test tube.

32 In this reaction, ammonium ions are oxidized by dichromate(VI) ions to nitrogen gas. To what are the dichromate(VI) ions reduced?

33 Write an equation for the reaction.

34 Why do ammonia and ammonium ions not undergo disproportionation reactions?

PRACTICAL 29
The preparation of sodium thiosulphate and an investigation of some of its properties

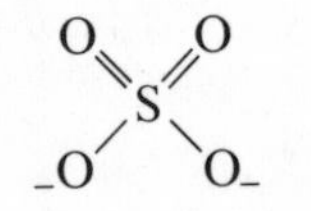

Sulphate ion, SO_4^{2-}

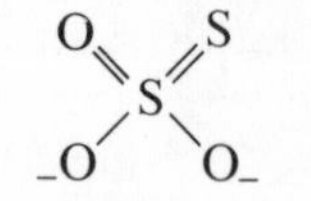

Thiosulphate ion, $S_2O_3^{2-}$

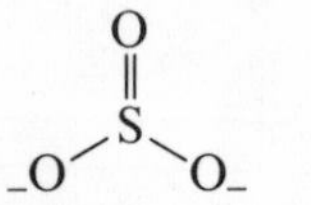

Sulphite ion, SO_3^{2-}

Figure 1 Structures of the sulphate, thiosulphate and sulphite ions

Introduction

This practical provides an opportunity to prepare a crystalline inorganic compound in as high a yield as possible. **Note:** the preparation involves refluxing for about two hours, so it may be necessary to start the practical during one lesson and complete it in the next.

The structure of the thiosulphate ion is shown in figure 1. Compare it with the sulphate ion, also shown in figure 1.

1 What is the oxidation number of sulphur in **a** $S_2O_3^{2-}$, **b** SO_4^{2-}, **c** SO_3^{2-}?

2 Would you expect the thiosulphate ion to act as an oxidizer, as a reducer, or as both? Explain your answer.

The thiosulphate ion can be formed by reacting sulphite ions with sulphur. A glance at the structure of the sulphite ion in figure 1 should make the chemistry of this reaction clear.

$$S_8 + 8SO_3^{2-} \longrightarrow 8S_2O_3^{2-}$$

In this practical you will be making crystals of sodium thiosulphate-5-water, $Na_2S_2O_3.5H_2O$, and investigating some chemical properties of the compound.

Procedure

Preparation of sodium thiosulphate-5-water
Fit a 100 cm^3 round-bottomed flask with a reflux condenser and put in it 10 g of hydrated sodium sulphite crystals, $Na_2SO_3.7H_2O$. Add 50 cm^3 of distilled water and 1·5 g of powdered roll sulphur. Arrange the apparatus over a tripod and gauze and reflux for 1–2 hours, after which time no further sulphur should dissolve.

3 How many moles of **a** $Na_2SO_3.7H_2O$, **b** sulphur have you used? Which reagent is in excess?

4 What is the maximum mass of $Na_2S_2O_3.5H_2O$ that can be formed in this preparation?

Allow the solution to cool, then filter to remove excess sulphur. Put the filtrate into an evaporating basin and evaporate until the volume of solution is about 10 cm^3. Leave to cool.

Sodium thiosulphate readily forms a **supersaturated solution** in water. This is a solution in which more solute is present than the solvent can normally hold, yet crystallization of the excess solute does not occur until the solution is 'seeded'. Your solution will probably be supersaturated and will not crystallize until the side of the evaporating basin is gently scratched with a glass rod to 'seed' the crystals.

When crystallization is complete, dry the crystals by gently pressing them between two sheets of filter paper and then weigh the dry crystals.

5 What is your percentage yield of sodium thiosulphate-5-water?

6 Reactions between inorganic compounds, unlike organic reactions, often occur very rapidly. Suggest a reason why this particular reaction is relatively slow and requires prolonged heating.

Reactions of sodium thiosulphate
In the following experiments it is instructive to compare the reactions of the thiosulphate ion with those of the sulphate ion which, as we have seen, is structurally similar.

EXPERIMENT 1 Action of heat

Heat a few crystals of $Na_2S_2O_3.5H_2O$ in a hard-glass test tube, gently at first then more strongly. Repeat using hydrated sodium sulphate crystals, $Na_2SO_4.10H_2O$.

7 Does the thiosulphate decompose on heating? If so, what is formed?

8 Compare the thermal stability of the thiosulphate ion with that of the sulphate ion. Explain your answer using the bond energies given in table 1.

EXPERIMENT 2 Reaction with iodine

Prepare a solution of sodium thiosulphate by dissolving about 2 g of crystals in about 20 cm^3 of water.

Put 2–3 cm^3 of a solution of iodine in potassium iodide into a test tube and add an excess of your thiosulphate solution.

9 You will probably have met this reaction in volumetric analysis. To what has the iodine been converted? To what has the thiosulphate been converted?

10 How is the thiosulphate behaving in this reaction?

11 Write a balanced equation for the reaction.

12 Would you expect iodine to react with sulphate ions? (Try the experiment if you are not sure.) Give a reason for your answer.

EXPERIMENT 3 Reaction with chlorine

Add an excess of chlorine water to 2–3 cm^3 of sodium thiosulphate solution. Observe the result, then add dilute hydrochloric acid followed by barium chloride solution.

13 To what has the thiosulphate been converted in this reaction?

14 How is the thiosulphate behaving here?

15 What differences are there between the reaction of thiosulphate with chlorine and its reaction with iodine?

16 Attempt to write a balanced equation for the reaction.

EXPERIMENT 4 Effect of dilute acid

Put 3 cm^3 of sodium thiosulphate solution in a test tube and add an equal volume of dilute hydrochloric acid. Observe the contents of the tube over the course of several minutes. Cautiously smell the solution after a few minutes.

17 What two substances can you identify as products of this reaction?

18 What is the oxidation number of sulphur in
a thiosulphate
b the two products of this reaction?

19 Write a balanced ionic equation for the reaction.

20 What type of reaction is this?

21 Would you expect sulphate ions to behave similarly? Give a reason for your answer.

22 The reaction of thiosulphate with acid is fairly slow. Suggest a reason why. Refer back to your answer to question 6. Can you see a link between this reaction and the reaction used to prepare sodium thiosulphate?

Bond	Bond energy /kJ mole^{-1}
S = O	523
S = S	431

Table 1 Average bond energies

Requirements

Each student or pair of students will require:

- Round-bottomed flask, with reflux condenser to fit (*Quickfit* apparatus is ideal).
- Filter funnel
- Filter paper
- Evaporating basin
- Glass rod
- 4 hard-glass test tubes
- Sodium sulphite-7-water, $Na_2SO_3.7H_2O$
- Powdered roll sulphur (**not** flowers of sulphur)
- Distilled water
- Sodium sulphate-10-water, $Na_2SO_4.10H_2O$

The following solutions at usual bench concentration:
- Iodine in potassium iodide
- Chlorine water
- Dilute hydrochloric acid
- Barium chloride

- Bunsen burner
- Tripod
- Gauze
- Access to balance

Time required 2 double periods

PRACTICAL 30
Alkanes

Requirements

Each student, or pair of students, will require:

- Rack and 4 test tubes
- Boiling tube
- Beaker (250 cm³)
- Small glass trough or large beaker
- Hard-glass watch glass
- Boiling tubes fitted with bungs and glass bends as in figure 1.
- Aluminium foil
- Glass wool
- Splints
- Access to a bright light source (ideally sunlight, but a high intensity lamp or even a fluorescent light will do)
- Hexane or cyclohexane
- Paraffin oil (medicinal paraffin)
- Paraffin wax
- Bromine. Sodium thiosulphate solution (approximately molar) should also be available for dealing with bromine spills.
- Concentrated sulphuric acid
- Concentrated ammonia solution

The following solutions at usual bench concentration:

- Bromine water
- Potassium manganate(VII) (potassium permanganate)
- Dilute sodium hydroxide
- Access to a fume cupboard

Time required 1–2 double periods

Introduction

The alkanes form the simplest family of hydrocarbons and can be regarded as the basic skeletons to which functional groups are attached in more complex organic compounds. They are in many ways the most important homologous series, because they are the principle constituents of crude oil, from which most organic chemical products are derived.

In this practical you will investigate some of the physical and chemical properties of alkanes. You will be working with a number of different alkanes or mixtures of alkanes.

Methane, CH_4	Natural gas from the laboratory gas taps is a good source. (North Sea gas is about 95% methane, most of the remainder being higher alkanes.)
Hexane, C_6H_{14}	Cyclohexane is a good substitute if hexane is not available. Cycloalkanes have very similar properties to straight-chain alkanes.
Medicinal paraffin	A liquid mixture of long-chain alkanes.
Paraffin wax	A solid mixture of long-chain alkanes.

Physical properties

A Volatility
Use a data book to answer the following questions.

1   What are the boiling points of methane and hexane?

2   Which is the first straight-chain alkane to be liquid at room temperature and pressure?

3   Which alcohol has a boiling point closest to that of hexane? Comment on your answer.

B Solubility
Put 2–3 cm³ of hexane in a test tube and add about twice this volume of water. Shake, then stand the tube in a rack.

4   Does hexane dissolve in water? Suggest a reason why the two liquids behave in this way.

5   Is hexane more or less dense than water?

6   Use a data book to discover how the densities of straight-chain alkanes change as their relative molecular masses increase.

Chemical properties

A Action of some common reagents on hexane
Investigate the reaction of hexane (a typical alkane) with
   a bromine water,
   b potassium manganate(VII) solution,
   c sodium hydroxide solution,
   d concentrated sulphuric acid.   **CARE Eye protection must be worn.**

Use about 2 cm³ of hexane with 2 cm³ of the reagent in each case, shake and look for any signs of a chemical reaction having occurred.

7   Does hexane react with any of the above substances under normal laboratory conditions?

8 The alkanes used to be called the **paraffins**. Suggest why this name was used.

9 In which substance is bromine more soluble – hexane or water? Why?

10 In which substance is potassium manganate(VII) more soluble – hexane or water? Why?

The experiments you have carried out suggest that with common reagents under ordinary laboratory conditions, alkanes are very unreactive. Alkanes are not however, completely inert. If they were they would not be very useful to man.

B Reaction of hexane with bromine
Put 2 cm^3 of hexane in each of two test tubes. Add **one drop** of bromine (not bromine water) to each tube.

CARE Bromine is corrosive and toxic. Wear eye protection and gloves and work in a fume cupboard when adding the bromine.

Wrap aluminium foil around one of the tubes so that it is light-proof, then stand the two tubes side by side in a test tube rack. Leave the rack in bright sunlight or near to a bright light source for 5–10 minutes, then examine the appearance of each tube. Blow a little ammonia across the top of each tube.

11 Under what conditions does hexane react with bromine?

12 What inorganic product is formed in this reaction?

13 Write a structural formula for one possible organic product of the reaction.

14 What type of reaction is this?

C Combustion of alkanes
Fill a boiling tube with methane from the gas tap. Cork the tube, stand it in a rack and apply a lighted splint to the mouth of the tube.

15 Write a balanced equation for the reaction which occurs.

16 Would you expect the same result if you held the tube upside down and lighted the gas? Explain your answer.

Put a small piece of paraffin wax on a hard-glass watch glass and apply a lighted splint to it.

17 Can the wax be easily ignited?

18 Why is wax harder to ignite than methane, even though they both contain alkanes?

19 Why does a candle have a wick?

D Cracking of alkanes
At high temperatures, the atoms in an alkane molecule vibrate rapidly. If the temperature is high enough, the vibration becomes sufficiently vigorous for chemical bonds to break. This breakage of the C—C bonds in alkanes leads to the formation of smaller hydrocarbon fragments, and is called **cracking**. The temperature required for cracking can be reduced by the use of solid catalysts.

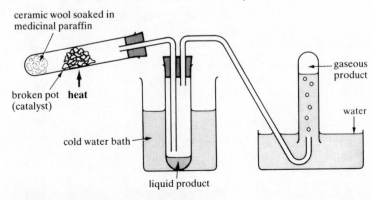

Figure 1 Cracking medicinal paraffin.

Set up the apparatus shown in figure 1. Wearing eye protection, heat the broken pot strongly and warm the medicinal paraffin gently so that a steady stream of vapour passes over the hot catalyst. (Beware of melting the rubber bung.) Continue until a few cm^3 of liquid product have collected, by which time you should have collected several tubes of gas. Shake some of the gas with a few drops of bromine water.

20 What happens? What structural feature does the gaseous product's molecule contain?

Test the gaseous product with a burning splint.
Compare the appearance and viscosity of the liquid product with that of the paraffin oil. Shake a little of the liquid product with an equal volume of bromine water.

21 What evidence is there that the liquid product contains smaller molecules than the original paraffin oil?

22 Eicosane (C$_{20}$H$_{42}$) is a typical alkane present in paraffin oil. Write an equation (using structural formulae) to represent one possible outcome of the cracking of eicosane.

23 Why are cracking reactions of this sort important in the petrochemical industry?

24 What would you expect to happen if you repeated the cracking experiment using the liquid product instead of paraffin oil? Test your prediction if you have time.

25 Summarize the important physical and chemical properties of alkanes that you have encountered in this practical. Where appropriate give an example of the industrial, social or environmental importance of that property.

PRACTICAL 31
Alkenes

Introduction

The intention of this practical is to prepare a gaseous and a liquid alkene and to look at some of their important reactions.

Preparing alkenes

Although large quantities of ethene and propene are obtained industrially from cracking reactions, the most convenient method of preparing small amounts of any alkene is by the dehydration of the corresponding alcohol. In this experiment you will prepare ethene and cyclohexene from ethanol and cyclohexanol respectively.

1   What is meant by the term dehydration?

2   Write equations for the preparation of **a** ethene and **b** cyclohexene by dehydration of their corresponding alcohols.

3   What is the most commonly used dehydrating agent?
    (We could use this reagent or concentrated phosphoric(V) acid to dehydrate alcohols, but in preparing ethene from ethanol we shall use a catalytic dehydrating agent such as aluminium oxide, silicon(IV) oxide or porous pot.)

4   Porous pot is made from clay. What elements will it contain?

Preparing ethene from ethanol

Arrange the apparatus as in figure 1. Heat the porous pot with a small flame. **CARE Eye protection must be worn.** Do **not** heat the ceramic wool directly and remember to remove the delivery tube from the water when you are not heating. Beware of melting the rubber bung. Discard the first test tube of gas and then collect three or four test tubes of ethene.

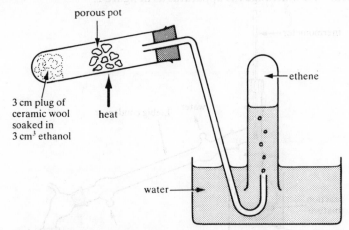

Figure 1 Preparing ethene from ethanol.

5   Why is the ceramic wool not heated directly?

6   Why must the delivery tube be removed from the water when you are not heating?

7   Why should you discard the first test tube of gas?

Requirements

For the preparation and reactions of ethene, each student or pair of students will require:
- Ceramic wool
- Ethanol
- Teat pipette
- Bunsen burner
- Porous pot
- Hard-glass test tube fitted with bung and delivery tube
- 4 test tubes
- Beaker (250 or 400 cm³)
- Splint
- Dilute bromine water
- Dilute potassium manganate(VII)
- Dilute sulphuric acid
- Sodium carbonate solution

For the preparation and reactions of cyclohexene, each student or pair of students will require:
- Round-bottomed flask or pear shaped flask (50 or 100 cm³)
- Still-head
- Thermometer (0–110°C) and holder
- Liebig condenser
- Collecting (receiving) vessel
- Beaker (250 or 400 cm³)
- Cyclohexanol
- Concentrated phosphoric(V) acid
- Measuring cylinder (10cm³)
- Separating funnel
- Rack and 3 test tubes
- Small conical flask
- Small crucible
- Teat pipette
- Bunsen burner
- Splints
- Saturated sodium chloride solution
- Anhydrous calcium chloride
- Ceramic wool
- Bromine water
- Dilute potassium manganate(VII)
- Dilute sulphuric acid
- Conc. sulphuric acid
- Cyclohexane

Time required 2–3 double periods

Reactions of ethene

A Burning
Ignite the ethene in one test tube.   CARE Eye protection must be worn.

8   Describe the flame and write an equation for the combustion.

B Reaction with bromine
Add 1 cm^3 of dilute bromine water to a test tube of ethene and shake.

9   Describe what happens, write an equation for the reaction and name the product.

C Reaction with potassium manganate(VII)
Add 1 cm^3 of dilute sulphuric acid and 3 drops of potassium manganate(VII) solution to a test tube of ethene and shake.

10   Describe what happens.
The potassium manganate(VII) oxidizes ethene to ethane-1,2-diol and other products.

11   To what has the $KMnO_4$ been converted?
Repeat this test using 1 cm^3 of sodium carbonate in place of dilute sulphuric acid.

12   What happens this time?

13   Is the $KMnO_4$ reduced to the same product as before? Explain your answer.

Preparing cyclohexene from cyclohexanol

The aims of this experiment are:

A   to prepare cyclohexene by dehydrating cyclohexanol using concentrated phosphoric(V) acid.
B   to introduce the method of purification for an organic liquid.

14   Concentrated phosphoric(V) acid is preferred to concentrated sulphuric acid as a dehydrating agent for alcohols because it gives a higher yield of alkene. Why does concentrated sulphuric acid give a lower yield of alkene? (Hint: What side reactions occur with concentrated sulphuric acid?)

Preparation
Put 10 cm^3 of cyclohexanol in a small round-bottomed flask. *Slowly* add 4 cm^3 of concentrated phosphoric(V) acid and mix thoroughly.   CARE Eye protection must be worn.   Now arrange the apparatus as in figure 2.

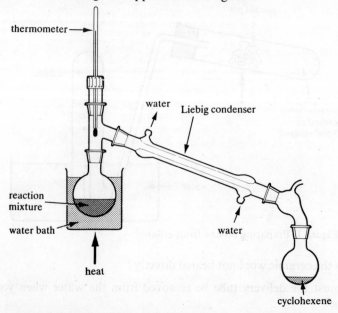

Figure 2 Preparing cyclohexene from cyclohexanol.

Heat the mixture in a water bath at 70°C for 15 minutes. Raise the temperature and distil very slowly, collecting the distillate which comes over between 70°C and 90°C.

Purification

Pour the distillate into a separating funnel and shake it with an equal volume of saturated sodium chloride solution.

Allow the two layers to separate and then run off the lower aqueous layer. Transfer the top layer (cyclohexene) to a small conical flask and add a few pieces of anhydrous calcium chloride. Mix the contents of the flask until the liquid is clear.

Finally, decant the alkene into a clean flask and re-distil it, collecting the liquid which distils between 81 and 85°C. (The boiling point of cyclohexene is 83°C.) Use your sample of cyclohexene in the experiments below.

15 Name two impurities present in the cyclohexene before purification.

16 What impurities are removed from the impure cyclohexene when it is shaken with sodium chloride solution? (Sodium chloride solution is used rather than water because it is denser than water and will separate from the cyclohexene more rapidly after shaking.)

17 Why is the cyclohexene mixed with anhydrous calcium chloride?

18 What are the three key stages in purifying an organic liquid?

19 What mass of cyclohexene did you obtain?
    Use the equation for the dehydration of cyclohexanol to calculate the theoretical yield of cyclohexene. (Assume that the phosphoric(V) acid is present in excess.)

20 What is your percentage yield of cyclohexene?

Reactions of cyclohexene

A Burning

Burn 1 *drop* of cyclohexene on a small tuft of ceramic wool in a crucible.

21 Suggest four products formed when the cyclohexene burns.

22 Write an equation for the complete combustion of cyclohexene (C_6H_{10}) in excess oxygen.

B Reaction with bromine

Add 1 cm^3 of bromine water to 3 drops of cyclohexene and shake.

23 Describe what happens, write an equation for the reaction and name the product.

C Reaction with potassium manganate(VII)

Add 1 cm^3 of dilute sulphuric acid and 3 drops of potassium manganate(VII) solution to 3 drops of cyclohexene and shake.

24 Describe and explain what happens.

D Reaction with concentrated sulphuric acid

Add 3 drops of concentrated sulphuric acid to 3 drops of cyclohexene.
CARE Eye protection must be worn.

25 Describe what happens, write an equation and name the product.

26 Many of the reactions of alkenes involve addition. What do you understand by the term 'addition reaction'?

27 Why do alkenes readily undergo addition reactions?

Further reactions

If time permits, carry out the reactions again, using cyclohexane in place of cyclohexene.

28 Account for the differences in reactivity between cyclohexene (a typical alkene) and cyclohexane (a typical alkane).

PRACTICAL 32

The preparation and properties of phenylethyne – a liquid alkyne

Requirements

For the preparation of phenylethyne each student, or pair of students, will require:

Stage A
- Dropping funnel (50 or 100 cm^3)
- Round-bottomed flask or pear-shaped flask (100 cm^3)
- Phenylethene (4 cm^3)
- 5% (v/v) solution of bromine in ethane-1,2-diol (ethylene glycol) – about 40 cm^3. (Working in a fume cupboard, put 15 cm^3 of bromine in 285 cm^3 of ethane-1,2-diol.)
- Access to a fume cupboard.

Stage B
- Potassium hydroxide (8 g)
- Antibumping granules (a few)
- Round-bottomed flask or pear-shaped flask (100 cm^3)
- Still-head
- Thermometer (0–250°C)
- Liebig condenser
- Collecting (receiving) vessel
- Measuring cylinder (10 cm^3)
- Bunsen burner

For the purification of phenylethyne each student or pair of students will require:
- Measuring cylinder (10 cm^3)
- Teat pipette
- 2 small flasks
- Anhydrous calcium chloride (a few granules)
- Access to a balance

For the purification of phenylethyne each student, or pair of students, will require:
- Rack + 3 test tubes
- Teat pipette
- Ammoniacal copper(I) chloride solution (3 cm^3). (Dissolve 5 g of copper(I) chloride in 50 cm^3 of dilute ammonia solution.)
- 5% (w/v) silver nitrate solution (3 cm^3)
- Dilute ammonia solution (a few drops)
- Bromine water (3 cm^3)
- Dilute sulphuric acid (2 cm^3)

Introduction

The first member of the homologous series of alkynes is ethyne (acetylene). This is a gas at room temperature and pressure, and so it is awkward to handle on a small scale.

On the other hand, phenylethyne (C_6H_5 . C≡CH) is a liquid which can be used to investigate the reactions of alkynes more conveniently.

The objectives of this experiment are to prepare phenylethyne and then examine its properties. The phenylethyne can be prepared by a relatively simple two-stage process:

A the bromination of phenylethene (styrene) forming 1-phenyl-1,2-dibromoethane, followed by

B the elimination of two molecules of hydrogen bromide from 1-phenyl-1,2-dibromoethane by reaction with alcoholic potassium hydroxide to form phenylethyne.

1 Write equations for stages **A** and **B** in the preparation of phenylethyne from phenylethene.

2 When two molecules of hydrogen bromide are removed from one molecule of a dibromocompound, the product is either an alkyne or a diene. Write the structural formulae and the names of all possible alkynes and dienes that could be obtained by the removal of two molecules of hydrogen bromide from one molecule of
   a 1,1-dibromobutane,
   b 1,2-dibromobutane,
   c 2,3-dibromobutane.

3 Why is there only one possible product when two molecules of hydrogen bromide are eliminated from one molecule of 1-phenyl-1,2-dibromoethane?

Preparation of phenylethyne

Stage A   Bromination of phenylethene
CARE Eye protection must be worn. Work in a fume cupboard – both bromine and phenylethene are irritants.

Put 50 cm^3 of the 5% solution of bromine in ethane-1,2-diol into a dropping funnel. Add the bromine solution a few cm^3 at a time to 4 cm^3 of phenylethene in a small round-bottomed flask. Swirl the flask thoroughly after each addition. Continue to add the bromine solution until some of it does not react and the yellow–orange colour remains after thorough mixing.

4 Why is a 5% solution of bromine used in this stage rather than pure bromine?

Stage B   Elimination of hydrogen bromide from 1-phenyl-1,2-dibromoethane
CARE Potassium hydroxide is very caustic. Avoid skin contact.

Weigh out 8 g of potassium hydroxide pellets (excess) and add them to the solution of 1-phenyl-1,2-dibromoethane obtained in stage A, a few at a time.

Mix the contents of the flask after each addition to ensure that the potassium hydroxide dissolves and reacts. If the reaction becomes too vigorous, cool the flask in cold water.

Add a few anti-bumping granules to the flask and arrange the apparatus for simple distillation. Warm the flask *gently* for a few minutes to ensure the reaction is complete. During this time a solid should appear. Now distil the mixture in the flask and collect the distillate (mainly water and phenylethyne) that comes over below 160°C in a 10 cm^3 measuring cylinder.

86

5 Why is the colour of the 1-phenyl-1,2-dibromoethane solution discharged during reaction with potassium hydroxide?

6 Suggest a possible mechanism for the reaction of potassium hydroxide with 1-phenyl-1,2-dibromoethane.

7 What is the solid which appears in the flask during stage B?

8 Draw a diagram of the apparatus used for distillation.

Purification of the phenylethyne

The distillate from stage B will collect as two layers in the measuring cylinder. The upper organic layer is impure phenylethyne. Separate the impure phenylethyne from the lower aqueous layer using a clean teat pipette. Transfer the phenylethyne to a small flask containing a little anhydrous calcium chloride. Allow the organic layer to stand in contact with anhydrous calcium chloride for some time (preferably overnight) and then decant the liquid into a small weighed flask.

Weigh the flask plus phenylethyne.

If time permits, re-distil the phenylethyne, collecting the fraction which boils between 141°C and 145°C. (The boiling point of phenylethyne is 143°C.)

9 What impurities in the phenylethyne would be removed by re-distilling it and collecting only the fraction that boils between 141°C and 145°C?

10 List the key stages in the purification of phenylethyne.

11 Calculate the maximum theoretical yield of phenylethyne. (Assume the density of phenylethene is 0·91 g cm^{-3}.)

12 What is your percentage yield of phenylethyne?

13 Why is your percentage yield less than 100?

Reactions of phenylethyne

Shake 3 drops of your phenylethyne with separate samples of each of the following reagents:

A 3 cm^3 of aqueous diamminosilver ions (ammoniacal silver nitrate). This can be made by adding drops of dilute ammonia solution to silver nitrate solution until the brown precipitate just dissolves.

B 3 cm^3 of aqueous diamminocopper(I) ions (ammoniacal copper(I) chloride).

C 3 cm^3 of bromine water.

D 2 cm^3 of dilute sulphuric acid plus 5 drops of potassium manganate(VII) solution.

E 2 cm^3 of dilute sodium carbonate plus 5 drops of potassium manganate(VII) solution.

14 Describe your observations for each test and explain the reactions which take place. Write equations wherever possible.

15 Would you expect the reactions of phenylethyne to be typical of alkynes? Explain your answer.

- Dilute sodium carbonate solution (2 cm^3)
- Potassium manganate(VII) solution (10 drops)

Time required 2 double periods

Requirements

Each student, or pair of students, will require:

Part 1:
- Benzene
- Cyclohexane
- Cyclohexene
- Methylbenzene
- Bromine water
- Dilute sulphuric acid
- Dilute potassium manganate(VII) solution
- Rack with 4 test tubes
- Access to a fume cupboard

Part 2:
- Methyl benzoate
- Concentrated sulphuric acid
- Concentrated nitric acid
- Ice
- Ethanol
- Conical flask (100 cm³)
- Measuring cylinder (10 cm³ or 25 cm³)
- Beaker (250 cm³ or 400 cm³)
- 2 small beakers (100 or 150 cm³)
- Rack with 2 test tubes
- Teat pipette
- Thermometer (0–110°C)
- Buchner funnel and flask
- Filter paper
- Melting point tube
- Access to balance
- Access to melting point apparatus
- Access to suction pump

Time required 2 double periods

Introduction

The most important, and the prototype, aromatic compound is benzene. Indeed, aromatic compounds are normally considered to be those compounds with chemical properties similar to benzene.

1 Write down the full structural formulae of benzene, cyclohexane, cyclohexene and methylbenzene.

2 Which of these compounds are
   a alkanes?
   b aromatic compounds?
   c cyclic compounds?
   d unsaturated hydrocarbons?

3 What reactions would you expect an unsaturated hydrocarbon to undergo with
   a bromine water,   b dilute acidified potassium manganate(VII)?

During the last few years there has been much concern over the use in schools of benzene and certain other aromatic compounds because of their toxic properties.

In the following tests, all reactions involving benzene should be carried out in an efficient fume cupboard.

1 Addition reactions

A Reaction with bromine water
Add 1 cm³ of bromine water to separate 1 cm³ samples of benzene, cyclohexane, cyclohexene and methylbenzene. Shake all four mixtures thoroughly.

4 Make a table of your results. Describe what happens in each case, write equations where appropriate and name the products of any reactions.

B Reaction with acidified potassium manganate(VII)
Add 1 cm³ of dilute sulphuric acid and 1 cm³ of dilute potassium manganate(VII) solution to separate 1 cm³ samples of benzene, cyclohexane, cyclohexene and methylbenzene. Shake all four mixtures thoroughly.

5 Make a table of your results. Describe what happens in each case, write equations where appropriate and name the products of any reactions.

6 Benzene is an unsaturated compound yet it does not react with bromine water or with dilute acidified potassium manganate(VII). Why is this so?

The characteristic reactions of benzene and other aromatic hydrocarbons involve substitution, because this type of reaction retains the delocalized π-electron system of the aromatic nucleus (see *Chemistry in Context*, sections 27.6 and 27.7). Addition reactions involving disruption of the aromatic ring do, however, occur (see *Chemistry in Context*, section 27.8).

7 State the conditions, name the products and write equations for the addition reactions between benzene and
   a hydrogen,
   b chlorine.
   What are the important uses of the products of these two reactions?

2 Substitution reactions

One of the most important substitution reactions of aromatic compounds is nitration. The intention of this experiment is:

A   to illustrate the technique of nitration.
B   to demonstrate the method of purifying an organic solid.

The compound chosen for nitration is methyl benzoate, a relatively simple derivative of benzene, of low toxicity, which is used in commercial heat meters. Methyl benzoate can be nitrated readily with the usual nitric acid/sulphuric acid nitrating mixture. The product is solid methyl 3-nitrobenzoate.

$$\text{C}_6\text{H}_5\text{COOCH}_3 + HNO_3 \longrightarrow \text{(methyl 3-nitrobenzoate)} + H_2O$$

A   Preparation of methyl 3-nitrobenzoate
CARE Eye protection must be worn.

Weigh out 2·7 g of methyl benzoate into a small conical flask and then dissolve it in 5 cm³ of concentrated sulphuric acid. When the solid has dissolved, cool the mixture in ice.

Prepare the nitrating mixture by carefully adding 2 cm³ of concentrated sulphuric acid to 2 cm³ of concentrated nitric acid and then cool this mixture in ice as well.

Now add the nitrating mixture drop by drop from a teat pipette to the solution of methyl benzoate. (Do not allow the nitrating mixture to get into the rubber teat.) Stir the mixture with a thermometer and keep the temperature below 10°C. When the addition is complete, allow the mixture to stand at room temperature for another 15 minutes.

After this time, pour the reaction mixture on to about 25 g of crushed ice and stir until all the ice has melted and crystalline methyl 3-nitrobenzoate has formed.

B   Purification of the methyl 3-nitrobenzoate
Filter the crystals using a Buchner funnel, wash them thoroughly with cold water and then transfer them to a small beaker.

Now, recrystallize the product from the minimum volume of hot ethanol. Warm 15 cm³ of ethanol to about 50°C by immersing it in a beaker of hot water away from all naked flames. Dissolve all the crystals in the minimum volume of this hot ethanol. Allow the solution to cool to room temperature, then immerse the beaker in iced water to complete the crystallization of methyl 3-nitrobenzoate.

Filter the crystals, dry them between filter papers and then weigh them. Record the mass obtained. Finally, obtain the melting point of the crystals. (Pure methyl 3-nitrobenzoate melts at 77·5°C.)

8   What conditions employed during the preparation help to prevent further nitration of the product to a dinitroderivative?

9   Give the names and structural formulae of two nitro-compounds that are very likely to contaminate the impure crystals of methyl 3-nitrobenzoate.

10   How do the conditions for nitration of methyl benzoate differ from those for the nitration of benzene?

11   Why were the crystals washed with water before recrystallization?

12   What happens during a recrystallization to
a impurities,   b the main product?

13   How is loss of the product kept to a minimum during recrystallization?

14   What are the usual stages in the purification of an organic solid?

15   How many moles of methyl benzoate and nitric acid were used in the preparation of methyl 3-nitrobenzoate? (Assume that concentrated nitric acid is pure HNO_3 and that its density is 1·5 g cm⁻³.) Which reactant is present in excess?

16   What is the theoretical yield of methyl 3-nitrobenzoate?

17   What is your percentage yield?

PRACTICAL 34
Bromobutane

Requirements

Each student, or pair of students, will require:

- Pear-shaped flask, 50 cm³
- Round flask 50 cm³
- Condenser
- Still-head
- Tap funnel to fit still-head (doubles as separating funnel)
- Thermometer, 0–110°C, and holder
- Measuring cylinder, 25 cm³
- Teat pipette
- Rack with 3 test tubes
- Beaker, 250 cm³
- Butan-1-ol
- Sodium bromide
- Concentrated sulphuric acid
- Anhydrous sodium sulphate
- Concentrated hydrochloric acid
- Ethanol
- Dilute nitric acid
- Dilute sodium hydroxide
- Silver nitrate solution, approximately 0·1 M (4 g in 250 cm³)
- Sodium bromide solution, approximately 0·1 M (2·5 g in 250 cm³)

Time required 3 double periods

Introduction

In this practical you will prepare 1-bromobutane in as high a yield and as pure a state as possible. You will meet several of the experimental techniques commonly used in the preparation of an organic liquid. Having prepared a sample of 1-bromobutane, you will investigate some of its chemical reactions.

Principle

Halogenoalkanes are normally prepared from their corresponding alcohol by a nucleophilic substitution reaction. The reagents most frequently used are hydrogen halides or phosphorus halides (e.g. PCl_5, PCl_3).

1 Draw the structure of the alcohol from which 1-bromobutane can be made, and indicate which bond must be broken.

Hydrogen halides and phosphorus halides show acid properties. In the presence of an acid, the OH group of an alcohol becomes protonated, by virtue of the lone pairs of electrons on the oxygen atom:

$$R-\ddot{O}H + H^+ \longrightarrow R-\overset{+}{\underset{H}{\ddot{O}}}-H$$

2 Explain why acid conditions favour the required bond cleavage.

3 Outline the mechanism of the reaction between hydrogen bromide and an alcohol to produce a bromoalkane.

4 The preparation of 2-bromo-2-methylpropane (an isomer of 1-bromobutane) by a similar method requires much less severe conditions than the preparation you will carry out here. Standing at room temperature for twenty minutes gives a good yield. Suggest a reason for this difference in the light of your proposed mechanism.

5 Give one example of an alternative preparation of monohalogenoalkanes that does not involve a substitution reaction. Could this method be readily applied to the preparation of 1-bromobutane?

You will be preparing 1-bromobutane by the reaction of butan-1-ol with hydrogen bromide, the latter being generated from the reaction of sodium bromide or potassium bromide with concentrated sulphuric acid. The 1-bromobutane prepared in this way will be impure, even after distilling it off from the reaction mixture, and a lot of your time will be spent in purifying the crude product. Impurities will include:

    unreacted butan-1-ol
    oxides and oxyacids of sulphur
    bromine
    hydrogen bromide
    water
    butoxybutane
    but-1-ene

Butan-1-ol is soluble in concentrated hydrochloric acid, so the first stage of purification is to shake with this reagent. Excess hydrochloric acid, and other acidic impurities are then removed by shaking with a base, sodium hydrogen carbonate. The 1-bromobutane is then dried using anhydrous sodium sulphate, and finally distilled to remove remaining impurities.

Procedure

Preparation of 1-bromobutane

Put 10 g sodium bromide, 7·5 cm³ (6 g) butan-1-ol and 10 cm³ water in a 50 cm³ pear-shaped flask. The mass of butan-1-ol used should be known so that you can calculate your final percentage yield. Fit the flask with a reflux condenser, and place a funnel in the top as shown in figure 1.

Surrounding the flask with a beaker of cold water to cool it during the addition of concentrated sulphuric acid. Put 10 cm³ of this acid in the tap funnel. **CARE Eye protection must be worn.** Gradually add it to the reaction mixture, shaking the flask from time to time to mix well.

Remove the funnel from the reflux condenser and the flask from the water bath. Set the apparatus over a tripod and gauze and heat gently over a low bunsen flame so that the mixture refluxes gently for 30–45 minutes. The flask now contains crude bromobutane, which must be distilled from the reaction mixture.

Allow the flask to cool and rearrange the apparatus for distillation as shown in figure 2. (The thermometer is not essential at this stage.) Boil the mixture in the flask and collect the distillate in a measuring cylinder. A form of 'steam distillation' takes place -- bromobutane and water distil simultaneously and form separate layers in the measuring cylinder. Continue the distillation until the upper organic layer in the flask has disappeared. Dismantle, then clean and dry your apparatus.

The distillate in the measuring cylinder has an organic and an aqueous layer. Before the bromobutane in the organic layer can be purified, the aqueous layer must be removed and discarded. In order to avoid discarding the wrong layer, it is essential to think carefully which layer is which. One way to decide is by using the densities given in the table below. If you are still not sure which is the aqueous layer, run in a few drops of water and note which layer they enter.

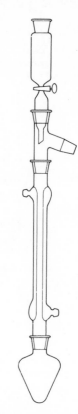

Figure 1 Apparatus for refluxing with addition.

Liquid	Density/g cm⁻³
water	1·0
butan-1-ol	0·81
1-bromobutane	1·3
concentrated hydrochloric acid	1·2

Use a teat pipette to remove the water layer. You now need to shake the crude product with concentrated hydrochloric acid to remove unreacted butanol. Put the organic layer in a small tap funnel then pour 10 cm³ of concentrated hydrochloric acid into the measuring cylinder you used to receive the distillate – this will wash out any traces of the crude product. Pour the acid into the tap funnel. Stopper the funnel and shake well, releasing the presssure occasionally. Allow the two layers to separate, decide which is the acid layer and discard it. Run the organic layer into a clean measuring cylinder.

You must now remove excess hydrogen chloride and other acid impurities by shaking with a base. Put the bromobutane in a clean tap funnel. Rinse out the measuring cylinder with 10 cm³ of sodium hydrogencarbonate solution, and pour this into the tap funnel. Stopper and shake the funnel cautiously, **regularly releasing pressure caused by the formation of carbon dioxide.** Allow the layers to separate, decide which is the bromobutane layer and run it into a 50 cm³ round flask.

The bromobutane is cloudy because of the presence of water and must now be dried. Add anhydrous sodium sulphate in small quantities, swirling after each addition, until the liquid is perfectly clear.

Finally the bromobutane must be distilled to remove any remaining organic impurities. Set up the apparatus for distillation and transfer the bromobutane to the distillation flask, making sure that the sodium sulphate remains behind. Distil the bromobutane, collecting the fraction with boiling range 100–104°C (the boiling point of 1-bromobutane is 102°C). Find the mass of bromobutane collected.

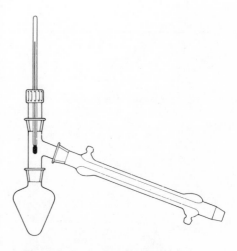

Figure 2 Apparatus for distillation.

6 Write a balanced equation for the overall reaction.

7 Calculate the maximum mass of 1-bromobutane that could be formed from the 6 g of butan-1-ol you used.

8 Calculate your percentage yield of 1-bromobutane.

9 Why are the reagents added in the particular order suggested?

10 Besides generating hydrogen bromide by reaction with sodium bromide, what other useful function does the concentrated sulphuric acid perform in this experiment?

11 Why is butan-1-ol much more soluble in concentrated hydrochloric acid than in water?

12 Why is sodium hydrogencarbonate used to remove acid impurities, rather than a stronger base like sodium hydroxide?

Reactions of 1-bromobutane

EXPERIMENT 1   Action of silver ions
Bromobutane is insoluble in water. In order to investigate its effect on aqueous silver ions, ethanol is used as a mutual solvent, since it dissolves both silver ions and bromobutane.

Add a few drops of bromobutane to 2 cm^3 of ethanol, then add 2 cm^3 of aqueous silver nitrate. Observe the reaction mixture for a few minutes. Repeat the experiment using sodium bromide solution instead of the solution of bromobutane in ethanol.

13 What does this experiment suggest about the way bromine is bonded in
   a bromobutane
   b sodium bromide?

14 What happens when the bromobutane is left to stand for a few minutes in the presence of aqueous silver ions? Why does this happen?

EXPERIMENT 2   Reaction with hydroxide ions
Put 2 cm^3 of dilute sodium hydroxide in a test tube. Add a few drops of bromobutane and warm very gently. Acidify with dilute nitric acid to neutralize the sodium hydroxide and then add 2 cm^3 of silver nitrate solution.

15 What happens? How do hydroxide ions react with 1-bromobutane?

16 Why must the sodium hydroxide be neutralized before adding silver nitrate solution?

17 What are the products of reactions between 1-bromobutane and:
   a ammonia
   b ethoxide ions, $CH_3CH_2O^-$
   c cyanide ions, CN^-?

18 What type of reactions are those referred to in questions **15** and **17**?

PRACTICAL 35
Nucleophilic substitution reactions of halogenoalkanes

Introduction

The carbon-halogen bond in halogenoalkanes is polarized:

$$-\overset{\delta+}{\underset{|}{C}}-\overset{\delta-}{Hal}$$

The positive charge on the carbon atom makes it attractive to negatively-charged groups or **nucleophiles**. Such groups may attack the carbon, displacing the halogen as a halide ion and resulting in a nucleophilic substitution reaction.

$$-\overset{\delta+}{\underset{|}{C}}-\overset{\delta-}{Hal} + {}^-OH \longrightarrow -\underset{|}{C}-OH + Hal^-$$

$$-\overset{\delta+}{\underset{|}{C}}-\overset{\delta-}{Hal} + \overset{H}{\underset{H}{\overset{|}{\underset{|}{O}}}} \longrightarrow -\underset{|}{C}-OH + H\,Hal$$

$$-\overset{\delta+}{\underset{|}{C}}-\overset{\delta-}{Br} + Cl^- \longrightarrow -\underset{|}{C}-Cl + Br^-$$

In this practical we will investigate the mechanisms of some nucleophilic substitution reactions by making simple qualitative comparisons of reaction rates.

Procedure

EXPERIMENT 1   Effect of the halogen atom on the rate of hydrolysis
In this experiment you will compare the rates of hydrolysis of 1-chlorobutane, 1-bromobutane, and 1-iodobutane.

$$CH_3CH_2CH_2CH_2Hal + H_2O \longrightarrow CH_3CH_2CH_2CH_2OH + H^+ + Hal^-$$

The rate of the reaction can be followed by carrying it out in the presence of silver ions. The halogenoalkanes, being covalently bonded, give no precipitate of silver halide, but as the reaction proceeds and halide ions are produced, a precipitate of silver halide gradually appears:

$$Ag^+(aq) + Hal^-(aq) \longrightarrow AgHal(s)$$

Halogenoalkanes are insoluble in water, so the reaction is carried out in the presence of ethanol, which acts as a mutual solvent for the halogenoalkane and silver ions.

Tube 1	Tube 2	Tube 3
1 cm^3 ethanol	1 cm^3 ethanol	1 cm^3 ethanol
2 drops 1-chlorobutane	2 drops 1-bromobutane	2 drops 1-iodobutane

Table 1 Tubes for experiment 1

Set up three labelled test tubes as described in table 1. Stand the tubes in a beaker of water at about 50°C and put a tube containing 5 cm^3 of 0·1 M silver nitrate solution in the same beaker. Leave the tubes for about 10 minutes so that they reach the temperature of the bath. Add 1 cm^3 of the silver nitrate solution to each of tubes 1, 2 and 3, working quickly and noting the time. Shake each tube to mix the contents and observe the tubes over the course of the next five minutes or so.

Requirements

Each student, or pair of students, will require:
- Rack and 9 test tubes
- Labels
- Beaker (250 cm^3)
- Thermometer (0–100°C)
- Measuring cylinder (10 cm^3)
- Stop-clock
- 1-Chlorobutane*
- 1-Bromobutane*
- 1-Iodobutane*
- 2-Bromobutane*
- 2-Bromo-2-methylpropane*
- Ethanol
- Silver nitrate solution, approximately 0·1 M (dissolve 4·25 g in 250 cm^3 of distilled water)
- Sodium iodide
- Propanone (acetone)

* With a separate dropping pipette for each bottle.

Time required 1 double period

93

Bond	Average bond energy /kJ mole^{-1}
C—Cl	339
C—Br	284
C—I	218

Table 2 Average bond energies for the carbon-halogen bonds in halogenoalkanes

1 Which halogenalkane undergoes hydrolysis fastest? Which is slowest?

2 Which halogenoalkane has the most polar carbon-halogen bond?

3 Is differing polarity the reason for the different rates of hydrolysis?

4 Suggest a possible explanation for the different rates of hydrolysis using the bond energies shown in table 2.

Theoretical background: the mechanism of nucleophilic substitution

Consider the substitution reaction:

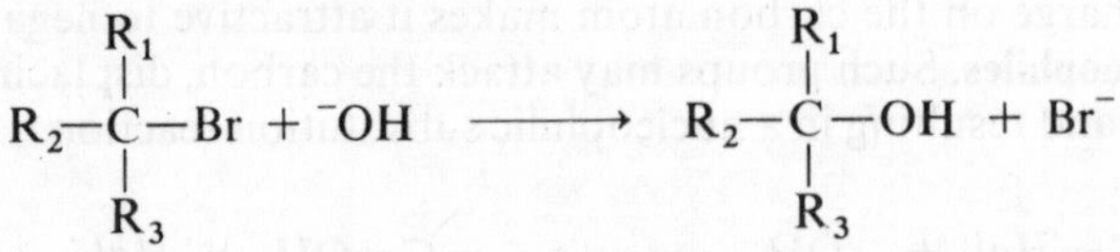

(R_1, R_2, and R_3 are alkyl groups or hydrogen atoms.) Studies of the kinetics of nucleophilic substitution suggest that these reactions can proceed by two possible mechanisms:

A Step by step mechanism
In this case, it is proposed that a two-step mechanism is involved:

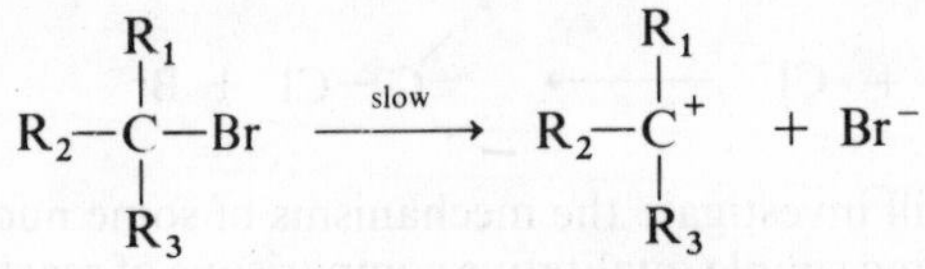

intermediate
carbonium ion

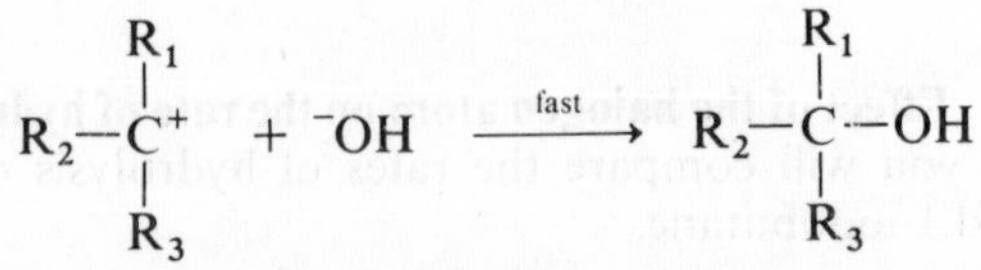

The first step, which is relatively slow, involves bond cleavage to form the intermediate carbonium ion. The carbonium ion is very unstable and reactive, so the second step is fast. The overall rate of the reaction is determined by the slow first step – the **rate-determining step**. (See *Chemistry in Context*, section 23.11 for further discussion of this.)

B Single-step mechanism
Here it is proposed that the reaction occurs in a single step. The OH$^-$ ion is attracted to the central carbon atom, and as it moves in, it repels the Br atom. At some 'middle' stage in the reaction, Br and OH are both partially bonded to the carbon, the OH on its way in and the Br on its way out. This is the **transition state** of the reaction. It is **not** a reaction intermediate which exists independently, but simply the middle stage in a continuous process during which the OH$^-$ moves in and the Br$^-$ moves out.

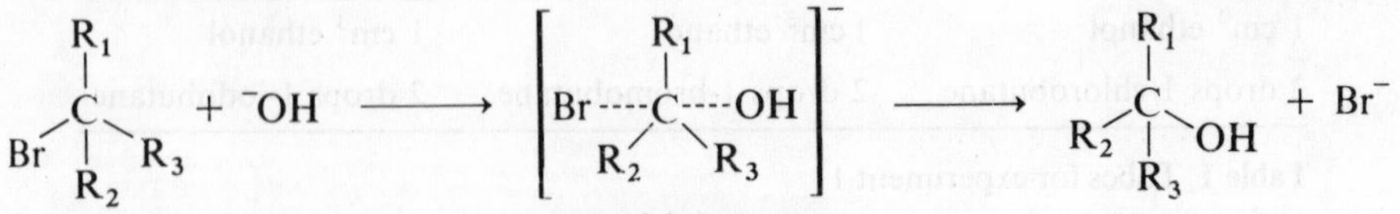

transition state

Factors determining which mechanism operates in practice
In many nucleophilic substitution reactions, both of the proposed mechanisms can and do operate, but in most cases one proceeds much faster than the other. The faster mechanism can often be taken as the mechanism for that reaction. In the experiments which follow you will investigate some of the factors which determine the particular mechanism for a given reaction.

Two of the most important factors are the structure of the halogenoalkane and the nature of the solvent.

A   Structure of the halogenoalkane

In the first, two-step mechanism, the rate of the reaction is determined by the ease with which the intermediate carbonium ion forms. If the substituent groups R_1, R_2, and R_3 are all alkyl groups rather than H atoms, they will tend to donate electrons, stabilizing the carbonium ion and favouring its formation.

$$
\begin{array}{ccc}
\underset{\underset{R_3}{|}}{\overset{\overset{R_1}{|}}{R_2-C-Br}} & \longrightarrow & \underset{\underset{R_3}{\uparrow}}{\overset{\overset{R_1}{\uparrow}}{R_2 \rightarrow C^+}} + Br^-
\end{array}
$$

If some or all of R_1, R_2, and R_3 are hydrogen atoms, the formation of the carbonium ion will be less favoured, so the rate of the overall reaction will be slower. Thus, the two-step mechanism is favoured by the presence of substituent alkyl groups. The reverse applies in the case of the single-step mechanism, which is favoured by the presence of substituent hydrogen atoms.

B   Nature of the solvent in which the reaction is carried out

Polar solvents, particularly water, favour ion formation and therefore the two-step mechanism. Conversely, the single step mechanism is favoured in non-polar solvents.

EXPERIMENT 2   Effect of the structure of the carbon skeleton on the reaction mechanism and reaction rate

The experimental method is similar to that in experiment 1. In this case, instead of varying the halogen atom for a given carbon skeleton, we will vary the structure of the alkane skeleton keeping the same halogen, bromine. The bromoalkanes used are

$$
\underset{\text{1-bromobutane}}{\underset{\underset{H}{|}}{\overset{\overset{H}{|}}{CH_3CH_2CH_2-C-Br}}} \qquad \underset{\text{2-bromobutane}}{\underset{\underset{H}{|}}{\overset{\overset{CH_3}{|}}{CH_3CH_2-C-Br}}} \qquad \underset{\text{2-bromo-2-methylpropane}}{\underset{\underset{CH_3}{|}}{\overset{\overset{CH_3}{|}}{CH_3-C-Br}}}
$$

The general reaction is the same as in experiment 1:

$$RBr + H_2O \longrightarrow ROH + H^+ + Br^-$$

(using RBr to represent the bromoalkane involved).

Tube 1	Tube 2	Tube 3
1 cm³ ethanol	1 cm³ ethanol	1 cm³ ethanol
2 drops 1-bromobutane	2 drops 2-bromobutane	2 drops 2-bromo-2-methylpropane

Table 3 Tubes for experiment 2

Set up three labelled test tubes as described in table 3, above. In this experiment it is satisfactory to work at room temperature.

Add 1 cm³ of 0·1 M silver nitrate solution to each tube, working quickly and noting the time. Shake each tube to mix the contents. Observe the tubes over the course of the next five minutes or so.

5   Which bromoalkane is hydrolysed fastest? Which is slowest?

6   Bearing in mind that the solvent in this experiment is a 1 : 1 mixture of ethanol and water (the water coming from the silver nitrate solution), which mechanism is likely to be favoured? Explain your answer.

7   Which bromoalkane would be most favoured by the mechanism you have proposed in your answer to question 6? Which would be least favoured? Explain your answer.

8   Explain the observed relative rates of hydrolysis of the three bromoalkanes.

EXPERIMENT 3 Effect of the solvent on the reaction mechanism and the reaction rate: the Finkelstein reaction

Experiments **1** and **2** have involved hydrolysis reactions in aqueous solution. It is, of course, impossible to carry out hydrolysis in a non-aqueous solvent, but by using a substitution reaction known as Finkelstein's reaction you can investigate the rate of a simple nucleophilic substitution reaction in a non-aqueous solvent, propanone. The reaction is the substitution of a bromide ion by an iodide ion:

$$RBr + I^- \longrightarrow RI + Br^-$$

The source of I^- is sodium iodide. Sodium iodide is soluble in propanone, but sodium bromide is not. As bromide ions are produced in the reaction, they combine with sodium ions to form a precipitate of sodium bromide.

$$RBr(Pr) + Na^+(Pr) + I^-(Pr) \longrightarrow RI(Pr) + NaBr(s)$$

[(Pr) indicates that the substance is in solution in propanone.]

This reaction can therefore be followed by timing the rate of appearance of the sodium bromide precipitate.

Dissolve 1 g of sodium iodide in 15 cm³ of propanone. This solution is called Finkelstein's Reagent. Put 5 cm³ of the reagent into each of three labelled test tubes and stand the tubes in a beaker of water at 35°C. Leave the tubes for 10 minutes to reach the temperature of the water bath.

Working quickly and noting the time, add 8 drops of 1-bromobutane to the first tube, 8 drops of 2-bromobutane to the second and 8 drops of 2-bromo-2-methylpropane to the third. Observe the three tubes over the course of the next 5 minutes or so.

9 Which bromoalkane undergoes nucleophilic substitution by iodide ions fastest? Which is slowest?

10 Bearing in mind that the solvent in this experiment is propanone, which reaction mechanism is likely to be favoured? Explain your answer.

11 Which bromoalkane would be most favoured by the mechanism you have proposed in your answer to question **10**? Which would be least favoured? Explain your answer.

12 Explain the observed order of reaction rate of the three bromoalkanes.

PRACTICAL 36
Alcohols

Introduction

Alcohols are organic compounds with the general formula:

$$R_2-\underset{\underset{R_3}{|}}{\overset{\overset{R_1}{|}}{C}}-OH$$

R_1, R_2 and R_3 may be hydrogen or any alkyl or aryl group. Most alcohols can be regarded as alkanes in which a hydrogen atom has been replaced by an —OH group. Examples are ethanol ($CH_3 . CH_2OH$), propan-2-ol ($CH_3 . CHOH . CH_3$) and cyclohexanol ($C_6H_{11}OH$).

The intention of this practical is to compare the physical properties of alcohols with those of other organic compounds of similar relative molecular mass and then to look at some of their typical chemical properties.

Comparing the physical properties of alcohols, alkanes and ethers

1 Copy out and complete table 1 below using a data book.

Compound	Relative Molecular mass	Structural Formula	Melting Point/°C	Boiling Point/°C	Relative solubility in water
Ethanol					
Methoxymethane					
Propane					

Table 1 Comparing the physical properties of ethanol, methoxymethane and propane.

2 Explain the relative volatilities of ethanol, methoxymethane and propane.

3 Explain the relative solubilities in water of ethanol, methoxymethane and propane.

4 Why has ethanol been compared with methoxymethane and propane rather than ethoxyethane and ethane?

Investigating the chemical properties of alcohols
A Reactions in which the O—H bond breaks

Reactions of alcohols involving cleavage of the O—H bond can be compared with similar reactions for water.

Reaction with sodium

CARE Eye protection must be worn.

Put 1 cm^3 of ethanol in a test tube and add one small piece of sodium. Try to identify the gas produced.

5 Describe what happens. Is the reaction of sodium with ethanol more or less vigorous than the reaction of sodium with water?

6 Describe and explain what happens when a piece of damp indicator paper is added to the sodium/ethanol mixture.

7 Write equations for the reactions of sodium with **a** water, **b** ethanol.

8 Explain the relative reactivities of ethanol and water with sodium.

Each student, or pair of students, will require:
- Rack with 4 test tubes
- Teat pipette
- Ethanol
- Methanol
- Propan-2-ol
- Full range indicator paper
- Sodium, cut into small pieces
- Small beaker (100cm³)
- Bunsen burner
- Splint
- Concentrated sulphuric acid
- Glacial ethanoic (acetic) acid
- 2-hydroxybenzoic (salicylic) acid
- Phosphorus pentachloride
- Silver nitrate solution
- Dilute sulphuric acid
- Dilute potassium dichromate(VI)
- About 20 cm of medium gauge copper wire wound into a spiral at one end.
- Access to a fume cupboard

Time required 2 double periods

Reaction with organic acids – Esterification

CARE Eye protection must be worn for this experiment.

Put 10 drops of ethanol in a test tube and add 10 drops of glacial ethanoic (acetic) acid. Carefully, add 6 drops of concentrated sulphuric acid, and then warm the mixture, without boiling, for 5 minutes. Now, pour the contents of the test tube into about 50 cm³ of water in a beaker, and smell cautiously.

9 Why must the mixture not be boiled?

10 The product, ethyl ethanoate (acetate) is an ester. Describe its smell.

11 Why does the smell of the ester become more prominent after the mixture is poured into water?

12 Write an equation for the reaction between ethanol and ethanoic acid, clearly showing the structure of the ester.

13 How does sulphuric acid catalyse this reaction?

14 What uses has ethyl ethanoate?

Repeat the experiment using 2-hydroxybenzoic acid (salicylic acid) in place of ethanoic acid. (Use approximately equal volumes of solid 2-hydroxybenzoic acid and ethanol.)

15 What is the name of the ester produced? What does it smell like?

16 This ester is a constituent of 'oil of wintergreen'. What is this oil used for?

17 How and why has isotopic labelling been used in the study of esterification reactions? (See *Chemistry in Context*, section 31.5.)

B Reactions in which the C—OH bond breaks

Reaction with phosphorus halides

CARE Eye protection must be worn. Work in a fume cupboard.

Add a *little* phosphorus pentachloride to 10 drops of ethanol. Collect some of the fumes evolved in a teat pipette and identify them by bubbling into silver nitrate solution.

When the fumes have subsided and the reaction is complete, pour the contents of the tube into 50 cm³ of water in a beaker and cautiously smell the product, which is an alkyl halide.

18 What gas is evolved when phosphorus pentachloride reacts with ethanol?
19 Write equations for the reactions of ethanol with:
   a phosphorus pentachloride,
   b phosphorus trichloride.

Reactions with concentrated hydrohalic acids
The reactions between alcohols and concentrated hydrohalic acids (or halide ions in concentrated sulphuric acid) are important in the preparation of alkyl halides. The preparation of 1-bromobutane in practical 34 uses this procedure.

20 Write an equation or equations for the preparation of 2-bromopropane using potassium bromide, concentrated sulphuric acid and an appropriate alcohol.

C Reactions of the —CH₂OH and >CHOH groups

Alcohols with a —CH₂OH group can be oxidized first to an aldehyde containing

the $-C\underset{H}{\overset{O}{\parallel}}$ group, and then to a carboxylic acid with the $-C\underset{OH}{\overset{O}{\parallel}}$ group.

$$\underset{\text{alcohol}}{-\overset{\displaystyle H}{\underset{\displaystyle H}{C}}-OH} \longrightarrow \underset{\text{aldehyde}}{-\overset{\displaystyle O}{\underset{\displaystyle H}{C}}} + 2H^+ + 2e^-$$

$$\underset{\text{}}{-\overset{\displaystyle O}{\underset{\displaystyle H}{C}}} + H_2O \longrightarrow \underset{\substack{\text{carboxylic}\\ \text{acid}}}{-\overset{\displaystyle O}{\underset{\displaystyle OH}{C}}} + 2H^+ + 2e^-$$

Alcohols with a $\rangle$CHOH group can be oxidized to ketones, containing the $\rangle$C=O group. Further oxidation is difficult.

$$\underset{\text{alcohol}}{\overset{\displaystyle H}{\underset{\displaystyle OH}{C}}} \longrightarrow \underset{\text{ketone}}{C=O} + 2H^+ + 2e^-$$

Oxidation with acidified potassium dichromate(VI)
Add 10 drops of dilute sulphuric acid and 5 drops of potassium dichromate(VI) solution to 5 drops of ethanol. Warm the mixture and smell cautiously.

21 Describe what happens and explain the colour changes.

22 Write equations (or half-equations) for the oxidation of ethanol.

23 What conditions and techniques would favour the oxidation of ethanol to
   a ethanal rather than ethanoic acid,
   b ethanoic acid rather than ethanal?

The preparation of ethanal and ethanoic acid from ethanol is considered in more detail in practical 37.

Repeat the experiment using first methanol and then propan-2-ol in place of ethanol.

24 Write equations (or half-equations) for the oxidation of methanol and propan-2-ol. Name the oxidation products. (Note: The final oxidation products from methanol are carbon dioxide and water.)

Catalytic oxidation of alcohols
Warm 10 drops of methanol in a test tube and then heat a clean copper spiral to red heat in a roaring bunsen flame.

25 What do you notice as the red-hot spiral is moved out of the hot bunsen flame? Write an equation for the reaction which occurs. Now introduce the red-hot spiral into the methanol vapour in the test tube.

26 Observe the copper spiral carefully and explain the changes in colour which take place on its surface. (Hint: methanol is oxidized to methanal (formaldehyde)).

27 Explain how the copper acts as a catalyst in the oxidation of methanol by air.

(Industrially, methanal is manufactured by passing a mixture of methanol vapour and air over a catalyst of silver, rather than copper, at 450–600°C.)

The influence of conditions and reacting quantities on the product of an organic reaction

The preparation of ethanal and ethanoic acid from ethanol

Introduction

When inorganic substances react there is usually only one possible set of products. The same cannot be said for reactions involving organic compounds where frequently two, or even more, possible products can be formed from a given set of reactants, depending on the conditions chosen for the reaction. One such example is the reaction between ethanol and concentrated sulphuric acid, which can produce ethene, ethyl hydrogensulphate or ethoxyethane (ether) depending on the conditions.

The intention of this practical is to illustrate how different conditions and different reacting quantities can influence the products of the oxidation of ethanol by acidified sodium dichromate(VI).

1 Write half-equations for the oxidation of
   a ethanol to ethanal,
   b ethanol to ethanoic acid.

2 Write a half-equation for the action of the dichromate(VI) ion as an oxidizing agent in acid solution.

3 What relative proportions of ethanol and dichromate would favour the formation of
   a ethanal rather than ethanoic acid,
   b ethanoic acid rather than ethanal?

1 The preparation of ethanal

CARE Eye protection must be worn.

Arrange the apparatus shown in figure 1. Put 5 cm³ of water in the flask and slowly add 2 cm³ of concentrated sulphuric acid. Agitate the flask to ensure the constituents are thoroughly mixed. Finally, add a few anti-bumping granules and replace the flask.

Dissolve 5 g of sodium dichromate(VI) in 5 cm³ of water in a small beaker, add 4 cm³ of ethanol and put this mixture in the dropping funnel.

Pass a stream of cold water through the condenser and surround the receiver with a mixture of ice and water. Heat the dilute acid in the flask until it begins to boil. Now remove the flame and allow the alcohol/dichromate(VI) mixture to drip into the flask slowly. The addition should take about 10 minutes. Towards the end of this time, it may be necessary to warm the flask in order to maintain *gentle* boiling. When the addition of the alcohol/dichromate(VI) solution is complete, a few cm³ of aqueous ethanal will have collected in the receiver. Smell the distillate cautiously and note the characteristic smell of ethanal (boiling point 21°C).

4 In many organic reactions, the materials are refluxed in order to increase the yield and improve the conversion of reactants to products. Why is the ethanol not refluxed with the oxidizing agent in this case?

5 Why is the ethanal distilled off as fast as it forms?

6 What happens when the alcohol/dichromate mixture is added to the hot dilute acid?

7 Why is the receiver cooled in a mixture of ice and water?

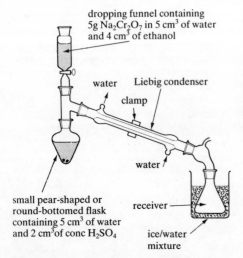

dropping funnel containing
5g Na₂Cr₂O₇ in 5 cm³ of water
and 4 cm³ of ethanol

water   Liebig condenser
   clamp

water

small pear-shaped or
round-bottomed flask
containing 5 cm³ of water
and 2 cm³ of conc H₂SO₄

receiver

ice/water
mixture

Figure 1 The preparation of ethanal.

8 Name three impurities which contaminate the ethanal distillate collected in the receiver and explain where these impurities come from. The purification of ethanal is a very lengthy operation as it is difficult to remove the last traces of water. In this experiment you are not expected to purify the ethanal.

Tests for ethanal in the distillate
Carry out the following two tests for ethanal in the distillate:
A   Reduction of diamminosilver(I) ions by ethanal
Put 3 cm^3 of silver nitrate solution in a test tube and add dilute ammonia solution drop by drop until the precipitate of silver oxide just dissolves. The clear solution contains $[Ag(NH_3)_2]^+$ ions. Now add 1 cm^3 of the distillate and warm the mixture. Pour the mixture down the sink with plenty of water immediately after use.

9 Describe and explain what happens. Write an equation or half-equations for the reaction which occurs.

B   Reduction of Fehling's solution by ethanal
Put 3 cm^3 of Fehling's solution in a test tube and add 1 cm^3 of the ethanal distillate. Warm the mixture and look for the formation of copper(I) oxide.   **CARE Eye protection must be worn.**

10 Describe and explain what happens. Write an equation or half-equations for the reaction which occurs.

2   The preparation of ethanoic acid

CARE Eye protection must be worn.

Arrange the apparatus shown in figure 2. Slowly add 6 cm^3 of concentrated sulphuric acid to 10 cm^3 of water in a small beaker. Make sure that mixing is complete after each addition. Now add 12 g of sodium dichromate(VI) to the diluted acid and stir the mixture until all the solid has dissolved. Put this acid/dichromate(VI) mixture in the flask and add a few anti-bumping granules.

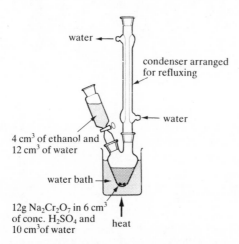

water ←
condenser arranged for refluxing
← water
4 cm^3 of ethanol and 12 cm^3 of water
water bath
12g Na$_2$Cr$_2$O$_7$ in 6 cm^3 of conc. H$_2$SO$_4$ and 10 cm^3 of water
heat

Figure 2 The preparation of ethanoic acid.

Place 4 cm^3 of ethanol and 12 cm^3 of water in the dropping funnel and start the flow of water through the condenser. Allow the alcohol/water mixture to drip into the flask slowly and with constant agitation.

When the addition of alcohol is complete and the initial reaction has subsided, heat the flask on a boiling water-bath for 10 minutes.

11 What conditions and relative proportions of the reactants are used in the above procedure to favour the oxidation of ethanol to ethanoic acid rather than to ethanal?

Assemble the apparatus for direct distillation through the Liebig condenser. Distil the mixture by direct heating with a small Bunsen flame until about 10 cm^3 of distillate have collected. Smell the distillate cautiously.

Requirements

Each student, or pair of students, will require:

Part 1
● Ethanol
● Conc. sulphuric acid
● Sodium dichromate(VI) (Na$_2$Cr$_2$O$_7$)
● Ice
● Dilute silver nitrate solution
● Dilute ammonia solution
● Fehling's solution (Mix equal volumes of solutions A and B. Store solutions A and B separately because the mixture will deteriorate.)
Solution A   Dissolve 17 g of CuSO$_4$. 5H$_2$O in 250 cm^3 of water.
Solution B   Dissolve 86 g of potassium sodium tartrate (Rochelle salt) and 30 g of NaOH in 250 cm^3 of water. Warming may be necessary.
● Pear-shaped or round-bottomed flask (50 or 100 cm^3)
● Still-head
● Dropping funnel (50 or 100 cm^3)
● Liebig condenser
● Receiver
● Adaptor to connect condenser and receiver
● Beaker (250 cm^3)
● Anti-bumping granules
● Measuring cylinder (10 cm^3)
● Small beaker (50 or 100 cm^3)
● Micro-burner (or Bunsen)
● Rack with 2 test tubes
● Teat pipette
● Access to balance

Time required (part 1) 1 double period

Part 2
● Ethanol
● Conc. sulphuric acid
● Sodium dichromate(VI)
● Small beaker (50 or 100 cm^3)
● Stirring rod
● Small two-necked flask (100 cm^3)
● Pear-shaped or round-bottomed flask (50 or 100 cm^3)
● Dropping funnel (50 or 100 cm^3)
● Liebig condenser
● Beaker (250 cm^3)
● Bunsen burner
● Anti-bumping granules
● Receiver
● Rack with 2 test tubes
● Teat pipette
● Full-range indicator paper or litmus paper
● Dilute ammonia solution
● Iron(III) chloride solution

Time required (part 2) 1 double period

12 Draw a diagram of the apparatus used for distillation.

13 Describe the smell of the distillate.

14 What impurities will the distillate contain? (The boiling point of CH_3COOH is 118°C.)

15 The sequence of operations below shows how the aqueous ethanoic acid distillate could be purified. Explain each stage of this purification and write equations where appropriate.

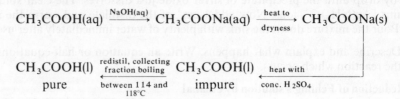

$$CH_3COOH(aq) \xrightarrow{\text{NaOH(aq)}} CH_3COONa(aq) \xrightarrow[\text{dryness}]{\text{heat to}} CH_3COONa(s)$$

$$\underset{\text{pure}}{CH_3COOH(l)} \xleftarrow[\substack{\text{between 114 and} \\ 118°C}]{\substack{\text{redistil, collecting} \\ \text{fraction boiling}}} \underset{\text{impure}}{CH_3COOH(l)} \xleftarrow[\text{conc. } H_2SO_4]{\text{heat with}}$$

Test for ethanoic acid in the aqueous distillate
Reaction with iron(III) chloride

The iron(III) chloride test must be carried out using neutral solutions. Add dilute ammonia solution drop by drop to 1 cm³ of the ethanoic acid distillate until the solution is just alkaline to indicator. Boil the solution to expel excess ammonia and then add 1 cm³ of iron(III) chloride. Ethanoate ions form a red complex ion with iron(III) ions.

16 Describe and explain what happens during the iron(III) chloride test.

PRACTICAL 38
Phenol

OH

Phenol

Introduction

Phenol (right) contains the same functional group as ethanol and other aliphatic alcohols. However, the benzene ring has a marked effect on the behaviour of the hydroxyl group, OH, so many of the properties of phenol are very different from those of ethanol. In this practical you will examine some of the differences between these two hydroxy-compounds. There will also be the opportunity to prepare phenyl benzoate and to use some of the standard techniques employed in purifying a solid organic compound and in testing its purity.

Procedure

CARE Phenol is toxic and highly corrosive. Avoid all skin contact. Eye protection must be worn throughout this practical.

EXPERIMENT 1   Combustion

Working in a fume cupboard, set fire to a small crystal of phenol on a piece of broken porcelain held in tongs.

1   With what sort of flame does phenol burn?

2   What does ethanol look like when it burns?

3   Explain the difference in the flames seen when ethanol and phenol burn.

EXPERIMENT 2   Acidic nature

Add a spatula-full of phenol to 5 cm³ of water in a test tube. Cork the tube and shake. Remove the cork and warm the tube in a beaker of hot water. Note what occurs, then cool the tube and again note what happens.

4   How does the solubility of phenol in water compare with that of ethanol in water?

To the cooled tube add a little sodium hydroxide solution and mix thoroughly.

5   Is phenol more soluble in aqueous sodium hydroxide than in water? If so, why?

Now add a little concentrated hydrochloric acid.

6   Describe and explain what happens.

Make a stock solution of phenol by putting a spatula-full of phenol into a boiling tube and half-filling with water.

Test a portion of the solution with universal indicator. For comparison, test a solution of ethanol in water with universal indicator.

7   Explain why ethanol and phenol differ in acidity.

EXPERIMENT 3   Reaction with sodium

Dissolve a few crystals of phenol in a few cm³ of ethanol. Put the solution in an evaporating basin and add a piece of sodium about the size of a grain of rice (**CARE**).   Observe the reaction and try to decide what is formed. Compare the reaction with that of a similar-sized piece of sodium with ethanol alone. Destroy any sodium left over, in excess ethanol.

8   Write an equation for the reaction of phenol with sodium.

9   Use your knowledge of the relative acid strengths of phenol and ethanol to explain why the reaction of sodium with the two compounds differs in vigour.

Each student, or pair of students, will
require:

- Rack and 4 test tubes (one fitted with a
  cork)
- Boiling tube
- Beaker (250 cm^3)
- Conical flask (250 cm^3), fitted with a
  bung
- Broken porcelain
- Tongs
- Spatula
- Evaporating basin
- Melting point tube and access to
  apparatus for melting point
  determination
- Glass rod
- Buchner funnel and flask
- Filter paper
- Access to suction pump
- Phenol
- Universal indicator paper
- Concentrated hydrochloric acid
- Concentrated sulphuric acid
- Ethanol
- Sodium metal, ready cut into pieces
  about the size of a grain of rice
- Benzoyl chloride (keep in fume
  cupboard)
- Bromine water
- Iron(III) chloride solution
- Benzene-1,2-dicarboxylic acid
  anhydride (phthalic anhydride)
- Access to a fume cupboard
- Protective gloves are desirable when
  using phenol.

Time required 2 double periods

EXPERIMENT 4   Esterification

Phenol differs from ethanol in the ease with which it forms esters. Ethyl esters are
readily formed by heating ethanol with a carboxylic acid in the presence of concentrated sulphuric acid, but phenyl benzoate, for example, cannot be prepared by
heating phenol with benzoic acid and concentrated sulphuric acid. The benzoic
acid must be activated by converting it to benzoyl chloride, which readily reacts
with phenol in basic conditions to give a good yield of phenyl benzoate.

This is sometimes called the Schotten-Baumann reaction.

10   Why is phenol more difficult to esterify than ethanol?

11   Why is benzoyl chloride effective in esterifying phenol?

In the experiment which follows you will prepare phenyl benzoate in this way
and, if time permits, purify your sample and check its purity by measuring its
melting point.

Put 5 g of phenol in a 250 cm^3 conical flask. Add 90 cm^3 of 2 M sodium
hydroxide solution. **Working in a fume cupboard**, carefully add 9 cm^3 of benzoyl
chloride.   **CARE Benzoyl chloride is harmful and its vapour irritates the eyes.
Avoid spilling the liquid or releasing the vapour outside the fume cupboard.**   Put a
bung in the flask and shake vigorously. Leave for 10 minutes, shaking from time to
time. Phenyl benzoate separates out as white crystals.

Filter off the crystals using a Buchner funnel, working in a fume cupboard or
near an open window because some unreacted benzoyl chloride may still be present. Wash the crystals several times with water, breaking up any lumps using a
glass rod.

The crystals can now be purified by recrystallization. Dissolve them in the
minimum possible quantity of hot ethanol in a boiling tube, heating the ethanol in
a water-bath at about 60°C and **not over a bunsen flame**. Now allow the solution to
cool so that the crystals separate out and can be filtered off, leaving impurities
behind in the solution.

The purity of your sample can be assessed by measuring its melting point. Seal a
melting point tube at one end, then put into it a few crystals of your sample. The
tube must now be heated slowly until the sample melts, at which point the temperature is noted. To do this you may use an electrical melting point apparatus or hot
oil – your teacher will show you the details of whichever method you are to use.
Remember to heat the tube very slowly, otherwise you may 'overshoot' and get an
inaccurate value for the melting point.

12   Record the melting point of your crystals. The melting point of pure phenol
benzoate is 71°C.

13   What can you say about the purity of your own sample?

EXPERIMENT 5   Complexing reactions

Add a few drops of iron(III) chloride solution to an aqueous solution of phenol.
Repeat the reaction using ethanol instead of phenol.

14   Describe the characteristic colour given by phenol and Fe^{3+}.

15   What is the cause of this colour?

EXPERIMENT 6   Substitution reactions of phenol

Add bromine water drop by drop to an aqueous solution of phenol, until it is
present in excess.

16   What happens?

17   Use a text book to find the formula of the product, and write an equation for
the reaction.

18 Ethanol has no reaction with bromine water. Explain the difference between the two compounds in this respect.

EXPERIMENT 7   Preparation of phenolphthalein

The indicator, phenolphthalein, can be prepared by the reaction between phenol and benzene-1,2-dicarboxylic anhydride (phthalic anhydride):

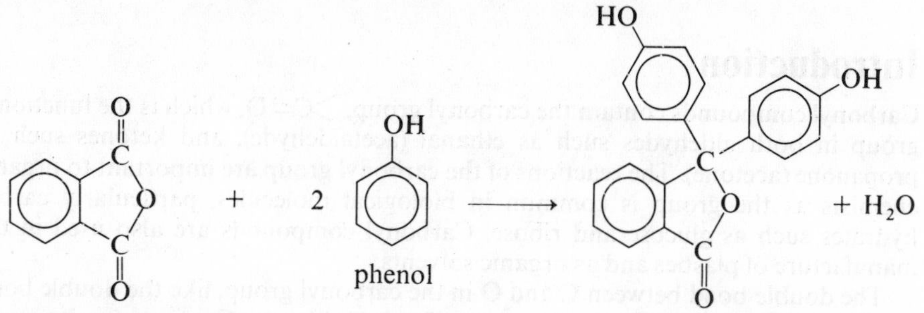

benzene-1,2-dicarboxylic
anhydride
(phthalic anhydride)

phenol

phenolphthalein

Mix 0·5 g of phenol and 0·5 g of benzene-1,2-dicarboxylic anhydride in a test tube and heat the mixture with 1 drop of concentrated sulphuric acid. Cool and then carefully add a little dilute sodium hydroxide solution. Investigate the colour change that occurs when the solution is acidified.

19 Phenolphthalein is a weak acid which dissociates in the following way:

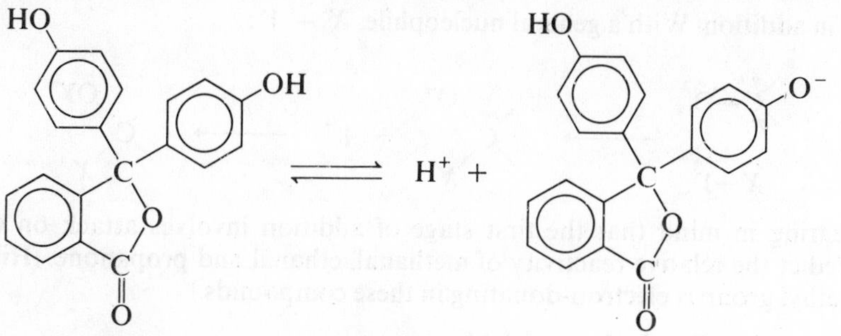

What are the colours of the undissociated and dissociated forms respectively?

20 Summarize the main differences between phenol and ethanol.

PRACTICAL 39
Carbonyl compounds

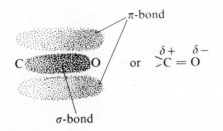

Figure 1 Electron density in the carbonyl bond

Introduction

Carbonyl compounds contain the carbonyl group, $>C=O$, which is the functional group in both aldehydes such as ethanal (acetaldehyde), and ketones such as propanone (acetone). The reactions of the carbonyl group are important to organic chemists as the group is common in biological molecules, particularly carbohydrates such as glucose and ribose. Carbonyl compounds are also used in the manufacture of plastics and as organic solvents.

The double bond between C and O in the carbonyl group, like the double bond in alkenes, consists of a σ-bond and a π-bond. Unlike the $C=C$ group, however, the carbonyl group does not have an even electron distribution between the two atoms and there is a greater electron density near the more electronegative oxygen atom (figure 1).

A   Addition reactions of carbonyl compounds

The electron distribution in the carbonyl bond makes the carbon atom attractive towards nucleophiles which attack and bond to it, breaking the π-bond and resulting in addition. With a general nucleophile, $\overset{\delta-}{X} - \overset{\delta+}{Y}$;

$$\underset{\underset{\overset{\delta-}{X}-\overset{\delta+}{Y}}{\uparrow}}{\overset{\delta+}{C}\overset{\delta-}{=}\overset{\delta-}{O}} \longrightarrow \underset{X}{\overset{O^-}{C}} + Y^+ \longrightarrow \underset{X}{\overset{OY}{C}}$$

1 Bearing in mind that the first stage of addition involves attack on the $\overset{\delta+}{C}$, predict the relative reactivity of methanal, ethanal and propanone. (Hint: the methyl group is electron-donating in these compounds.)

Reactions with sodium hydrogensulphite
Mix 2 cm³ of propanone with 1 cm³ of saturated sodium hydrogensulphite solution. Judge whether a reaction has taken place by seeing if heat is evolved and then cool the mixture in a stream of cold water.

2 Describe what happens during this reaction.

3 Draw a full structural formula for the hydrogensulphite ion.

4 Write an equation for the reaction of propanone with sodium hydrogensulphite.

Repeat the experiment using ethanal in place of propanone.

5 Why does the propanonehydrogensulphite crystallize more readily than the ethanalhydrogensulphite?

6 Why are these carbonylhydrogensulphite addition compounds fairly soluble in water?

7 Why are carbonylhydrogensulphate compounds more difficult to prepare than carbonylhydrogensulphite compounds?

B   Polymerization of carbonyl compounds

Carbonyl compounds, like alkenes, can undergo self addition (polymerization) by addition across the $C=O$ double bond. Ketones are not reactive enough to polymerize easily, but aldehydes can readily be converted to a variety of addition polymers.

Polymerization of methanal (formaldehyde)

Put 10 drops of methanal solution on a watch glass and warm this over a beaker of hot water. **CARE Work in a fume cupboard. Methanal causes irritation of skin and eyes.**

8   Describe what happens.

The residue which forms is poly(methanal) (below). A plastic made from poly(methanal) has been used to replace metal in machine parts such as gear wheels and clips.

$$-CH_2-O-CH_2-O-CH_2-O-CH_2-O$$

Working in a fume cupboard, scrape some of the poly(methanal) into a test tube and investigate the action of heat on it.

Polymerization of ethanal

9   Ethanal, unlike methanal, requires chemical reagents to induce polymerization. Is this consistent with the relative reactivities of the two compounds? Explain.

Ethanal resin

Warm 10 drops of ethanal with 3 cm³ of concentrated sodium hydroxide solution. **CARE Eye protection must be worn.**

10   Describe what happens and smell the product cautiously.

C   Condensation reactions of carbonyl compounds

In some cases, the addition reactions of carbonyl compounds are followed by the elimination of a molecule of water. Many of these elimination reactions involve derivatives of ammonia with the general formula $X-NH_2$.

This sort of reaction, in which two molecules combine with the elimination of water, is an example of a **condensation** or **addition-elimination** reaction.

The most useful reagent for low molecular mass carbonyl compounds is 2,4-dinitrophenylhydrazine:

The products of condensation reactions between 2,4-dinitrophenylhydrazine and carbonyl compounds are crystalline solids with well-defined melting points. They are therefore useful in identifying individual carbonyl compounds.

Preparation of propanone 2,4-dinitrophenylhydrazone

Add 5 cm³ of 2,4-dinitrophenylhydrazine solution (Brady's Reagent) to 3 drops of propanone. If crystals do not form, add a little 2 M sulphuric acid, warm the mixture and then cool in iced water.

If time permits, filter off the crystals, wash them with a little water, dry them between filter papers and then recrystallize them from ethanol. Determine their melting point and compare your value with the accepted value of 128°C for the melting point of propanone 2,4-dinitrophenylhydrazone.

11   Describe the appearance of propanone 2,4-dinitrophenylhydrazone crystals.

12   Write an equation for the reaction between propanone and 2,4-dinitrophenylhydrazine.

Other carbonyl compounds react in a similar manner to propanone.

Requirements

Each student, or pair of students, will require:

- Rack with 4 test tubes
- Ethanal (acetaldehyde)
- Aqueous methanal (40% solution – formalin)
- Propanone (acetone)
- Glucose
- Saturated sodium hydrogensulphite solution
- Concentrated sodium hydroxide solution (Dissolve 30 g of NaOH in 100 cm³ of water)
- Dry ether
- Ice
- Salt
- Hydrogen chloride generator (Add concentrated sulphuric acid to sodium chloride and dry the gas by passing through concentrated sulphuric acid.)
- 2,4-dinitrophenylhydrazine solution – Brady's reagent (Dissolve 2 g of 2,4-dinitrophenylhydrazine in 4 cm³ of concentrated sulphuric acid and add carefully, with cooling, 30 cm³ of methanol. Warm gently to dissolve any undissolved solid and then add 10 cm³ of water.)
- 2 M sulphuric acid
- Ethanol

The following solutions at usual bench concentration:

- Potassium dichromate(VI)
- Silver nitrate
- Dilute ammonia

- Fehling's solution (Mix equal volumes of solutions A and B. Store solutions A and B separately because the mixture will deteriorate.)

Solution A   Dissolve 17 g of CuSO₄ . 5H₂O in 250 cm³ of water.

Solution B   Dissolve 86 g of potassium sodium tartrate (Rochelle salt) and 30 g of NaOH in 250 cm³ of water. Warming may be necessary.

- Teat pipette
- Watch glass
- Small beaker (150 or 250 cm³)
- Bunsen burner
- Tripod
- Gauze
- Filter paper
- Melting point tubes
- Access to melting point apparatus
- Access to fume cupboard

Time required 2–3 double periods

D Oxidation of carbonyl compounds

Aldehydes possess a hydrogen atom attached to their carbonyl group. This hydrogen is activated by the carbonyl group and is readily oxidized to —OH. Aldehydes are therefore readily oxidized to carboxylic acids.

Ketones, however, have no hydrogen atom joined directly to the carbonyl group, so they are not readily oxidized.

13 What would you expect to be the order of ease of oxidation of methanal, ethanal and propanone?

1 Oxidation by acidified dichromate(VI)

Compare the ease of oxidation and hence the reducing power of methanal, ethanal and propanone by adding 5 drops of each substance *in turn* to 2 drops of potassium dichromate(VI) solution and 10 drops of dilute sulphuric acid. If nothing happens in the cold, warm gently.

14 Write the carbonyl compounds (methanal, ethanal and propanone) in order of increasing reducing power. Do your results agree with your answer to question **13**?

15 Write equations (or half-equations) for the reactions which have occurred.

2 Oxidation by diamminosilver(I) ions (Tollen's Reagent)

Put 3 cm³ of silver nitrate solution into a *clean* test tube and add dilute ammonia solution drop by drop until the precipitate of silver oxide just dissolves. The clear solution contains $[Ag(NH_3)_2]^+$ ions. Add 10 drops of ethanal and warm the resulting mixture in a beaker of hot water. Rinse out the test tube immediately afterwards.

16 Describe and explain what happens. Write an equation (or half-equations) for the reaction which occurs.

Repeat the experiment using first methanal and then propanone in place of ethanal.

17 How do the reactions of methanal and propanone compare with that of ethanal?

3 Oxidation by Fehling's solution (Copper(II) ions complexed with tartrate ions)
CARE Eye protection must be worn.

Put 3 cm³ of Fehling's solution in a test tube and add 10 drops of ethanal. Boil the mixture gently and note the formation of copper(I) oxide.

18 Fehling's solution contains Cu^{2+} ions in an alkaline solution of tartrate ions. The tartrate ions complex with Cu^{2+} ions and prevent the precipitation of copper(II) hydroxide. Describe and explain what happens when ethanal reacts with Fehling's solution. Write an equation (or half-equations) for the reaction which occurs. (In your equation represent the complexed copper(II) ion as Cu^{2+} for simplicity.)

E Sugars: naturally occurring carbonyl compounds

Sugars are sweet-tasting soluble carbohydrates.

The most obvious feature of their structures is the presence of large numbers of —OH groups.

19 What do you understand by the term 'carbohydrate'?

20 Why are carbohydrates very soluble in water?

As well as showing the properties of hydroxy compounds, sugars such as glucose show many properties that are typical of carbonyl compounds. The carbonyl properties of glucose arise from the fact that it can exist in an 'open-chain' form as well as its normal 'ring' form, as shown below. In aqueous solution about 1% of glucose exists in the open-chain form which carries the aldehyde group.

ring form   open-chain form

1 Condensation with 2,4-dinitrophenylhydrazine

Carry out the experiment in part **C**, using half a spatula measure of glucose in $1 cm^3$ of water in place of 3 drops of propanone.

21 Describe the appearance of the crystals which form and write an equation for the reaction between glucose and 2,4-dinitrophenylhydrazine. (Use the open-chain formula for glucose in this equation.)

2 Oxidation by diamminosilver(I) ions (Tollen's Reagent)

Carry out experiment 2, in part **D**, using half a spatula measure of glucose in $1 cm^3$ of water in place of 10 drops of ethanal. Rinse out the test tube immediately after the reaction.

22 Describe and explain what happens.

3 Oxidation by Fehling's solution

Fehling's test is used to detect sugar in the urine of people suffering from diabetes. The pancreas of these patients produces insufficient insulin to cope with the sugar in their diet. This means that sugar accumulates in the blood and, when it reaches a certain concentration, it is excreted by the kidneys and appears in the urine.

Carry out experiment 3, in part **D**, using half a spatula measure of glucose in $1 cm^3$ of water in place of 10 drops of ethanal.

23 Describe and explain what happens. Depending on the concentration of glucose, the solution may simply turn green, produce a fine yellow precipitate or give a dark red precipitate.

PRACTICAL 40
Carboxylic acids

Requirements

Each student, or pair of students, will require:

- Rack and 6 hard-glass test tubes
- 2 boiling tubes
- Evaporating basin
- Beaker (250 cm³)
- Dropping pipette
- Splints
- Universal indicator paper (pH 1 to 4 or full range 1 to 14) or, better, access to a pH meter
- Access to a refrigerator
- Access to a fume cupboard
- Concentrated (glacial) ethanoic (acetic) acid
- Concentrated methanoic (formic) acid
- 2 M ethanoic acid (dilute 116 cm³ of glacial ethanoic acid to 1 dm³ with distilled water)
- 0·1 M ethanoic acid (dilute 5·8 cm³ of glacial ethanoic acid to 1 dm³ with distilled water)
- 0·1 M ethanol in water (dilute 1·5 cm³ of ethanol to 250 cm³ with distilled water)
- 0·1 M hydrochloric acid (dilute 8·6 cm³ of concentrated acid to 1 dm³ with distilled water)
- Phosphorus pentachloride (keep in fume cupboard)
- Pentanol (amyl alcohol)
- Concentrated sulphuric acid
- Ethanol

The following solutions at usual bench concentration:

- Dilute sodium hydroxide
- Dilute hydrochloric acid
- Potassium manganate(VII) (potassium permanganate)
- Iron(III) chloride (ferric chloride)
- Dilute ammonia
- Sodium ethanoate

Time required 2–3 double periods

Introduction

Carboxylic acids contain the functional group 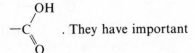 . They have important industrial use, and occur widely in nature, both free and combined as esters, particularly in fats and oils.

In this practical you will begin by considering some of the properties of ethanoic acid, a typical carboxylic acid and then look at some of the unusual properties of methanoic acid and ethanedioic acid. The structural formulae of these acids are shown in the table below.

Systematic name	Other name	Structural formula	
methanoic acid	formic acid	$H-C\overset{O}{\underset{OH}{}}$	(HCOOH)
ethanoic acid	acetic acid	$CH_3-C\overset{O}{\underset{OH}{}}$	(CH_3COOH)
ethanedioic acid	oxalic acid	$\overset{C-OH}{\underset{C-OH}{}}$	(($COOH)_2$)

Structural formulae of methanoic, ethanoic and ethanedioic acids

Procedure

A   The properties of ethanoic acid as a typical carboxylic acid

CARE Concentrated (glacial) ethanoic acid is corrosive. Eye protection must be worn throughout this practical, and any spills should be washed away immediately with plenty of water.

EXPERIMENT 1   Physical properties of glacial ethanoic acid
Put a test tube of glacial ethanoic acid in a refrigerator and observe after an hour.

1   What happens? Why is concentrated ethanoic acid called 'glacial'?

Cautiously smell the vapour from a bottle of glacial ethanoic acid.

2   Is the smell familiar? How did the original name 'acetic acid' come to be used for what we now call ethanoic acid?

Test the solubility of glacial ethanoic acid in water.

3   Is the acid soluble in water? Explain your answer in terms of the structure of the acid.

4   Would you expect octadecanoic acid, $C_{17}H_{35}COOH$ to be soluble in water? Explain your answer.

EXPERIMENT 2   Acid properties of ethanoic acid
Using universal indicator paper, or preferably a pH meter, measure the pH of
0·1 M solutions of
    a   ethanol
    b   ethanoic acid
    c   hydrochloric acid.

5   Record your results and list these compounds in order of increasing acid
strength.

6   Would you classify ethanoic acid as a strong or a weak acid?

7   What explanation can you give for the difference in acid strength of ethanol
and ethanoic acid?

Measure the pH of a 0·1 M solution of sodium ethanoate.

8   Why might sodium ethanoate be expected to be neutral?

9   Is it neutral in fact? If not, why not?

Put 10 cm^3 of 2 M sodium hydroxide solution in an evaporating basin. Add
5 cm^3 of 2 M ethanoic acid and smell the resulting mixture.

10   Can ethanoic acid still be smelled? If not, why not?

Add a further 5 cm^3 or so of 2 M ethanoic acid to the evaporating basin, until
the acid is just in excess. (How can you test whether it is in excess?) Evaporate the
contents of the evaporating basin over a beaker of boiling water until all the water
has left the basin and white crystals remain. (This may take some time. Get on with
another part of the practical while you are waiting.)

11   Identify the white crystals and write an equation for the reaction that has
occurred.

Put a few of the crystals in a test tube and add a few cm^3 of dilute hydrochloric
acid. Gently warm the tube and cautiously smell the vapour coming off.

12   What has been formed? Explain the reaction that has occurred in terms of
your answer to question **5**.

EXPERIMENT 3   Reaction of ethanoic acid with phosphorus pentachloride
**CARE Phosphorus pentachloride is corrosive and reacts violently with water. Do
not let it come into contact with water, and avoid spills. Stopper the bottle immediately after use.**
Working in a fume cupboard, put 2 cm^3 of **glacial** ethanoic acid in a boiling
tube. Carefully add about 0·5 g of phosphorus pentachloride.

13   What evidence is there that a reaction has occurred?

14   Give the formula of the organic product of this reaction and write an equation. (Consult a text book if necessary.)

15   How does phosphorus pentachloride react with ethanol? How does this reaction resemble the reaction of phosphorus pentachloride with ethanoic acid?

**EXPERIMENT 4   Reaction of ethanoic acid and ethanoates with neutral
iron(III) chloride**
Fe^{3+} ions give a characteristic colour with ethanoate ions, but only in neutral
solution.
Prepare a neutral solution of iron(III) chloride (which is often acidic) as follows.
Add ammonia solution dropwise to a few cm^3 of iron(III) chloride solution, until a
faint precipitate just begins to form. Filter or decant the solution from the precipitate and then, working in a fume cupboard, gently warm the solution to drive off
excess ammonia. You now have a neutral solution of iron(III) chloride.
Put 2–3 cm^3 of sodium ethanoate solution in a test tube and add a few drops of
neutral iron(III) chloride solution. (Sodium ethanoate is used here instead of ethanoic acid because it is nearly neutral.)

16   What colour is given by ethanoate ions with neutral iron(III) chloride?

17   What other organic compounds give a characteristic colour with iron(III)
chloride? What colour do they give?

EXPERIMENT 5   Esterification

Put 2 cm^3 of pentanol (amyl alcohol) in a boiling tube and add 1 cm^3 of concentrated sulphuric acid. Add a few drops of ethanoic acid and warm the tube gently.   CARE Allow the tube to cool a little, then pour the contents into a beaker containing about 50 cm^3 of cold water. Cautiously smell the vapour.

18   Describe the smell of the vapour.

19   The pentanol you used was probably a mixture of isomers of formula $C_5H_{11}OH$. Using pentan-1-ol as a representative example, name and write the structural formula of the ester that has been formed.

20   Write an equation for the reaction.

21   What is the purpose of the sulphuric acid in this reaction?

22   Why is the reaction mixture poured into cold water before smelling?

23   Give the names and structural formulae of the esters that would have been formed if
   a   ethanol had been used instead of pentanol.
   b   propanoic acid had been used instead of ethanoic acid.

EXPERIMENT 6   Oxidation

Put 2 cm^3 of glacial ethanoic acid in a test tube. Add a few drops of potassium manganate(VII) solution and warm gently.

24   Are there any signs of reaction?

25   Is ethanoic acid easily oxidized?

EXPERIMENT 7   Dehydration

Put 2 cm^3 of concentrated sulphuric acid in a boiling tube and warm it gently.   CARE Now add a few drops of ethanoic acid to the warm sulphuric acid and observe closely to see if any gas is evolved. If a gas comes off, attempt to identify it.

26   Is ethanoic acid readily dehydrated by concentrated sulphuric acid?

The reactions of ethanoic acid that you have seen in the first part of this practical are typical of carboxylic acids in general. Methanoic acid and ethanedioic acid are not typical, however, and you will now look at some of their properties.

B   Some properties of methanoic acid

CARE Methanoic acid is corrosive and its vapour is irritant. Eye protection must be worn.

EXPERIMENT 8   Physical properties

Investigate the smell and the solubility of methanoic acid.

27   Are these properties similar to those of ethanoic acid?

EXPERIMENT 9   Esterification

Repeat the esterification experiment, as in experiment 5, using ethanol instead of pentanol and methanoic acid instead of ethanoic acid.

28   Describe the smell of the vapour.

29   Name and write the structural formula of the ester that has been formed.

30   Does methanoic acid behave in a similar way to ethanoic acid in this reaction?

EXPERIMENT 10   Oxidation

Repeat the oxidation experiment, as in experiment 6, using methanoic acid instead of ethanoic acid.

31   Are there signs of reaction?

32   Is methanoic acid easily oxidized? If so, to what?

EXPERIMENT 11   Dehydration

Repeat the dehydration experiment, as in experiment **7**, using methanoic acid instead of ethanoic acid.

33   Is methanoic acid readily dehydrated? If so, what is formed?

C   Some properties of ethanedioic acid

CARE Ethanedioic acid is poisonous. Although it is less corrosive than other acids, you should continue to take the same precautions.

EXPERIMENT 12   Physical properties

Investigate the smell and the solubility of ethanedioic acid.

34   Compare the volatility of ethanedioic acid with that of methanoic and ethanoic acids.

35   Relate any difference in volatility to the structure of ethanedioic acid.

36   Compare the solubility of ethanedioic acid with that of the other two acids. Attempt to explain any differences.

EXPERIMENT 13   Esterification

Repeat the esterification experiment as in experiment **5**, using ethanol instead of pentanol and ethanedioic acid instead of ethanoic acid.

37   Describe the smell of the vapour.

38   Name and write the structural formula of the ester that has been formed.

39   Is the volatility of this ester similar to the volatility of the ethanoate and methanoate esters you prepared?

40   Explain your answer to question **39** in view of the difference in volatility between ethanedioic acid and the other two acids.

EXPERIMENT 14   Oxidation

Repeat the oxidation experiment, as in experiment **6**, using an aqueous solution of ethanedioic acid instead of ethanoic acid.

41   Are there signs of reaction?

42   Is ethanedioic acid easily oxidized? If so, to what?

EXPERIMENT 15   Dehydration

Repeat the dehydration experiment, as in experiment **7**, using ethanedioic acid instead of ethanoic acid.

43   Is ethanedioic acid readily dehydrated? If so, what is formed?

44   Draw up a table summarizing the reactions of carboxylic acids and show the results for each of the three acids used in this practical.

PRACTICAL 41
Amines

Requirements

Each student, or pair of students, will require:

- Rack with 4 test tubes
- Teat pipette
- 2 small beakers (100 cm³)
- Universal indicator paper
- Boiling tube
- Thermometer (0–110°C)
- Butylamine
- Phenylamine
- Ethanoyl chloride (acetyl chloride)
- Ammonium chloride
- Sodium nitrite
- Concentrated hydrochloric acid
- Phenol
- 2-Naphthol
- Bromine water
- Ice

The following solutions at usual bench concentration:

- Ammonia
- Hydrochloric acid
- Sodium hydroxide
- Copper(II) sulphate

Access to a fume cupboard

Time required 2 double periods

Introduction

Amines are one of the most important groups of organic nitrogen containing compounds. They can be regarded as derivatives of ammonia in which one or more of the hydrogen atoms in NH_3 is replaced by aryl or alkyl groups. If only one of the hydrogen atoms in NH_3 is substituted, we get a compound of the form RNH_2, called a primary amine. The simplest primary amines are CH_3NH_2, methylamine, and $CH_3CH_2NH_2$, ethylamine.

1   Write full structural formulae for ammonia, butylamine and phenylamine (aniline).

2   Which simple substance would you expect amines to resemble?

3   Predict two properties of butylamine.

Free amines are relatively rare in nature, but they do occur in decomposing protein such as meat and fish. Normally the $-NH_2$ group is associated with other functional groups and in this respect it forms an important part of proteins. Compounds containing the $-NH_2$ group are important in the manufacture of drugs, dyes and nylon.

The intention of this practical is to consider the properties of an alkylamine (butylamine) and an arylamine (phenylamine) and to compare their properties with those of ammonia. Butylamine is chosen because it is a liquid and not too volatile. Phenylamine is also a liquid, colourless when pure, but the sample you use is likely to be dark-coloured due to atmospheric oxidation.

1   Reactions of amines as bases

CARE Phenylamine is toxic and harmful because of skin absorption. Avoid all skin contact. Butylamine is an irritant and is extremely flammable. Work in a fume cupboard if possible.

A   Reaction with water and indicators
Shake 2 drops of butylamine and 2 drops of phenylamine separately with 2 cm³ of water. Now test each of these solutions and a solution of ammonia with universal indicator paper.

4   Comment on the solubility of the amines in water and account for any differences in solubility.

5   Comment on the basic strength of the amines relative to ammonia and account for any differences in their strength as bases.

B   Reaction with dilute hydrochloric acid
Shake 2 drops of butylamine and 2 drops of phenylamine separately with 2 cm³ of dilute hydrochloric acid.

6   Describe what happens.

7   Write equations for the reactions which have occurred.

8   Explain the different solubility of phenylamine in water and in dilute hydrochloric acid.

Add dilute sodium hydroxide to the solution of phenylamine in hydrochloric acid until the mixture is alkaline.

9   Describe and explain what happens.

2 Reactions of amines as ligands

Reaction with copper(II) sulphate solution
Add ammonia solution dropwise to 2 cm^3 of copper(II) sulphate solution until the ammonia is present in excess.

10 Describe and explain what happens. Write equations for the reactions which occur.

Repeat the experiment using first butylamine and then phenylamine in place of ammonia solution.

11 Describe and explain what happens. Do the amines react in a similar fashion to ammonia?

3 Reactions of amines as nucleophiles

A Reaction with ethanoyl chloride and benzoyl chloride
CARE Work in a fume cupboard. Eye protection must be worn.

Ammonia and amines can act as nucleophiles in attacking the positive centres in molecules such as ethanoyl chloride, CH_3COCl and benzoyl chloride, C_6H_5COCl. For example, ammonia reacts very vigorously with ethanoyl chloride forming ethanamide.

ethanamide

12 Would you expect phenylamine to react with ethanoyl chloride more or less vigorously than ammonia? Explain your answer.

Add 1 cm^3 of ethanoyl chloride dropwise to an equal volume of phenylamine in a boiling tube, shaking after the addition of each drop.

13 Describe what happens.

14 Write an equation for the reaction which occurs. Name the solid organic product.

B Reaction with nitrous acid
Dissolve a full spatula measure of ammonium chloride in 3 cm^3 of hot water. Now add an equal amount of solid sodium nitrite.

15 Describe and explain what happens. (The reaction can be regarded as one between ammonia and nitrous acid, these two substances being formed by the reaction between ammonium ions and nitrite ions:

$$NH_4^+ + NO_2^- \longrightarrow NH_3 + HNO_2 .)$$

Put 3 drops of butylamine in a test tube and add concentrated hydrochloric acid dropwise until a clear solution is formed. Dilute the mixture to 3 cm^3 with water, add a spatula measure of sodium nitrite and warm gently. Repeat the experiment using phenylamine in place of butylamine.

16 Describe and explain what happens. Write equations for the reactions which occur.

17 What organic product can you smell in the reaction of phenylamine with nitrous acid on warming?

The reactive group in nitrous acid is thought to be the nitrosyl cation, NO^+.

When amines react with nitrous acid, they first form the diazonium ion, $\overset{+}{R}-N\equiv N$.

$$RNH_2 + NO^+ \longrightarrow R-\overset{+}{N}\equiv N + H_2O$$

Unless the diazonium ion is stabilized in some way, it decomposes forming nitrogen and the new carbonium ion, R^+ which takes part in a variety of further reactions.

$$R-\overset{+}{N}\equiv N \longrightarrow R^+ + N_2$$

In the case of arylamines, such as $C_6H_5NH_2$, stabilization of the diazonium ion can occur by delocalization from the benzene ring, provided the temperature of the reactants is kept below 10°C. The diazonium salts that form are useful reactive intermediates, as the following experiment shows.

Make a solution of 2 drops of phenylamine in $2\ cm^3$ of dilute hydrochloric acid and add it to about $5\ cm^3$ of a crushed ice/water mixture at 5–10°C. Add to this mixture a spatula measure of sodium nitrite and stir well to ensure that the solid dissolves. This solution contains the benzenediazonium ion, $C_6H_5-\overset{+}{N}\equiv N$.

Use the solution to prepare azo dyes as follows:

A Dissolve a spatula measure of phenol **(CARE Corrosive – avoid skin contact and wear eye protection)** in $2\ cm^3$ of dilute sodium hydroxide solution. Cool this solution to below 5°C and add the diazonium solution drop by drop.

B Repeat the test using 2-naphthol in place of phenol.

18 Describe what happens in each case. Write equations for the reactions which occur. (See *Chemistry in Context*, section 32.7.)

19 Why is it so important to keep the solutions below 10°C in these reactions?

4 Reactions of the aromatic ring in phenylamine

Put 3 drops of phenylamine in a test tube and add concentrated hydrochloric acid drop by drop until the phenylamine dissolves. Now add bromine water dropwise until no further change occurs.

20 Describe what happens. Write an equation for the reaction which occurs.

21 Why can phenylamine undergo substitution with bromine so much easier than benzene?

PRACTICAL 42
Polymers

Introduction

This practical is in two parts. In the first part you will prepare some representative polymers. In the second part you will investigate the way in which the molecular structure of a polymer can influence its physical properties. You may find it helpful to refer to *Chemistry in Context*, sections 26.7 and 31.5, during this practical and in answering the questions.

A   The preparation of some polymers

Polymers can be classified in a number of ways. One way is to group them according to the type of chemical reaction by which they are prepared. Thus, there are **addition polymers** and **condensation polymers**.

1   Explain the terms 'addition' and 'condensation' as applied to chemical reactions.

Polymers differ in the way in which they behave when heated. **Thermoplastic** materials soften on heating and harden again on cooling. **Thermosetting** materials set hard on heating and cannot be softened or melted. Strong heating only tends to decompose them.

2   Name one article made from a thermoplastic material and one made from a thermosetting material.

The strength and elasticity of polymers varies widely. Highly elastic polymers are called **elastomers**. **Fibres** have low elasticity and high tensile strength. Intermediate between elastomers and fibres are **plastics**.

3   Name one elastomer, one plastic and one fibre.

The preparation of an addition polymer – polystyrene

Polystyrene, systematically called poly(phenylethene) is made by polymerizing phenylethene (styrene), whose formula is shown below. A catalyst, di(dodecanoyl) (lauroyl) peroxide is needed to speed up the polymerization, which is otherwise very slow.

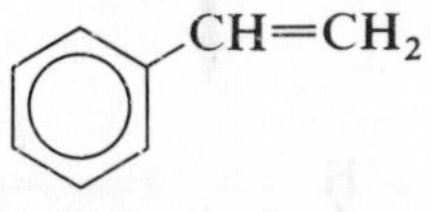

Working in a fume cupboard, put 10 cm^3 of phenylethene in a boiling tube. Add about 0·2 g of di(dodecanoyl) peroxide and shake the tube until the catalyst has dissolved. Plug the tube with cotton wool and heat it in a water bath at about 100°C for 20 minutes, then examine the contents and compare with the original phenylethene.

4   What evidence is there that polymerization is taking place?

5   Draw the structure of a section of the polymer chain.

Polymerization can be completed by heating the tube in an oven or water bath at about 50°C until a solid block is obtained. Of course, the glass tube would have to be broken to remove the polymer. Alternatively, the semi-polymerized material can be used as a casting or embedding resin by pouring it into a mould which has been lightly smeared with liquid detergent.

6   Classify polystyrene as
a   thermoplastic or thermosetting
b   elastomer, plastic or fibre.

Requirements

Each student, or pair of students, will require:

- Rack and 4 test tubes
- 2 boiling tubes
- Beaker (400 cm^3)
- Beaker (100 cm^3)
- Glass rod
- Tweezers
- 2 disposable polystyrene cups (normal, not expanded polystyrene)
- Cotton wool
- Broken porcelain
- Tongs
- Sharp knife or scissors
- Access to a fume cupboard
- Access to an oven
- Oil bath*
- Thermometer, $0-250°C$*
- * these materials may only be required for a teacher demonstration.
- Phenylethene (styrene)
- Di(dodecanoyl peroxide (lauroyl peroxide)
- 5% solution (freshly prepared) of decanedioyl dichloride (sebacoyl chloride) in 1,1,1-trichloroethane (10 cm^3)
- 5% solution of 1,6-diaminohexane in water (10 cm^3)
- 40% aqueous solution of methanal (formalin)
- Urea
- Polystyrene, chips, or fragments of a polystyrene cup, toy etc.
- Natural rubber, small pieces
- Urea-methanal plastic, small pieces (pieces of a broken white electrical fitting such as a plug or a socket would do)
- 1,1,1-Trichloroethane
- Benzene-1,2-dicarboxylic anhydride (phthalic anhydride)
- Ethane-1,2-diol (ethylene glycol)
- Butanedioic acid (succinic acid)
- Polythene film

Time required 2–3 double periods

The preparation of a condensation polymer – nylon 6.10

Nylon 6.10 is prepared by a condensation reaction between decanedioyl dichloride (sebacoyl chloride) and 1,6-diaminohexane.

$$Cl-C-CH_2CH_2CH_2CH_2CH_2CH_2CH_2CH_2C-Cl$$
$$\quad\;\; O \qquad\qquad\qquad\qquad\qquad\qquad\qquad O$$

decanedioyl dichloride

$$H_2NCH_2CH_2CH_2CH_2CH_2CH_2NH_2$$

1,6-diaminohexane

Working in a fume cupboard, put 10 cm^3 of a 5% solution of decanedioyl dichloride in 1,1,1-trichloroethane into a 100 cm^3 beaker.
CARE The vapour from this solution is harmful.
Very carefully, using a dropping pipette, add 10 cm^3 of a 5% solution of 1,6-diaminohexane in water. Do **not** allow the two solutions to mix – try to float one layer on top of the other so that the nylon forms at the interface. Use tweezers to pull the interface film out of the liquid and wind it on to a glass rod. The glass rod can then be rotated, pulling a continuous filament of nylon from the interface and winding it on to the rod. **Do not touch this filament by hand.**

7 Draw the structure of a section of the nylon 6.10 chain.

8 What other product is formed in this reaction, apart from nylon?

9 What reagents would you use if you were making nylon 6.6?

10 Classify nylon as
   a thermoplastic or thermosetting
   b elastomer, plastic or fibre.

The preparation of a thermosetting plastic – urea-methanal resin
Urea and methanal react together to form polymer chains as shown in figure 1.

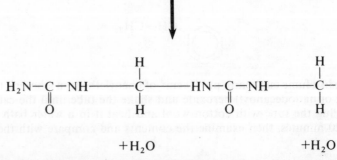

Figure 1 Polymerization of urea and methanal

These chains can then cross-link with other chains via further molecules of methanal (figure 2). The extensive cross-linking in this polymer makes it set hard on formation and prevents it from softening on heating. (See part 2 of this practical.)

Working in a fume cupboard put 20 cm^3 of a 40% solution of methanal in water into a disposable plastic cup.   **CARE Irritant vapour.**   Add about 10 g of urea and stir thoroughly. Add a few drops of concentrated sulphuric acid, stirring all the time.
CARE Eye protection must be worn.
When polymerization is complete it should be possible to remove the solid mass of polymer intact from the container. Wash the polymer thoroughly before handling.

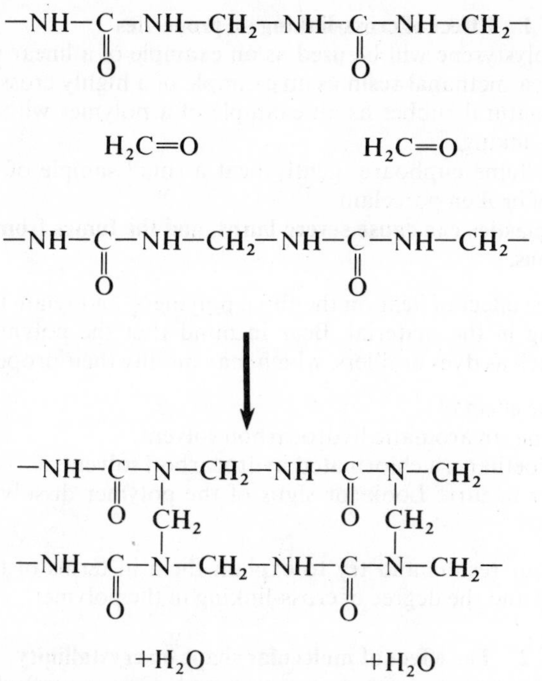

$$-NH-\overset{\overset{\displaystyle O}{\|}}{C}-NH-CH_2-NH-\overset{\overset{\displaystyle O}{\|}}{C}-NH-CH_2-$$

$$H_2C=O \qquad\qquad H_2C=O$$

$$-NH-\overset{\overset{\displaystyle O}{\|}}{C}-NH-CH_2-NH-\overset{\overset{\displaystyle O}{\|}}{C}-NH-CH_2-$$

↓

$$-NH-\overset{\overset{\displaystyle O}{\|}}{C}-\underset{\underset{\displaystyle CH_2}{|}}{N}-CH_2-NH-\overset{\overset{\displaystyle O}{\|}}{C}-\underset{\underset{\displaystyle CH_2}{|}}{N}-CH_2-$$
$$-NH-\overset{\overset{\displaystyle O}{\|}}{C}-\underset{|}{N}-CH_2-NH-\overset{\overset{\displaystyle O}{\|}}{C}-\underset{|}{N}-CH_2-$$

$$+H_2O \qquad\qquad +H_2O$$

Figure 2 Cross-linking between different urea-methanal chains

11 Describe the appearance of the polymer.

12 Is this a condensation polymer or an addition polymer?

13 What was the purpose of the concentrated sulphuric acid?

14 This polymer is often used to make electrical fittings such as plugs. How could this polymer be made into the shape of an electrical plug, bearing in mind that it is a thermosetting material?

B The properties of some polymers

The properties of materials can usually be related to their molecular structure, and polymers are no exception. A number of molecular properties are particularly important in deciding the strength, elasticity, solubility and softening point of polymers.

a Chain length
In general, increasing the chain length increases the strength of a polymer.

b Cross-linking
A high degree of cross-linking reduces the extent to which the individual chains can move, reducing elasticity and increasing hardness and softening point (figure 3).

c Crystallinity
No polymer is completely crystalline – in some regions the chains are regularly arranged and in others they are at random (figure 4). A polymer will therefore have both crystalline and amorphous (non-crystalline) regions, although the proportion of one to the other will vary according to the polymer. A high degree of crystallinity increases the density, the softening point and the tensile strength of the polymer. Crystalline polymers are usually opaque, while non-crystalline ones are transparent and glassy. Fibres have very high crystallinity, but in elastomers crystallinity is very low. Two factors are of major importance in determining the degree of crystallinity:

i Intermolecular forces between chains.   A polymer carrying highly polar groups along its chain will have strong interactions between chains and this will tend to increase crystallinity.

ii Shape of the chains.   Linear chains pack together well, giving high crystallinity. Chains carrying many branches, or with irregularly arranged side-groups, pack together less well, giving lower crystallinity.

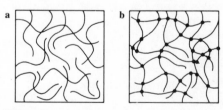

Figure 3
a Polymer with no cross-linking
b Polymer with high degree of cross-linking

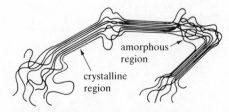

Figure 4 Crystalline and amorphous regions in a polymer.

119

EXPERIMENT 1 Effect of cross-linking on properties

In these tests, polystyrene will be used as an example of a linear polymer with no cross-linking, urea-methanal resin as an example of a highly cross-linked polymer, and vulcanized natural rubber as an example of a polymer with an intermediate amount of cross-linking.

Working in a fume cupboard, gently heat a small sample of each polymer in turn on a piece of broken porcelain.

CARE Molten plastics can cause severe burns, and the fumes from burning plastics are often poisonous.

15 Compare the effect of heat on the three polymers, and relate it to the degree of cross-linking in the material. Bear in mind that the polymers may contain additives such as dyes or fillers, which may modify their properties.

Investigate the effect of:

a methylbenzene, an aromatic hydrocarbon solvent,

b 1,1,1-trichloroethane, a chlorinated hydrocarbon solvent,

on each polymer in turn. Look for signs of the polymer dissolving, softening or swelling.

16 Tabulate your results and try to explain them in terms of the nature of the solvent used and the degree of cross-linking in the polymer.

EXPERIMENT 2 The effect of molecular shape on crystallinity

In this experiment two polyesters are prepared. One is made by condensation polymerization between butanedioic acid and ethane-1,2-diol (figure 5a), and the other by a similar reaction between benzene-1,2-dicarboxylic acid and ethane-1,2-diol (figure 5b).

a $HOOCCH_2CH_2COOH$ $HOCH_2CH_2OH$
   butanedioic acid   ethane-1,2-diol

b 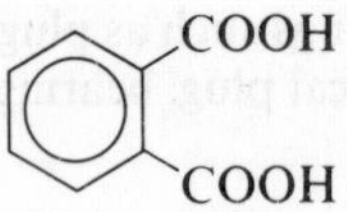
   benzene-1,2-dicarboxylic acid $HOCH_2CH_2OH$
                                                ethane-1,2-diol

Figure 5

17 Why are these polymers called polyesters?

18 Draw a representative section of the chain of each polyester showing at least two repeating units. If possible, build a ball-and-stick model of a section of each polymer.

19 Which polymer chain is more linear?

20 Which polymer has chains which will pack together more regularly?

21 Which polymer is likely to be more crystalline?

The experiment involves heating the monomers at 180°C using an oil bath.

CARE This operation can be hazardous and should be carried out in a fume cupboard. Precautions must be taken to ensure the oil bath cannot be knocked over.

Your teacher may prefer to demonstrate the reaction rather than carry it out as a class practical.

Label two boiling tubes A and B. Into tube A put 5·6 cm³ of ethane-1,2-diol and 11·8 g of butanedioic acid. Into tube B put 5·6 cm³ of ethane-1,2-diol and 14·8 g of benzene-1,2-dicarboxylic anhydride (phthalic anhydride). These quantities represent 0·1 mole in each case. Heat both tubes in an oil bath at 180°C for one hour, then leave the tubes overnight.

22 What evidence of crystallinity is there in each polymer? Does this evidence agree with your answer to question 21?

EXPERIMENT 3   The effect of 'cold drawing' on the crystallinity of a polymer: the effect of stretching polythene film

With a pair of sharp scissors or a sharp knife cut out a strip of polythene film. Make sure the edges of the strip are clean-cut. Stretch the film lengthways by pulling it slowly and steadily – avoid sudden jerks. Record any changes that occur in the film as you stretch, particularly the width of the strip, its appearance, its elasticity and its strength, as indicated by the force required to stretch it.

23   What changes occur in the crystallinity of the polythene as it is stretched?

24   If the polymer chains are distributed as shown in figure 6 before stretching, indicate by means of a sketch how you think they may be arranged after stretching.

25   The operation you have just carried out is called 'cold drawing'. What industrial importance does this process have?

Repeat the experiment using a fresh strip of polythene, but this time stop pulling when you think the crystallinity of the sample has reached a maximum, but before it breaks. Now pull the strip across its width, that is at right angles to the original direction of stretching.

26   What happens? Suggest an explanation in terms of the arrangement of the polymer chains.

Take a polystyrene drinking cup (made of normal, not expanded polystyrene) and try to tear it, pulling in the first direction shown in figure 7. Now try to tear it at right angles, pulling in the second direction shown in figure 7.

27   Suggest a reason for any differences in the ease with which the polystyrene can be torn in the different directions.

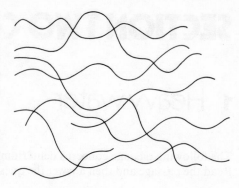

Figure 6 Polythene chains before cold drawing.

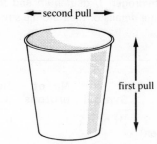

← second pull →

first pull

Figure 7 Tearing a polystyrene cup.

121

SECTION TWO Comprehension questions

1 Heavy water

The following passage concerns deuterium oxide, D_2O, often called 'heavy water'. Read the passage and then answer the questions following it.

Isotopes of hydrogen

Three isotopes of hydrogen are known: protium (usually just called 'hydrogen'), deuterium and tritium. The table below gives some details of their atomic structure and relative abundance.

Name	Symbol	No. of protons	No. of neutrons	Abundance in natural hydrogen
protium (hydrogen)	$_1^1H$ or H	1	0	99·984%
deuterium (heavy hydrogen)	$_1^2H$ or D	1	1	0·015%
tritium	$_1^3H$ or T	1	2	very rare— 1 part in 10^{17} (radioactive and unstable)

Table 1 Isotopes of hydrogen

The electron structure of these atoms is of course identical, so the isotopes are chemically very similar. Any differences in their chemical behaviour are due to the different masses of the atoms – deuterium atoms are twice as heavy as normal hydrogen atoms. As the ratio of the masses of the isotopes is large compared with the isotopes of other elements (for example 1·06 for $^{37}Cl : {}^{35}Cl$ and 1·01 for $^{238}U : {}^{235}U$), differences in reactivity between deuterium and hydrogen are more marked than for other isotope pairs. In spite of this, chemical differences between the two isotopes are small, although biological systems, which are highly sensitive to changes in reaction conditions, sometimes show noticeably different behaviour when deuterium atoms are substituted for hydrogen atoms. In general, deuterium is slightly less reactive than hydrogen, probably because of the higher bond energy of the D—D bond (440 kJ mole^{-1} compared with 436 kJ mole^{-1} for the H—H bond).

Heavy Water

Most of the hydrogen atoms in the earth's crust occur combined as water. Ordinary water contains H_2O, HDO and D_2O molecules, the proportion of D atoms being about 0·015 per cent, or about 1 in 6000. Although the chemical properties of H_2O and D_2O are very similar (D_2O can for example be drunk

quite safely), some of their physical properties differ quite noticeably, as the table below shows.

	H_2O	D_2O
melting point/°C	0	3·82
boiling point/°C	100	101·4
density at 20°C/g cm^{-3}	0·9982	1·1059
solubility of NaCl at 25°C/g per 100 g	3·6	3·0
viscosity/10^{-4} kg m^{-1} s^{-1}	10·9	12·6
dielectric constant at 25°C	82	80·5

Table 2 A comparison of some physical properties of H_2O and D_2O

Because of its higher density D_2O is often called 'heavy water'.

Heavy water has many important uses. It is employed as a source of deuterium atoms in the preparation of deuterated compounds (e.g. NaOD, C_2H_5OD) and in isotopic tracer experiments. Perhaps its best known use is as a moderator in nuclear reactors. The fission of ^{235}U is triggered by neutrons, but only 'slow' neutrons are effective in causing the uranium nucleus to split. The function of a moderator is to slow down the neutrons, and D_2O is particularly effective for this purpose because it does not absorb neutrons but merely reduces their kinetic energy.

The manufacture of heavy water

The separation of isotopes is extremely difficult because of the very small differences in their physical and chemical properties. The isotopes of hydrogen are easier to separate than those of any other element, but their separation is nevertheless a difficult and tedious process. Deuterium is manufactured by separating D_2O from natural water. There are two ways of doing this, either by fractional distillation, using the small difference in boiling point between H_2O and D_2O, or by electrolysis. During the electrolysis of natural water, hydrogen is liberated at the cathode six times faster than deuterium, so the residual water becomes progressively richer in deuterium. Heavy water is therefore manufactured by the repeated electrolysis of an aqueous electrolyte. One cubic centimetre of 99 per cent pure D_2O can be obtained from about 35 dm^3 of electrolyte from old electrolysis cells.

The drama of Telemark

The phenomenon of nuclear fission was known well before the Second World War, and both the Allies and the Nazis realized its potential and the possibility of building an atomic bomb. By 1942, the Allies were working hard on the problem but were still far from a solution, when it became clear that German scientists too were trying to develop atomic fission with a view to making a bomb. Crucial to their work was the heavy water needed as a moderator for the fission reaction. For many years, the major world production of heavy water had been carried out at Rjukan, in the Telemark mountains of Norway. The site was near a large hydro-electric generating station. By 1942 the Nazis had occupied Norway and it was clear that to stop the Germans developing an atomic bomb, the Rjukan factory must be destroyed.

In early 1942 a group of Norwegian commandos moved into the Telemark mountains to prepare the ground and provide Britain with intelligence about heavy water production. In March an attempt was made to parachute an airborne British sabotage group into the area, but it failed when their gliders crashed. The Norwegian reception committee withdrew into the mountains where they spent the next winter, eking out what was left of their food with wild reindeer. In February 1943, a new group was dropped into the area and this time they successfully linked up with the Norwegian commandos and made their way to the Rjukan factory. Using wire-clippers they entered the plant and placed explosives on the heavy water containers. In the explosion which followed, the containers were largely destroyed and the saboteurs escaped.

However, the Germans rebuilt the plant, and to counter any further threats the Americans decided to bomb the factory. In November 1943, 140 Flying Fortresses bombed the factory,

The Vermork factory near Rjukan, Telemark where heavy water was produced.

and the Nazis abandoned all plans for heavy water production. There remained however, some stocks of partially processed heavy water, and the Nazis decided to remove these to the safety of Germany. Part of the journey involved transporting the heavy water across Lake Tinnsjø by ferry. The Allies received intelligence reports of the Germans' intention to move the water, and the night before the ferry was due to sail, a Norwegian saboteur group placed a time charge on the ferry. On 20 February 1944, the ferry blew up and sank, taking with it the containers of heavy water. Four German guards and fourteen Norwegian civilians were killed, but the battle for heavy water in Norway was over.

Questions

1 Assuming that naturally-occurring hydrogen contains 1 deuterium atom in 6000, and that these proportions also apply to the hydrogen atoms in natural water, calculate the ratio of $D_2O : HDO : H_2O$ molecules in natural water.

2 Why are the isotopes of hydrogen 'easier to separate than those of any other element'?

3 Starting with D_2O as your only source of deuterium atoms, how could you prepare **a** NaOD **b** DCl **c** ND_3 **d** D_2SO_4 **e** CH_3COOD **f** CH_3CH_2OD?

4 Suggest an explanation for the fact that 'in the electrolysis of natural water, hydrogen is liberated at the cathode six times faster than deuterium' even when H_2O and D_2O are present in equal concentrations.

5 Explain, using an example, what is meant by an 'isotopic tracer experiment'.

6 Why might biological experiments in which deuterium is used as a tracer give unreliable results?

7 Why were the Allies anxious to prevent the Nazis producing and stockpiling heavy water?

8 Why was the heavy water factory situated at this particular place in the Telemark mountains?

9 Was it right for the Allies to prevent the production of a German atomic bomb and yet go ahead themselves in producing and eventually using such a bomb?

2 Nuclear alert at Three Mile Island

This article from the *Daily Mail* of 4 April 1979 refers to a crisis that occurred at a nuclear power station in Harrisburg, Pennsylvania, and led to the evacuation of civilians who lived nearby. Eventually the reactor was made safe. Read the article and then answer the questions which follow it.

THAT NUCLEAR NIGHTMARE ON THREE MILE ISLAND

Harrisburg power station, where disaster so nearly struck

Why it couldn't happen here..!

BY ANGUS MACPHERSON

AS the engineers on Three Mile Island grew more confident yesterday that they had their rogue

actively considered as things went wrong in Harrisburg. 'I suppose we can forget about

and valves and it is clear that the Harrisburg cliff-hanger began with either a leak or a burst that spilled water out of the

throughout the world — nearly 300 of them the water-cooled type, similar to Harrisburg.

Water reactors are easier and more economic to operate than gas ones, so most of the developing countries and many in Europe, including France, have plumped for the Harrisburg type.

This is why most of the world has bought its reactors from America rather than

As the engineers on Three Mile Island grew more confident yesterday that they had their rogue nuclear reactor back under control, a major disaster seemed to have been avoided. But as they slowly shut down the atomic heart of the Harrisburg power station, how much of the world's controversial nuclear power programme were they shutting down with it?

It was not some obscure pitfall of sub-atomic structure that went wrong at Harrisburg, it was, essentially, bad plumbing – something that anyone can understand. This is likely to magnify the world-wide impact of the inevitable looming US-style post-mortem.

This particular knife-edge incident could *not* have happened at any of Britain's thirteen operational atomic power stations. We have chosen a path in design that has set us apart from most of the rest of the world, including America. But the British nuclear industry has been under immense pressure – particularly from the Central Electricity Generating Board – to adopt Harrisburg-type designs. This radical switch was still being actively considered as things went wrong in Harrisburg. 'I suppose we can forget about that now,' said one Electricity Board official grimly. 'At least we shall have to have another very hard think.'

A nuclear reactor is a highly complex variation on a domestic heating system. At its heart the rods of uranium fuel produce heat and intense radioactivity inside a pressure vessel of eight-inch steel surrounded by a 'bunker' of 4 ft thick concrete. Something has to go in to extract the heat and transfer it to the steam turbines of the electricity generators.

The British gas reactor design uses carbon dioxide gas circulating around the system to do this. Most American power stations – and the majority in the world – use circulating water. Experts wrangle endlessly about the respective merits of these designs. But one difference is that, to keep the water from boiling off into steam, the American water design has to operate at a pressure close to one ton per square inch; anything up to seven times as high as in a British gas-type reactor. This enormously increases the problems of making leak-free joints and valves and it is clear that the Harrisburg cliff-hanger began with either a leak or a burst that spilled water out of the cooling system. The pile then began to overheat. What's more, some of the remaining water in the system was decomposed by the heat to give off a huge bubble of hydrogen. As the temperature inside the pile continued to mount, the station technicians faced a double possibility of disaster.

The hydrogen bubble might explode, which would scatter nuclear fall-out over a huge area. Alternatively, the metal pile of the reactor might melt, and literally bore its way out through the floor in a white-hot radioactive soup. The exact scale of the human catastrophe that either of these nuclear escapes could cause is angrily debated. But if either broke on an unsuspecting community it is likely that hundreds or thousands would die and several thousands more would be made seriously ill by radiation sickness.

The seriousness of the risk was underscored by President Carter's personal mission to the plant and the plans to evacuate up to 600,000 people from within a 20 mile radius of the plant.

There are more than 500 nuclear reactors operating throughout the world – nearly 300 of them the water-cooled type, similar to Harrisburg. Water reactors are easier and more economic to operate than gas ones so most of the developing countries and many in Europe, including France, have plumped for the Harrisburg type. This is why most of the world has bought its reactors from America rather than Britain over the past decade. However, this could dramatically change in the wake of Harrisburg. The next British Government is going to hesitate a long while before it allows the C.E.G.B. to switch to American-style reactors.

No one can be 100 per cent sure that unsuspected hazards do not lurk in the design of gas reactors as well. But it does seem, in the light of the past few days, that the British models score significantly over the Americans on some safety points – and this could lead to a new interest from the rest of the world. Our 13 home-designed reactors contributing 13 per cent of our electricity are all of the gas type, and are in comparatively remote spots like Dungeness, Sizewell, on the Suffolk coast, or Wylfa, in Anglesey. They have not themselves been free from accident. There have been at least two serious fires. In the most notorious incident, at Windscale in 1957, considerable radio-activity was released into the air. But they do not run the risk of the appalling hazard of an explosive gas 'bomb' that has been revealed at Harrisburg.

The nuclear engineers may with some justice claim that they did, in the end, avert the disaster. The safety system proved strong enough to cope with an entirely unforeseen hazard. But it is conventional engineering, not nuclear technology, that is in the dock.

The atomic genie has not revealed any unsuspected power or viciousness in the past few days. But are we clever or careful enough to build a bottle that can hold him?

Angus MacPherson, Why it couldn't happen here!,
Daily Mail, April 4th, 1979

Questions

1 Outline very briefly how uranium 'fuel' produces heat in a reactor.

2 The Harrisburg crisis began when the nuclear pile began to overheat. Why did this overheating occur, and why is it less likely to happen in the type of power station used in Britain?

3 Despite the overheating, there was never any danger of a nuclear explosion, that is, of the uranium exploding like an atomic bomb. Explain why not.

4 The main danger was caused by a large bubble of hydrogen in the reactor. Write an equation to show how this hydrogen was formed. Why was there a chance that the hydrogen bubble might explode?

5 The formation of hydrogen must have been accompanied by oxygen formation. What happened to this oxygen?

6 Why would the explosion of the hydrogen bubble have been catastrophic, even though the civilian population had been evacuated?

7 With fossil fuel reserves becoming rapidly depleted, nuclear fission is an increasingly attractive source of energy. Yet there are risks associated with its use, as this article shows. Should we continue to develop nuclear power, or should we stop? Justify your view.

3 The quality of air

The following passage by Professor D. Bryce-Smith, which first appeared in *Chemical Age International*, outlines the major air pollutants in the United Kingdom. Read the passage and then answer the questions which follow it.

It is interesting to note that the quality of the air was recognized as an important factor in health by Hippocrates about 2½ millenia ago.

The atmosphere of a city such as London has been contaminated for centuries, mainly with soot and sulphur dioxide resulting from the burning of coal. These, especially the latter, still constitute a serious threat to health, augmented in more recent times by numerous other toxic materials, including carbon monoxide, asbestos, and organic and inorganic lead compounds.

Sulphur dioxide results mainly from the combustion of coal and fuel oil, with a relatively small contribution from petrol. There are no legal limits on the sulphur content of coal and oil, but the Central Electricity Generating Board have in recent times, been using fuels in London with average sulphur contents of 1·6 per cent in coal and 2·5 per cent in oil, although in 1970 some of the coal they used contained up to 4·3 per cent sulphur. The ratio of SO_2 to SO_3 in flue gases is typically about 30 : 1, but further oxidation of SO_2 to the trioxide can occur in the atmosphere, in part through oxides of nitrogen

acting as oxygen-transfer agents. Sulphuric acid in rainfall is therefore one of the end products, and one which is almost certainly costly to society both by its effect on health and through its corrosive effects on fabrics, masonry, metalwork, etc. The removal of sulphur dioxide from chimney fumes is technically feasible, and might even appear economically attractive if the real cost of the present policy to public health, etc. were taken into account.

Carbon monoxide. Most of this comes from motor car exhaust fumes (comparatively little from diesel fumes). Its toxic effects result from deactivation of haemoglobin in the bloodstream. Long-term levels in the vicinity of moderately heavy traffic tend to range between 10 and 20 parts per million, although short-term peaks more than ten times this have been recorded. Continuous exposure of adults to levels above 50 parts per million can definitely be harmful, and lower levels are suspected of affecting mental processes. The effects of mild exposure of healthy people to carbon monoxide are probably only temporary as it is not a cumulative poison.

The polluted atmosphere of Sydney, Australia.

126

Oxides of nitrogen. About 30–40 per cent of the oxides of nitrogen in the atmosphere result from the limited degree of nitrogen fixation which occurs in the internal combustion engine. Present atmospheric levels are low relative to levels known to produce adverse physiological effects. Oxides of nitrogen (mainly NO and NO_2) are known to be an important factor in the production of 'photochemical smog'; but this is not yet a major problem in the U.K.

Asbestos is present in the atmosphere as minute airborne fibres from building and insulating materials, brake drums, etc. There are several forms, the most dangerous of which is crocidolite, which is known to cause fibrositis and lung cancer. C.H.Um (British Medical Journal 1971) has shown a major increase in the proportion of 'asbestos bodies' in the lungs of Londoners. The percentages of cases in which these 'bodies' were found in the years 1936, 1946, 1956 and 1966 were 0, 3, 14 and 20 per cent respectively.

Lead. This is a major contaminant of city air, being present in both inorganic and organic forms. All the organic lead and at least 90 per cent of the inorganic lead in city air are now recognized as originating from organo-lead anti-knock additives in petrol such as tetraethyl-lead. According to recent studies in London and Stockholm, the ratio of inorganic to organic lead near traffic is about 10 : 1. Organo-lead compounds, being fat soluble, are more readily able to cross the blood-brain barrier than inorganic lead.

Most of the organo-lead compounds in petrol end up as an aerosol of inorganic lead compounds in the exhaust gases. The danger to health from inhalation of airborne inorganic lead should be considered in the context of the total exposure from all sources including diet and water. (Lead pipes, for example in the Glasgow area cause the average level of lead in the drinking water to be 160 micrograms per litre, in comparison with the upper safe level of 100 micrograms per litre set by the World Health Organisation.) Children are also liable to ingest significant amounts of lead from chewing and sucking lead-painted articles and from general city dust. City dust commonly contains between 1000 and 10000 parts per million of lead, and values up to nearly 50000 parts per million have been recorded.

The contribution which inhalation of airborne inorganic lead can make to the total amount absorbed daily from all sources has been much debated. A recent authoritative report from the U.S. National Academy of Sciences has concluded that the inhalation route can provide up to half the amount which comes from diet, but sometimes much less than this, and occasionally up to twice the amount from diet. The contribution from airborne lead can therefore be a major one, although its precise extent depends upon the habits, location and physiology of the individual.

Inorganic lead is toxic to the nervous system, including the brain, but the effects on young children are now recognized to be particularly serious. A high percentage of young children who experience even a mild attack of lead poisoning suffer permanent brain damage. The effects are manifest years later as educational disorders of the dyslexia type and/or emotional and behavioural disturbances, including a tendency to violent and impulsive behaviour.

Abstracted by Michael Vokins in *School Science Review*, September, 1974, from D. Bryce-Smith, The Quality of Air, *Chemical Age International*, December, 1972

Questions

1 Explain briefly in your own words how sulphuric acid comes to be present in rainfall in industrial areas.

2 What is meant by 'the limited degree of nitrogen fixation which occurs in the internal combustion engine', and why is it responsible for the presence of NO and NO_2 in the atmosphere?

3 How does lead come to be present in the atmosphere?

4 What are the other major sources of exposure to lead, apart from the atmosphere?

5 Why have smoke and soot become less serious pollutants in city air in recent times, while pollution by carbon monoxide, asbestos and lead have become more serious?

6 What steps could be taken to reduce the level of carbon monoxide in city air? What are the drawbacks (in terms of expense, infringement of liberty, etc.) of the steps you propose?

7 Answer question **6** with reference to sulphur dioxide instead of carbon monoxide.

8 What other harmful air pollutants, not mentioned in the passage, are likely to be present in the air in urban areas?

9 Is pollution an inevitable result of industrial activity? Discuss.

4 Compounds of the noble gases

This passage concerns the chemical compounds of the noble gases. Read the passage, and then answer the questions following it.

Until 1962, no one had succeeded in making genuine chemical compounds of the noble gases. Some attempts had been made to combine the gases with fluorine, but these had been unsuccessful. In any case, the octet rule, proposed by Kossel and Lewis in 1916, had led chemists to believe that the noble gases must be totally inert.

The breakthrough came in the early 1960's when Neil Bartlett was experimenting with platinum hexafluoride, PtF_6, a very powerful oxidizing agent. Bartlett found that this compound was capable of oxidizing oxygen molecules to give a stable compound:

$$O_2 + PtF_6 \longrightarrow O_2^+ PtF_6^-$$

Bartlett reasoned that since the first ionization energy of xenon (1170 kJ mole^{-1}) was very close to that of molecular oxygen (1180 kJ mole^{-1}), a similar reaction might take place between xenon and platinum hexafluoride. Furthermore, as the O_2^+ ion and the Xe^+ ion are approximately the same size, the lattice energies of the two compounds should be similar. When Bartlett mixed xenon with gaseous platinum hexafluoride there was an immediate reaction, forming xenon hexafluoroplatinate, $Xe^+PtF_6^-$, a yellow solid. This discovery led to great experimental activity, and within a year, several other compounds of the noble gases had been prepared. These included three fluorides of xenon, made by direct combination under different conditions:

$$Xe + 3F_2 \longrightarrow XeF_6$$

$$Xe + 2F_2 \longrightarrow XeF_4$$

$$Xe + F_2 \longrightarrow XeF_2$$

Xenon tetrafluoride crystals

Hydrolysis of XeF_6 gave two oxygen-containing compounds: XeO_3 and $XeOF_4$.

$$XeF_6 + 3H_2O \longrightarrow 6HF + XeO_3$$

$$XeF_6 + H_2O \longrightarrow 2HF + XeOF_4$$

Since these initial discoveries, many other compounds of xenon have been prepared, all containing xenon bonded to the three most electronegative elements, fluorine, oxygen and chlorine. Krypton difluoride, KrF_2, has also been made, though it is less stable than XeF_2, and radon difluoride has been prepared, despite the difficulties involved in working with a radioactive noble gas. No compounds of helium, neon and argon have yet been made.

Questions

1 Why did the discovery of the reaction between platinum hexafluoride and molecular oxygen open the way to the preparation of the first noble gas compound?

2 Why was it significant that 'the lattice energies of the two compounds (O_2PtF_6 and $XePtF_6$) should be similar'?

3 Draw 'dot-cross' diagrams to show the electronic structures of **a** XeF_2 **b** XeF_4 **c** XeF_6. Show outer-shell electrons only.

4 Why did Kossel and Lewis' octet rule lead chemists to believe that the noble gases were totally inert?

5 Use the electron pair repulsion theory to predict the shapes of the molecules XeF_2 and XeF_4.

6 Why do all the known xenon compounds contain the element bonded to fluorine, oxygen or chlorine?

7 Explain why KrF_2 is less stable than XeF_2, and why no compounds of helium, neon and argon have yet been prepared.

8 Why was xenon the first noble gas to be shown to form chemical compounds? Why are there far more compounds known for xenon than for any other noble gas?

9 Despite the discovery that noble gases can be made to combine with other elements, they are nevertheless a very unreactive group of elements. What uses of the noble gases can you think of that depend on their lack of chemical reactivity?

5 Hydrogen storage in metal hydrides

The following passage by J. J. Reilly and Gary D. Sandrock is taken from *Scientific American*. It concerns the use of metal hydrides as a possible storage medium for hydrogen fuel in hydrogen-powered vehicles of the future. Read the passage and then answer the questions which follow it.

What fuel will power the motor vehicles of the world when petroleum is no longer an economic source? Hydrogen is high on the list of candidates, but the problem of storing it safely and compactly has seemed to stand in the way. Now a solution to that problem is at hand in the form of metal hydrides: chemical compounds of hydrogen and metals. Already hydrogen stored in this way has served as a fuel for buses operated experimentally in the U.S. and West Germany, and for an experimental automobile. Metal hydrides have also served on an experimental basis as energy-storage compounds for levelling peak demands on an electric-power system. In addition they hold promise for such applications as refrigeration, heat pumps and heat engines.

Hydrogen is by far the most abundant element in the universe. It is the raw material from which all the other elements have been made in the interior of stars. Moreover, it is chemically unique in that it can behave either like an alkali metal or like a halogen, that is, in forming a chemical bond it can either donate an electron (as an alkali metal does) or accept one (as a halogen does). This property is useful in processes that make chemical compounds by combining hydrogen with one or more elements.

Since hydrogen on the earth is almost always combined with another element or elements in a compound such as water, it must be separated in order to serve human purposes. The annual worldwide production of pure hydrogen is about 10 trillion cubic feet. The main consumer is the chemical industry, which utilizes hydrogen as a raw material in the manufacture of a large number of products ranging from plastics to fertilizers.

Hydrogen is attractive as a fuel because it has the highest density of energy per unit of weight of any chemical fuel, is essentially nonpolluting (the main by-product of burning it is water) and can serve in a variety of energy converters ranging from internal-combustion engines to fuel cells. In the near future hydrogen could be extracted relatively cheaply from coal. In the more distant future, when fossil fuels are no longer economic, hydrogen could be separated from water by the process of electrolysis, driven by nuclear, solar or other forms of energy.

Hydrogen powered vehicles.

The present methods of storing hydrogen are suitable and safe for the present industrial uses of hydrogen, but they would never do for moving vehicles or for special applications where compactness is required. For example, hydrogen stored as a compressed gas calls for large and heavy vessels. At a typical pressure of 136 atmospheres hydrogen gas in a steel container weighs about 30 times more than an equivalent amount of gasoline, and 99 per cent of the weight is in the container. The same container takes up about 24 times more space than a container holding the equivalent amount of gasoline. Hydrogen as a liquid is useful in certain circumstances, but the energy consumed in the liquefaction process is a major fraction of what could be generated by burning the hydrogen. Moreover, liquid hydrogen would present a serious and probably insoluble safety problem if it were to be considered as a common fuel for use in motor vehicles. Liquid hydrogen is extremely cold (it boils at 20 K, −253°C), and it is highly volatile if it is spilled. Metal hydrides, in contrast, store hydrogen compactly and safely at ambient temperatures.

Most elemental metals will form metal hydrides. In many cases the reaction is simple and direct, consisting merely of bringing gaseous hydrogen (H_2) in contact with the metal (M). In chemical shorthand a typical reaction can be written $M + H_2 \rightleftharpoons MH_2$. The arrows point in two directions, which means that the reaction is reversible. Its direction is determined by the pressure of the hydrogen gas. If the pressure is above a certain level (termed the equilibrium pressure), the reaction proceeds to the right to form the metal hydride; if it is below that level, the metal hydride decomposes into the metal and gaseous hydrogen. The metal is in the form of particles in order to provide a large surface area for reaction with the gas.

The primary reason metal hydrides have been proposed for the storage of hydrogen as an energy carrier is that they accommodate an extremely high density of hydrogen. Indeed, it is possible to pack more hydrogen into a metal hydride than into the same volume of liquid hydrogen. A consideration of the mechanism by which a metal hydride is formed reveals why such a high packing density is possible.

When gaseous hydrogen is brought in contact with a metal that forms a hydride, hydrogen molecules (H_2) are adsorbed onto the surface of the metal. Some of the molecules dissociate into hydrogen atoms (H), which then enter the crystal lattice of the metal and occupy specific sites among the metal atoms. Such locations are called interstitial sites. They must have a certain minimum volume in order to easily accommodate the hydrogen atom. (See figure 1.)

As the pressure of the gas is increased a limited number of hydrogen atoms are forced into the crystal. Usually at some critical concentration and pressure the metal becomes saturated with hydrogen and goes into a new phase: the metal-hydride phase. If the hydrogen pressure is now slightly increased further, much greater amounts of hydrogen are absorbed. Ultimately all the original hydrogen-saturated metal phase will be converted into the metal-hydride phase. Since metal crystals have many interstitial sites, it is possible for them to accommodate large amounts of hydrogen in a highly compact manner. In many hydrides the number of hydrogen atoms in the crystal will be two or three times the number of metal atoms.

Although the density of hydrogen by volume in a hydride is high, the density by weight is less satisfactory (compared with pure hydrogen) because of the weight of the associated metal. It is only the high density of energy by weight of hydrogen as a fuel that makes metal hydrides feasible for the storage of energy. Table 1 compares the energy density of some hydride storage systems with the energy densities of other power sources.

Table 1 Energy densities characteristic of various automotive power sources either already in existence or proposed are indicated in this table. As the figures in the column at the right show, metal hydrides lag far behind gasoline in terms of energy density, but they are competitive with electric batteries in this respect. (No allowance has been made in these calculations for the weight of the container holding either the metal hydrides or the gasoline.) The figures for the metal hydrides are based only on the available hydrogen in each case. The particular magnesium hydride tested contained an additive of nickel amounting to about 10 percent by weight.

Power source	Energy density (watt-hours per kilogram)	Conversion efficiency (percent)	Net energy density (watt-hours per kilogram)
Lead-acid battery	30–50	70	21–35
Lithium-metal sulphide battery	150	70	105
Iron-titanium hydride ($FeTiH_{1.7}$)	510	30	153
Magnesium-nickel hydride (Mg_2NiH_4)	1110	30	333
Magnesium hydride (MgH_2)	2332	30	700
Gasoline	12880	23	2962

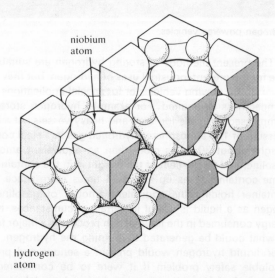

Figure 1 Interstitial sites in the body-centred cubic structure of the metal niobium. Two unit cells are shown, each with a niobium atom at the centre. Interstitial sites are shown fully occupied by hydrogen atoms.

niobium atom

hydrogen atom

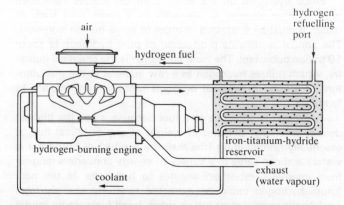

air

hydrogen fuel

hydrogen-burning engine

coolant

hydrogen refuelling port

iron-titanium-hydride reservoir

exhaust (water vapour)

Figure 2 Hydrogen-storage system based on the use of a metal hyride is shown in this schematic diagram, coupled with a standard internal-combustion engine that has been adapted to burn hydrogen gas.

Another fundamental property of metal hydrides is their heat of formation, the heat given off when the hydride is formed by the reaction of hydrogen with the metal. In order to decompose the hydride back into its original constituents, metal and hydrogen gas, the same amount of heat must be added to the system; it is termed the heat of decomposition. The heat effect can be quite large. It is roughly proportional to the stability of the hydride, that is, to the ease or difficulty of taking hydrogen out of the system. The stabler the hydride is, the higher the temperature and heat of decomposition will be. The fact that heat is evolved when hydrogen is stored in a metal hydride and is required when hydrogen is released is of great practical consequence. It is a primary consideration in the design of systems for storing hydrogen in metal hydrides. Figure 2 shows one way a hydride storage system can be coupled with an internal combustion engine.

From *Hydrogen Storage in Metal Hydrides* by J. J. Reilly and Gary D. Sandrock. © February, 1980, *Scientific American*, Inc.

Questions

1 Outline the advantages and disadvantages of hydrogen as a fuel for internal combustion engines.

2 Explain why 'hydrogen has the highest density of energy per unit of weight of any chemical fuel'. (Paragraph 4).

4 The passage gives the annual worldwide consumption of hydrogen as about 10 trillion cubic feet. Convert this figure to kg. State any assumptions you make. (1 foot = 0·304 m; 1 trillion = 10^{12})

4 Explain briefly in your own words why metal hydrides allow such high packing density of hydrogen to be achieved.

5 Use the data below to calculate the mass of hydrogen stored in 1·0 cm^3 of magnesium hydride, MgH$_2$, assuming no volume change occurs when magnesium is converted to the hydride. Compare this value with the mass of hydrogen in 1·0 cm^3 of liquid hydrogen, and discuss the significance of your answer in relation to the use of magnesium hydride as a hydrogen storage medium.
Mg = 24, H = 1
Density of magnesium = 1·74 g cm^{-3}
Density of liquid hydrogen = 0·07 g cm^{-3} (at 20 K)

6 Interstitial metal hydrides of the type described in this article have variable composition. For example, iron titanium hydride is often represented as FeTiH$_x$ where x can have a value up to about two. Explain why such hydrides have variable composition, in contrast to most other chemical compounds.

7 Consider the equation M + H$_2$ $\rightleftharpoons$ MH$_2$ referred to in the passage.
a Rewrite the equation showing state symbols and a conventional sign to indicate an equilibrium reaction.
b Write an expression for the equilibrium constant for this reaction.
c Use your answer to b to explain why 'If the pressure is above a certain level (termed the equilibrium pressure), the reaction proceeds to the right to form the metal hydride; if it is below that level, the metal hydride decomposes into the metal and gaseous hydrogen'.

8 How might metal hydrides be used as 'energy storage compounds for levelling peak demands on an electric-power system'?

9 Study *figure 2* carefully.
a Why does the engine coolant circulate through the iron-titanium hydride reservoir?
b What procedure would be used to recharge the iron-titanium hydride reservoir?

6 The framework of the planet: Silicates

The following passage, taken from *Chemistry, Matter and the Universe* by R. E. Dickerson and I. Geis, with illustrations copyright by I. Geis, concerns silicate minerals. Read the passage, then answer the questions following it.

In the centre of the third row of the periodic table sits silicon (Si 2,8,4), with four outer electrons like carbon (C 2,4). The rocks of our planet are derived from silicon dioxide, SiO_2, and are surrounded by an atmosphere composed in part of carbon dioxide gas, CO_2. This does not seem remarkable until we recognise that silicon and carbon, which have the same outer electronic structure, should have similar chemical properties. Why the gross difference in properties in their oxides?

The difference in properties arises because a silicon atom is larger than a carbon atom. Sharing two electron pairs with another atom in a double bond requires a closer approach of atoms than for a single bond. Silicon, with an inner core of ten electrons, cannot get close enough. Carbon, with a two-electron inner core, is smaller and can make C=O bonds. Two such double bonds build a CO_2 molecule, O=C=O. Rather than making double bonds to two oxygen atoms, it is easier for silicon to make single bonds to four oxygens, arranged around the Si atom at the corners of a tetrahedron (figure 1). Each of these oxygen atoms can bridge two silicon atoms, and the result is an endless three-dimensional lattice of silicate tetrahedra as in **quartz**, shown in figure 1. If silicon were smaller and could make double bonds to oxygen, there would be no reason not to expect discrete molecules of O=Si=O. Quartz would be a gas instead of a very hard mineral, and the history of our planet would be vastly different.

Pure quartz, with the overall composition of SiO_2, is an endless framework of Si and O atoms. Each Si is surrounded by four O atoms at the corners of a tetrahedron, and each O atom is shared between two adjacent tetrahedra (figure 2). One silicon atom has 'half a share' in each of four oxygen atoms around it, so the number of O atoms per Si atom is $\frac{1}{2} + \frac{1}{2} + \frac{1}{2} + \frac{1}{2} = 2$, thereby accounting for the overall composition of SiO_2. In other types of silicates, one or more of the four oxygens around a silicon atom may not be shared with other silicons, and the unshared oxygen atoms each carry one negative charge.

The smallest freestanding unit of silicon and oxygen is the silicate ion, SiO_4^{4-}, with none of the four oxygen atoms shared, and with each of them negatively charged (figure 3). These silicate tetrahedra can exist separately in three-dimensional ionic structures, or they can be linked together by sharing oxygens to form one-dimensional chains, two-dimensional double chains (ladders) and sheets, and three-dimensional frameworks such as quartz. These one-, two-, and three dimensional structures are the basis for all silicate minerals. Any negative charges arising from unshared oxygen atoms are balanced by positive metal ions inserted alongside the chains or between the layers.

Minerals containing discrete SiO_4^{4-} ions have the general formula X_2SiO_4, where X is a metal ion carrying two positive charges, needed to balance the negative charges of the SiO_4^{4-} ions. The mantle of the earth, the 3000 km deep layer lying between the earth's crust and its core is made of minerals of this type, called **olivines**, in which X is Mg^{2+} or Fe^{2+}, in any

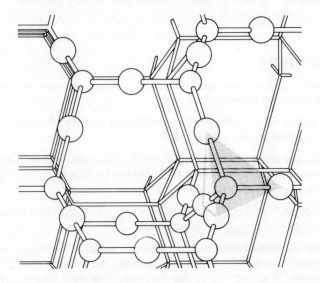

Figure 1 Silicate tetrahedra in quartz.

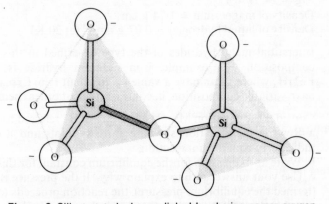

Figure 2 Silicate tetrahedra are linked by sharing corner oxygen atoms. Any unshared oxygens are left with a negative charge.

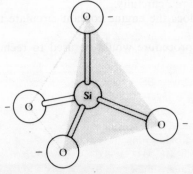

Figure 3 The free SiO^{4-} ion.

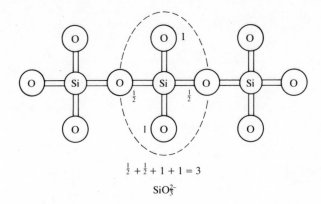

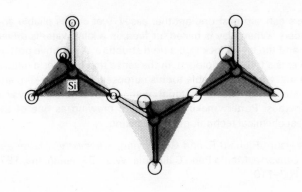

$$\tfrac{1}{2} + \tfrac{1}{2} + 1 + 1 = 3$$

$$SiO_3^{2-}$$

Figure 4 Chain silicates.

proportions. Minerals like olivine are very dense and hard.

Less dense silicate minerals are produced if the SiO_4^{4-} tetrahedra are linked into long chains. Each Si then shares two of its O atoms with two other Si atoms, leaving it with a net of $1 + 1 + \tfrac{1}{2} + \tfrac{1}{2} = 3$ oxygens, and two negative charges on its two wholly owned oxygen atoms. The overall (empirical) composition of these chain silicates is SiO_3^{2-} (figure 4). These chains have only half the negative charge per Si atom that olivine has, and hence need fewer metal ions to counterbalance the charge. Because of this, and also because of the open way in which the chains are packed in the mineral, chain silicates are less dense than olivine. Many of them floated to the top of the mantle when the earth's interior was molten and helped build the crust, the outer layer of the planet, which is about 33 km thick under the continents but only 5 km thick beneath the ocean basins. **Pyroxenes** (figure 5a) are single-chain silicates, with the chains held together by positive ions. **Amphiboles** (figure 5b) are double-chain, or ladder, structures. All these minerals cleave easily along the chain direction, but the covalent Si—O bonds within a chain are not easily broken. This is why **asbestos** (an amphibole) is fibrous and stringy.

Silicate tetrahedra also can be linked into endless sheets, with three of the four oxygen atoms shared, and only one O atom per Si left with a negative charge (figure 5c). This negative O is fully owned by one Si, while the other three are shared; thus the overall ratio of O to Si is $1 + \tfrac{1}{2} + \tfrac{1}{2} + \tfrac{1}{2} = 2\tfrac{1}{2}$ to one. The silicate sheet has the composition $SiO_{5/2}^-$ or $Si_2O_5^{2-}$. Even fewer metal ions are required to balance the negative charges than in olivines or pyroxenes, so **micas** and **clays** with sheet structures are lighter yet. They are believed to be present only in the crust of the Earth. The familiar flaking of mica arises because it is easy to separate silicate sheets, but much harder to break bonds within the sheets. (Recall the similar behaviour of graphite.)

Kaolinite is a typical clay mineral. It has one Al^{3+} associated with the negative charge on each silicate tetrahedron in the sheet, and two OH^- ions balance the other two charges on Al^{3+}. It is a layer structure, with the negatively charged oxygen atoms all pointing out to one side of the sheet of tetrahedra, aluminium atoms coordinated to these negative oxygens, and hydroxide ions on the other side of the Al^{3+}. This sandwich of silicate, Al^{3+}, and OH^- is stacked in layers to build up the three-dimensional structure. Water and other small molecules can get between the layers of kaolinite. Because the

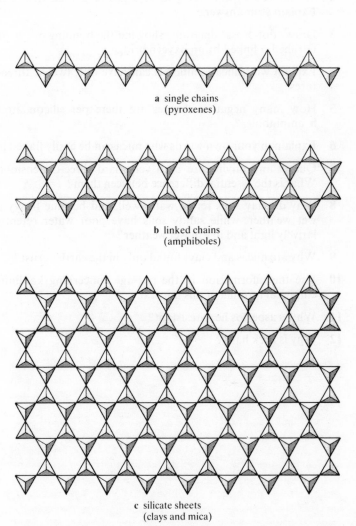

a single chains
(pyroxenes)

b linked chains
(amphiboles)

c silicate sheets
(clays and mica)

Figure 5 Chains and sheets of silicate tetrahedra

layers can slip past one another easily, wet clay is pliable and slippery. When clay is baked, or fired, in a kiln, water is driven out and the layers lock into a rigid structure. A primitive pottery maker is a true technologist, in the sense that he takes a natural material that is unsuitable for his purposes, fires it in a kiln, and transforms it into a material with quite different physical properties. Pottery making ranks with brewing as one of the oldest chemical technologies of mankind.

Dickerson, Richard E. and Geis, Irving, *Chemistry*, *Matter and the Universe*, Menlo Park, California: W. A. Benjamin, Inc. 1976 pp. 107–110.

Questions

1 Explain in your own words why CO_2 is a gas, but SiO_2 is a solid with a high melting point.

2 Would you expect GeO_2 to be a gas or a solid with a high melting point? Explain your answer.

3 Draw 'dot-cross' diagrams showing the bonding in **a** SiO_4^{4-} **b** two SiO_4 tetrahedra linked by an oxygen bridge.

4 Explain why silicate minerals can have one, two or three-dimensional structures.

5 How many negative charges are there per silicon atom in **a** pyroxenes **b** amphiboles?

6 Explain in your own words why mica can be easily flaked into sheets.

7 Quartz and olivine are both very hard three-dimensional silicate minerals. What is the essential difference between them?

8 Why do clay soils retain water well, and become heavy and waterlogged in wet weather, while sandy soils have poor water retention and remain relatively light and dry in wet weather?

9 Why are micas and clays found only in the earth's crust?

10 Use the information in the passage concerning kaolinite to work out the empirical formula of this mineral.

11 Why is asbestos heat-resistant?

12 Why is rock hard?

7 Laser photochemistry

The following passage, taken from *Esso Magazine*, is part of an article on some research activities being carried out by Esso's research scientists. The passage concerns the use of lasers to influence the products of a chemical reaction. Read it carefully and then answer the questions that follow.

I would like to discuss a fascinating new area of chemistry, namely laser photochemistry, which we hope will find wide application in a number of important areas. By way of introduction, let me review some basics.

In order to promote a chemical reaction, energy must be supplied to the reactant molecules. This energy is required to overcome an energy barrier to reaction. Conventionally, this energy is supplied in the form of heat, which is distributed randomly throughout the molecule. As the molecule heats up, it will eventually reach a point where the internal energy exceeds that of the lowest energy barrier. A chemical bond is then broken and the reaction is initiated.

What does it mean, however, when we speak of 'heating up a molecule'? Molecules consist of atoms held together by chemical bonds. These bonds behave much like springs, and the atoms oscillate in what are termed molecular vibrations. Each of these vibrations is characterized by a frequency which depends on the mass of the atoms and the strength of the bond. According to classical molecular physics, heating a molecule involves an equal sharing of the thermal energy among all the vibrational modes in the molecule.

There is a possible way, however, of getting round this so-called 'law of equipartition'. We can make use of the laser, with its ability to generate precisely controlled, single-frequency energy to excite a single vibrational mode (figure 1).

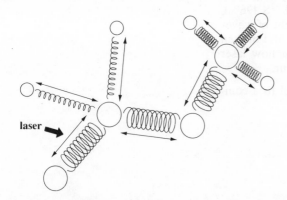

Figure 1 The vibrational modes of a molecule represented by showing the bonds as springs.

With the laser, we can perform experiments to answer a very critical question: does the energy deposited by the laser in a given vibrational mode remain localized long enough to break a specific bond? If it does, we then have a powerful means of controlling chemical reactions, which, in turn could have a profound influence on processing technology. I would like to share with you the results of two experiments performed in our laboratories which are very significant in the study of energy localization.

Consider a molecule of cyclopropane, the simplest cyclic organic structure, consisting of three carbon and six hydrogen

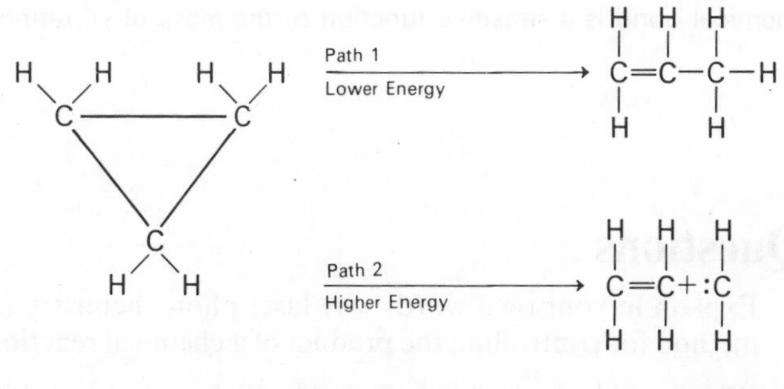

Figure 2 Cyclopropane reactions.

atoms. There are two reaction pathways available to this molecule. The first is a rather low energy pathway, in which the ring opens to give propene. The second is a higher energy pathway in which the cyclopropane breaks into ethene and a CH_2 fragment (figure 2).

We have studied these two pathways by using tuned laser energy to excite either of two vibrational modes. The first mode involves the so-called 'carbon-hydrogen stretch' in which the hydrogen atoms alternately approach the carbon, very much like a semaphore. Laser excitation of this mode resulted in the formation of propene with a selectivity greater than 99 per cent.

The second vibrational mode involves the 'wagging' motion of the plane of the CH_2 group. When this mode was excited with a laser of the proper frequency radiation, the reaction products were roughly a 50 : 50 mixture of propene and fragmentation products.

These results indicate a dramatic shift in product distribution depending on which vibrational mode is excited, and provide compelling evidence for mode-specific energy localization.

The second experiment I'd like to mention involves a much larger molecule than cyclopropane. In this case we are studying the effect on a reaction of varying the location in the molecule, rather than the vibrational mode, in which the laser energy is absorbed. Here we are dealing with a complex molecule in which various groups are loosely attached, or co-ordinated, to a core of a uranium and two oxygen atoms.

We studied the dissociation of this molecule, which involves breaking a bond between the UO_2 core and one of the co-ordinating groups. The molecule is excited with laser energy of a frequency corresponding to the vibrational frequency of either one of two bonds at different locations. The first of these is the asymmetric stretching frequency of the uranium-oxygen bond in the UO_2 core. The second is the stretching frequency of the carbon-oxygen bonds in one of the co-ordinating groups, spatially removed from the core.

The results of these experiments indicate that the degree of dissociation of the particular uranium-oxygen bond depends quite strongly on which of the two bonds is 'pumped'. There is less dissociation when the more remote carbon-oxygen bond is activated.

We believe that this experiment is extremely significant. It illustrates that laser photochemistry can achieve selective 'heating' of *part* of a molecule, something which was not previously possible. Such a process is inherently an extremely efficient way of activating a chemical reaction. We are depositing energy only in that portion of the molecule which is involved in a given reaction. In the classical method of molecular activation which I discussed earlier, energy is supplied to the entire molecule, a much more inefficient approach.

There is another important aspect of our laser chemistry research. As I mentioned earlier, the vibrational frequency of a chemical bond is a sensitive function of the mass of vibrating atoms. We can use the inherently narrow spectral distribution of a laser to dissociate preferentially molecules containing a particular isotope of a given element. In fact, we have already demonstrated such isotope selectivity for a variety of elements.

Laser photochemistry is providing us with an immensely powerful tool for probing complex molecular structures such as coal, and for achieving selective chemical reactions including isotope separation.

Dr. Peter J. Lucchesi (Vice-president of Corporate Research at Exxon Research and Engineering Company), Foundations for the Future, *Esso Magazine*, Autumn, 1979.

Questions

1   Explain in your own words why laser photochemistry may provide a powerful method for controlling the product of a chemical reaction.

2   Why can this method only be used with laser light, and not with ordinary light?

3   In what ways is the analogy between chemical bonds and springs an accurate one? When does the analogy break down?

4   What is meant by 'mode-specific energy localization' in the eighth paragraph?

5   Draw diagrams to illustrate the two different vibrational modes described for cyclopropane.

6   Which spectroscopic method, used to investigate the structures of organic molecules, employs the fact that specific bonds in a molecule vibrate at specific frequencies?

7   Why are Esso, an oil company, interested in research methods that are useful for 'probing complex molecular structures such as coal, and achieving ... isotope separation'?

8   Naturally-occurring sulphur contains two isotopes, ^{34}S (4%) and ^{32}S (96%). The ^{34}S—F bond in $^{34}SF_6$ can be excited and broken by laser light of wavelength $1 \cdot 082 \times 10^{-5}$ m, while the ^{32}S—F bond in $^{32}SF_6$ is excited and broken by light of wavelength $1 \cdot 061 \times 10^{-5}$ m. Using this data, explain how laser photochemistry could be used to separate the two isotopes of sulphur.

9   Suggest some ways in which laser photochemistry might be useful to a company like Esso.

8 Fuel and fuel cells

The following passage is from *Chemical Principles* by R. E. Dickerson, Gray and Haight. Read the passage, then answer the questions following it.

For thousands of years, man has known from experience that when you want energy, you burn something. The first fuel was wood, and then later came coal and some of the oils and gases that seep out of swampy places. The first Industrial Revolution in England was based on an energy source – coal – that had been commonplace since the beginning of recorded history. Man also could burn fats and tallow from animals to get energy, but only in the last two hundred years did he appreciate that this was what the animals were doing with them as well. The energy for the operation of almost all living creatures is obtained by combustion; that is, by the burning of some energy-rich compound in the presence of oxygen. On this particular planet, which thus far has had a sufficient oxygen supply, this is a reasonable thing to do.

There are two main problems in extracting useful energy from chemical substances: finding the substances with the most energy, and finding ways to get the maximum energy out in a usable form. Both of these problems become important when the supply of possible fuels is short, when their use harms the environment with waste products, when the supply of oxygen is not unlimited, or when we have to transport the fuels far from their natural source, as in space exploration.

The most thoroughly tested fuel systems on earth are those of living organisms. The tests have been going on for three billion years or more, and the penalty for a bad experiment is extinction. If we look at the table below, we see that living organisms have not done too badly. The most efficient combustion energy source, in terms of kilojoules of heat per gram of fuel, is hydrogen gas. Floating gasbag organisms have never evolved. Aside from the logistics problem, natural sources of hydrogen gas are not that abundant. Hydrocarbons such as gasoline yield one third as much energy per gram, and have the advantage of being in the more compact, liquid phase. Animals store most of their energy in fats, which are esters of long, gasoline-like fatty acids such as stearic acid. As the table below shows, these fats are very nearly as good energy-storage compounds as gasoline on a weight basis. The esterification helps to make them solids instead of liquids, an additional practical advantage for a creature that must move about.

Proteins (represented by the data for alanine, an amino acid) and carbohydrates (represented by glucose) are only half as efficient in storing energy. But for plants, which do not move about, the energy yield per gram of fuel is really not very important. A redwood tree does not need to be weight-conscious. It happens that the biochemistry required to store energy in fatty acid molecules and fats, and to retrieve it again when needed, is complicated. In contrast, carbohydrate synthesis and breakdown are simpler and faster. Plants give up the unimportant energy-per-gram advantage of fats, for the significant advantage of easier carbohydrate biochemistry. In green plants, energy is stored as starches rather than fats. Even animals use the fast-access storage advantage of carbohydrates. We store a limited amount of energy as glycogen, a starch-like molecule, as a buffer against sudden energy needs.

When we leave our natal planet and begin exploring space, the energy-per-gram problem of all animals becomes even more serious. Gasoline then becomes too energy-poor, and we turn to direct combustion of hydrogen for the first stage of the Saturn rockets. We also look for newer and better ways of generating electrical energy to operate the spacecraft systems. One approach is not to attempt to carry energy from our home planet on the mini-planet of a spacecraft, but to generate electricity directly from the energy of the sun by means of solar batteries. As yet, we do not have the technology to make this a source of large power supplies. But if we have to use terrestrial fuel, then it should at least be the best fuel on a weight basis. The answer is the generation of electrical power from a fuel cell that uses the combustion of hydrogen as its energy source:

$$H_2(g) + \tfrac{1}{2}O_2(g) \longrightarrow H_2O(l) \qquad \Delta H = -287 \text{ kJ}$$

About the worst thing that we could do with hydrogen in our spacecraft would be to burn it like we do most fuels on earth, and use the heat to generate electricity. We would be lucky to convert as much as 25% of the energy available in the

Fuel	Physical state	Relative molecular mass	Heat of combustion/ kJ mole^{-1}	Heat of combustion per gram/kJ g^{-1}
H_2	g	2	−287	−144
C_8H_{18} (octane, a typical component of gasoline)	l	114	−5473	−48
$C_{17}H_{35}COOH$ (stearic acid, a typical fatty acid)	s	285	−11390	−40
$H_2NCH(CH_3)COOH$ (alanine, a typical amino acid)	s	89	−1630	−18
$C_6H_{12}O_6$ (glucose, a typical carbohydrate)	s	180	−2827	−16

Table 1 Energy obtained per gram from some fuels during combustion with oxygen

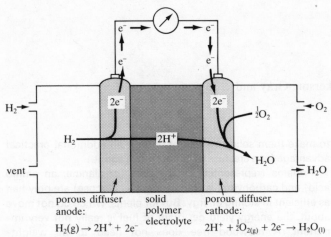

anode:
$$H_2(g) \rightarrow 2H^+ + 2e^-$$

cathode:
$$2H^+ + \tfrac{1}{2}O_{2(g)} + 2e^- \rightarrow H_2O_{(l)}$$

porous diffuser solid
                 polymer
                 electrolyte

porous diffuser

Figure 1 A hydrogen-oxygen fuel cell. Hydrogen gas introduced at the left dissociates in the porous, conducting barrier at the left side. Hydrogen ions migrate to the right through the electrolyte (for spacecraft applications, a solid polymer), and the electrons flow through the external circuit. At the right diffuser, these electrons, hydrogen ions from the electrolyte, and molecular oxygen combine to produce water. In a gravity field, the product water will run off into a storage tank; in free space, it can be drawn away from the diffuser cathode by a wick. In either event, the resulting water is available for human use. The standard e.m.f. of the hydrogen fuel cell is calculated to be 1.23 volts.

reaction into electricity. And nobody wants a furnace, boiler, and power plant operating within the close confines of a space capsule.

There is a better way. Why convert chemical energy into heat energy and then into electrical energy, if the heat step can be omitted? A **fuel cell**, which makes a direct chemical–electrical conversion, is shown in the figure above. The overall reaction is combustion of hydrogen. As with any other electrical cell, the half-reactions have been separated at two electrodes, and the electrons made to flow through an external circuit from anode to cathode. This cell is no different in principle from more familiar electric cells. The innovation is that new reactants (H_2 and O_2) are fed constantly into the cell, and

the product (H_2O) is drawn off so that the voltage of the cell remains constant and the power output is uninterrupted. The great advantage of all fuel cells is that they convert chemical energy directly into electrical energy without the intermediate conversion through heat energy. The thermodynamic limitations of a heat engine are thereby avoided.

Methods that are tolerable on board a spacecraft, operating on a NASA budget, do not necessarily work economically on earth. Why do we persist with antiquated coal-burning power stations, which pollute our skies with smoke and our rivers with warm water? Why not react coal, or at least readily available methane gas, directly in a fuel cell? The problems are all practical, and not theoretical. There is no known reversible electrode suitable for the methane half-reaction,

$$CH_4(g) + 2H_2O(l) \longrightarrow CO_2(g) + 8H^+ + 8e^-$$

If there were, we immediately could set up a methane fuel cell similar to the one shown, and a revolution in energy production would be at hand. Many hundreds of man-years have been spent in seeking suitable reversible electrodes for hydrocarbon fuels or coal, but so far with only limited success.

So, in spite of all our knowledge about electrochemical cells, the one key answer that could transform our use of fuels eludes us. The old fuels are good, but what we do with them in the process

Bond energy $\longrightarrow$ Heat $\longrightarrow$ Electrical energy $\longrightarrow$ Other uses

is sloppy, inefficient and polluting. We cannot even escape this thermodynamic trap with our glorious new atomic energy. For after developing this new source of energy from nuclear reactions, what do we do with it as the first step in generating electricity? We use it to boil water! James Watt and Ernest Rutherford, wherever they are, must be laughing.

R. E. Dickerson, H. B. Gray and G. P. Haight Jr.,
Chemical Principles,
Menlo Park, California, The Benjamin/Cummings Publishing
Company, Inc. 1974. pp. 756–760.

Questions

1 Explain in your own words why green plants store their energy in carbohydrates rather than fats, even though on a weight basis the latter are a more compact energy store.

2 Explain why hydrocarbons like octane release more energy per gram on combustion than carbohydrates like glucose.

3 Is it completely accurate to say that 'the energy for the operation of almost all living creatures is obtained by combustion'?

4 What are the principle differences between a fuel cell like the one described here and a conventional electrochemical cell like the Daniell cell?

5 What material might be used to construct the porous diffuser anode and cathode in the fuel cell? Give a reason for your choice.

6 Write half-equations for the reactions that would occur at the anode and cathode in a fuel cell powered by methane. Use these half-equations to produce an overall equation showing the net reaction occurring in the cell.

7 Outline the advantages that would be gained by generating electricity in fuel cells rather than in the thermal power stations used at present.

8 Do you consider that fuel cells are likely to replace thermal power stations to a significant extent in the next 20 years or so? Give a reason for your answer.

9 Even if fuel cells can be developed to the point where they can meet industrial and domestic demands for electricity, they will always depend on the availability of fossil fuel resources, which we know are limited. In view of this, is it worth spending money on research and development of fuel cells?

9 Acids and bases in the home

The reactions of acids and bases are of great interest to the chemist, not least because of their many applications in industry and the home. There are many important domestic uses of acids and bases, and the following passage summarizes a few of them. Read the passage and then answer the questions following it.

Acids used in the home

Cleaners

Acids react with and dissolve bases, and this makes them useful for a number of applications. Metals tarnish because they react with the air to form a layer of oxide or sulphide. Acids can be used to dissolve this layer. One of the most difficult oxide layers to remove is rust, Fe_2O_3; rust removers usually contain a strong mineral acid such as hydrochloric or phosphoric(V) acid. The latter has the advantage that it also forms a protective phosphate layer to help prevent further corrosion.

Such strong mineral acids can be hazardous, though, and for most domestic purposes weaker acids are used. Rust stains on cloth can be removed using ethanedioic (oxalic) acid (table 1), but this has the disadvantage of being toxic. The compound is probably effective because of the complexing of iron ions by ethanedioate ions.

Other metals can be cleaned using ethanoic or citric acid (table 1). Vinegar is a five per cent aqueous solution of

Systematic name	Common trivial name	Formula	Natural occurrence	K_a/mole dm^{-3}
hydrochloric acid	—	HCl	in the stomach	large
phosphoric(V) acid	—	H_3PO_4	—	large
methanoic acid	formic acid	H—C—OH $\parallel$ O	ants, nettles	1.6×10^{-4}
ethanoic acid	acetic acid	CH_3—C—OH $\parallel$ O	vinegar	1.8×10^{-5}
ethanedioic acid	oxalic acid	COOH $\mid$ COOH	some plants e.g. rhubarb	(1) 3.5×10^{-2} (2) 4.0×10^{-5}
2-hydroxypropane-1,2,3-tricarboxylic acid	citric acid	CH_2—COOH $\mid$ HO—CH—COOH $\mid$ CH_2—COOH	citrus fruit	(1) 8.4×10^{-4} (2) 1.8×10^{-5} (3) 4.0×10^{-6}
2,3-dihydroxy-butanedioic acid	tartaric acid	HO—CH_2—COOH $\mid$ HO—CH_2—COOH	grapes	(1) 1.0×10^{-3} (2) 4.6×10^{-5}

Table 1 Some acids commonly encountered in the home

Some acids and bases commonly found in the home.

ethanoic acid, and citric acid is present in varying amounts in lemons and other citrus fruit. An effective way of cleaning brass or copper is to rub it with a piece of lemon. Vinegar should be avoided with copper and its alloys because of the formation of poisonous copper ethanoate. In any case, metals should always be washed well with water or washing soda (see below) after cleaning with acid, in order to remove possible toxic products and prevent the retarnishing which occurs if the metal is left in contact with acid.

Hard water often forms undesirable deposits in kettles, baths and lavatory pans. These deposits are usually carbonates of calcium and magnesium (white) or iron (brown) which can be dissolved with acid. A weak organic acid is usually sufficient – vinegar or a slice of lemon will 'de-fur' a kettle. Commercial preparations containing methanoic acid (table 1) can be bought which do the same job rather more quickly.

Food additives

Acids have been used as food preservatives for centuries. Food-spoilage bacteria are inhibited or killed by low pH, so pickling in vinegar is an effective way of increasing the life of, say beetroot or herrings. Nowadays, we have other methods of preserving food which affect its flavour and texture less than pickling. Nevertheless, we continue to pickle food and to add acids to it because we like the sharp flavour which acids produce. Vinegar and lemon juice are frequently used to flavour food.

It is sometimes necessary to generate carbon dioxide in food so that the mixture will 'rise'. For example, baking powder is added to cakes to generate carbon dioxide bubbles and make them lighter. Baking powder is usually a mixture of sodium hydrogencarbonate and 2,3-dihydroxybutanedioic acid (tartaric acid – see table 1). Some sweets contain 'sherbet' – a mixture of sodium hydrogencarbonate and citric acid which generates carbon dioxide and a pleasant fizzing sensation when moistened in the mouth.

Bases used in the home

Cleaners

Many household cleaning problems involve grease-bound dirt. Fats, grease and oil bind dirt to a surface, making the dirt difficult to remove with water alone, because water and grease do not mix. The usual solution to this problem is to use a detergent, which helps grease and water to mix. For some very greasy surfaces, such as ovens and kitchen surfaces, detergent alone is not very effective. In such cases, an alkali cleaner can be used which reacts with the grease chemically.

The greases, fats and oils encountered in the kitchen are all of natural origin, and as such contain the triesters of long-chain carboxylic acids. Like all esters, they are hydrolysed (saponified) by alkalis.

$$
\begin{array}{ccc}
CH_2OOC-R_1 & CH_2OH & R_1COO^- \\
| & | & \\
CHOOC-R_2 + 3\,OH^- \longrightarrow & CHOH \quad + & R_2COO^- \\
| & | & \\
CH_2OOC-R_3 & CH_2OH & R_3COO^- \\
\text{fat} & \text{propane-1,2,3-} & \text{carboxylate} \\
(R_1, R_2 \text{ and } R_3 & \text{triol} & \text{anions} \\
\text{represent long} & & \\
\text{hydrocarbon chains)} & &
\end{array}
$$

Long-chain carboxylate anions of the type formed here have detergent properties because the product of saponification of a fat is **soap**. Thus, alkalis not only break down fats, but they turn them into soap. They are therefore very effective in removing fatty, greasy deposits. Oven cleaners, for example, usually contain sodium hydroxide (caustic soda). They are applied to the warm oven surface and left for half an hour or so. During this time the grease is partly saponified so that it can later be easily removed using warm water.

Some drain-cleaning preparations contain solid sodium hydroxide together with a little aluminium powder. When this mixture is poured down the drain it comes into contact with water and the following reaction occurs:

$$Al + OH^- + 3H_2O \longrightarrow Al(OH)_4^- + \tfrac{3}{2}H_2$$

This reaction is very exothermic and at the high temperature reached, the grease blocking the drain melts and the cleaning action of the sodium hydroxide is much more effective.

Ammonia and sodium carbonate (washing soda) are two more examples of alkaline cleaning agents. An aqueous solution of ammonia contains hydroxide ions because of the following equilibrium:

$$NH_3 + H_2O \rightleftharpoons NH_4^+ + OH^-$$

Ammonia is present in many liquid cleaners. It has the advantage of being a weaker alkali than sodium hydroxide, so it is less caustic. Household ammonia was first manufactured from coal, and this gave an impure cloudy solution. With the discovery of the Haber process, ammonia could be made very pure and clean, but manufacturers continued to make household ammonia cloudy by adding soap because consumers had become used to a cloudy product.

Sodium carbonate is another weak alkali:

$$CO_3^{2-} + H_2O \rightleftharpoons HCO_3^- + OH^-$$

It is often present in scouring powder.

Medicinal antacids

Acid indigestion is caused by an excess of hydrochloric acid in the stomach. Indigestion tablets contain weak bases, usually oxides, hydroxides or carbonates, which neutralize the excess acid. Table 2 gives a few examples. Carbonates have the advantage of being very weak bases which act quickly, but they have the disadvantage that carbon dioxide is produced when the carbonate is neutralized by an acid, and this can have embarrassing consequences. Some indigestion tablets are effervescent. They usually contain citric acid and sodium hydrogencarbonate (sodium bicarbonate), with the latter in excess. When water is added effervescence occurs, but the solution formed has an overall alkaline pH.

Carbonates	Hydroxides	Oxides
$NaHCO_3$	$Al(OH)_3$	Al_2O_3
$MgCO_3$	$Mg(OH)_2$	MgO
$CaCO_3$		

Table 2 Some bases present in indigestion tablets

Questions

1   Why do metals re-tarnish if left in contact with acid after cleaning?

2   A badly stained aluminium saucepan can often be cleaned by boiling in it some water containing apple peelings. Why is this?

3   Look at table 1. Why are *three* K_a values given for 2-hydroxypropane-1,2,3-tricarboxylic acid (citric acid)?

4   Write an equation for the reaction that occurs when vinegar is used to clean the surface oxide off copper. Why should vinegar not be used to clean copper cooking utensils?

5   Assuming that vinegar contains 5 g of ethanoic acid per 100 cm^3, use the information in table 1 to calculate   **a** the molarity of ethanoic acid in vinegar and   **b** the pH of vinegar.

6   Why does sherbet not fizz until water is added?

7   Suppose one of your family has spilled oven cleaner on their skin. What common household substance would be best to apply to the spill?

8   A method once used to clean a greasy oven involved warming the oven, putting a dish of ammonia solution inside, closing the door and leaving overnight. The next day the oven could be easily cleaned with warm water. How did this method work?

9   Could greasy, oily deposits on a car engine be removed using sodium hydroxide? Explain.

10   Why should a greasy aluminium frying pan never be cleaned with sodium carbonate (washing soda)?

11   Why is sodium hydrogencarbonate used to treat indigestion, but not sodium carbonate?

10 Anions of the alkali metals

You probably associate the Alkali Metals, Group I of the Periodic Table, with high reactivity and the formation of ions with a single positive charge. Recently, however, compounds of these metals have been made which contain **negatively** charged alkali metal ions, as this extract from *Scientific American* explains. Read the passage and then answer the questions that follow it.

Na⁺Na⁻ crystals.

The alkali metals are a group of elements whose most notable and most familiar chemical property is their eagerness to give up an electron. The alkali metal sodium, for example, readily donates an electron to chlorine, forming sodium chloride. Metallic sodium is so strongly disposed to get rid of an electron that it will even split a water molecule, displacing a hydrogen atom and forming sodium hydroxide; the reaction can be a violent one. Because the metallic form of sodium is so reactive it does not exist in nature; the element is found only as a positive ion, or cation, denoted Na⁺. The cation forms when the atom surrenders one of its electrons to some other chemical species.

These properties of the alkali metals have been known since the beginning of chemistry, and it has therefore come as a surprise to learn that alkali metals can also accept an electron, acting in a way which is precisely the reverse of their usual one. The addition of an electron to the neutral sodium atom, for example, forms the negatively charged anion, Na⁻. This anion has been known for some time as a stable species in gaseous sodium; what is more important, it has recently been discovered in solutions and even as a component of a crystalline salt. It appears that the anions of all the other alkali metals can also be prepared; those of potassium (K⁻), rubidium (Rb⁻) and cesium (Cs⁻) have already been observed. It may even be possible to prepare salts containing the simplest possible anion: the immobilized electron.

The key to the preparation of the alkali metal anions, curiously, is the trapping of the alkali **cations** in an organic molecule with a cagelike structure. Ordinarily, any negative metal ions present in a solution would quickly react with positive ions to yield neutral atoms of the metal. When the cations are sequestered in an organic cage molecule, the resulting complex is so stable that the 'backsliding' reaction is prevented. The negative ions and the crystals containing them are nonetheless highly reactive. They cannot be exposed to air or moisture, and they are stable for long periods only when stored at low temperature.

The alkali metals comprise six elements: lithium, sodium, potassium, rubidium, cesium and francium. (The last is a rare, radioactive species.) In the usual arrangement of the periodic table of the elements they are listed in the first column. This grouping reflects similarities in the chemical properties of all the alkali metals and in their underlying electronic structures: each has a solitary electron in its outermost, or valence, shell of electrons.

The alkali metals can be regarded as noble gases with one extra electron (and, of course, one extra proton in the nucleus). Lithium, for example, has the filled $1s$ orbital of a helium atom, and one additional electron half filling the $2s$ orbital. Sodium has filled $1s$, $2s$ and $2p$ orbitals, as in neon, and, in addition, one $3s$ electron. The remaining alkali metals are similar in structure; in each of them the outermost electron is alone in an s-type orbital.

The strong tendency of the alkali metals to lose an electron and form a cation can now be understood. With the removal of the valence electron, each of the alkali metals takes on the exceptionally stable structure of a noble gas. Even though the resulting ion has an unbalanced positive charge, in the presence of an electron acceptor it is stabler than the neutral atom.

Adding one electron to an alkali metal results in a configuration that is much less strongly favoured than the noble-gas structure of the cation. Nevertheless, the extra electron does fill an s-type orbital, and a filled orbital is somewhat stabler than a half-filled one. Under certain carefully contrived circumstances that slight gain in stability is enough to favour the existence of alkali metal anions.

From *Anions of the Alkali Metals* by James L. Dye.

Questions

1. Why are cations of alkali metals, for example Na^+, more stable than their anions, for example Na^-?

2. The passage explains that the anion Na^- has been known for some time as a stable species in gaseous sodium. Why is Na^- more likely to exist in the gas phase than in solution or as a component of a crystalline salt?

3. What normally happens when Na^+ and Na^- ions are in solution together? Write an equation for the reaction which occurs.

4. What conditions have been employed to prevent this happening, thus allowing the existence of Na^-?

5. How would you attempt to prove experimentally that a solution actually contained alkali metal anions?

6. Which of the alkali metals, Li to Cs, would form anions most readily? Explain your answer.

7. Would the Group II elements be as likely to form anions as the Group I elements? Explain your answer.

8. The elements of Group VII, the halogens, are normally thought of as only forming singly-charged anions, for example Cl^-. Is there any chance that compounds could be prepared containing halogen **cations**? Which of the halogens would be most likely to form cations? Explain your answer.

11 Knocking and the search for anti-knock agents

The following passage by C. B. Hunt is taken from the *School Science Review*. It concerns the phenomenon known as 'knocking' or 'pinking', which occurs under certain conditions in internal combustion engines. Read the passage, then answer the questions following it.

Knocking

When the piston reaches the top of its cycle in the cylinder, at the end of the compression stroke, a spark is passed from the spark plug; an explosion occurs in the combustion chamber and the flame front is steadily propagated through the cylinder, accompanying the expansion of the hot gas. As the flame front advances, the unburned mixture or 'end-gas' is compressed and the resulting rise in temperature ordinarily causes it to burn smoothly. However, it may sometimes burn explosively and the sudden, violent ignition gives rise to a metallic rattling or 'knocking' in the cylinder. This causes power loss in the engine and in extreme circumstances can cause cracking of the cylinder head. This phenomenon is known as **knocking** or **knock**. The main factors affecting knock are the composition of the fuel and the compression ratio of the engine, high compression ratios leading to more knock.

Knock in the internal combustion engine was investigated in America by Thomas Midgley Jnr., who was financed by General Motors Research Corporation. It was Midgley who was the first to produce a commercial anti-knock gasoline additive.

Midgley's search

Midgley first encountered the problem of knock in 1916 while he was investigating the substitution of kerosine for gasoline in small, portable lighting units which consisted of an internal combustion engine attached to a d.c. generator. Kerosine worsened the problem of knock very considerably. In the earliest stages of his work Midgley thought that the knock arose from the slow vaporization of kerosine and a completely false theory (that the absorption of radiant energy by a dye would promote vaporization and hence eliminate knock) led him to the discovery that iodine, when dissolved in the gasoline or kerosine, inhibited knock. The discovery was tantalizing since his original theory proved incorrect (dyes had no effect) and the mode of action of iodine remained unknown.

In the following five years or so, Midgley developed a theory to explain knock, and searched through thousands of chemicals for other anti-knock agents. His search was unsuccessful until he changed his tactics from trial and error to a systematic search. Midgley explained:

'... in the search for a material with which to control

knocking in an internal combustion engine, the following determinations were arrived at.

1 Iodine, dissolved in motor fuel in very small quantities, greatly enhanced the anti-knock characteristics of the fuel.

2 Oil-soluble iodine compounds had a similar, though modified, effect.

3 Aniline, its homologues, and some other nitrogenous compounds were effective, though their effectiveness varied over a wide range depending on the hydrocarbon radicals attached to the nitrogen atom.

4 Bromine, carbon tetrachloride, nitric acid, hydrochloric acid, nitrites, and nitro-compounds in general increased knocking when added to the fuel and air mixture.

5 Selenium oxychloride was extremely effective as an anti-knock material.

6 A large number of compounds of other elements had shown no effect.

With these facts before us, we adopted a correlational procedure based on the periodic table. What had seemed at times a hopeless quest, covering many years and costing a considerable amount of money, rapidly turned into a "fox hunt". Predictions began fulfilling themselves instead of fizzling out . . .'

Table 1 Properties of some anti-knock agents

Compound	% by vol. (see 1)	Approx. no. of molecules (see 2)
C_2H_5I	1·6	2 150
Xylidene	2·0	2 600
$(C_2H_5)_4Sn$	1·2	7 100
$(C_2H_5)_4Se$	0·4	11 750
$(C_2H_5)_4Te$	0·1	50 000
$(C_2H_5)_4Pb$	0·04	215 000

1 Percentage by volume of agent that must be mixed with kerosine to produce a given anti-knock effect.

2 Number of molecules of the fuel-air mixture on which one molecule of the anti-knock agent can act.

4	5	6	7
14 Si	15 P	16 S	17 Cl
32 Ge	33 As	34 Se	35 Br
50 Sn	51 Sb	52 Te	53 I
82 Pb	83 Bi	84 Po	85 At

Figure 1 Groups 4, 5, 6 and 7 of the Periodic Table.

The correlation of behaviour with chemical structure and the periodic table was the key to success. It was a simple, logical step to proceed through groups VII and VI to group IV and organic derivatives of tin and lead (figure 1), since anti-knock properties seemed to improve as a group was descended (iodine better than bromine, tellurium better than selenium) and as the table was traversed from right to left. Tetraethyl tin, although effective, caused pre-ignition but when, on 9 December 1921, in the General Motors research laboratories, Midgley started an engine fuelled with gasoline containing tetraethyl lead (TEL) he was rewarded by the sound of smoothly-functioning combustion with no audible trace of knock. The anti-knock properties of TEL were so superior to those of any previously investigated substance (see table 1) and so evident at very low concentrations that it was clear that the search for an anti-knock gasoline additive was over. Success had not come solely from the rational application of chemical theory; nor had it been stumbled on purely by chance. It came from a happy combination of both.

TEL in action

In conjunction with General Motors, Midgley filed a patent application which was granted in the United Kingdom in 1923. It is worth quoting a small section because Midgley emphasized the economic importance of his discovery, an importance which was to be dramatically re-emphasized fifty years later: 'The present tendency is to produce lower grades of gasoline in order to obtain a sufficient output for the increasing demand for motor fuels and to reduce the compressions of the engines

"She'll never say 'No' . . . . . on 'BP' Ethyl!"

FASTEST for cars

so that these lower grades of fuel may be used without knocking. As the lowering of engine compression reduces the efficiency of the engine, a still greater output of fuel is required to meet the increase in fuel required to operate larger and less efficient engines. The principal objects of the present invention are to overcome these difficulties and to provide a means for using either low or high grades of motor fuel more efficiently and so reduce the quantity of fuel used. . . . In particular we have found tetraethyl lead a most effective and convenient substance for our purpose . . .

For some commercial purposes it will probably be sufficient to use but one part of tetraethyl lead in 2000 parts of gasoline.'

The elaboration of a laboratory bench discovery into commercial or industrial production is seldom just a scaling-up job. TEL might be outstandingly effective but how was it to be manufactured? The newly-formed Ethyl Gasoline Corporation eventually adopted a route to TEL involving the reaction of chloroethane with a lead/sodium alloy:

$$4CH_3CH_2Cl(g) + 4Pb/Na(s) \longrightarrow$$

$$(CH_3CH_2)_4Pb(l) + 3Pb(s) + 4NaCl(s)$$

Since the combustion of gasoline containing TEL, deposited metallic lead and lead(II) oxide in the engine cylinders, a lead 'scavenger', such as 1,2-dibromoethane, was incorporated into the gasoline. This ensured the removal of lead and lead(II) oxide as volatile lead bromide or oxybromide in the exhaust gases. The blend of TEL and scavenger became known as 'ethyl fluid'. This innovation involved a considerable expansion in the industrial output of bromine and, in collaboration with the Dow Chemical Company, Ethyl Gasoline Corporation established a new chemical plant, extracting bromine from sea water in which it is to be found at concentrations of about $0·07$ g l^{-1}.

Gasoline containing TEL was sold for the first time on 2 February 1923, in Dayton, Ohio, and was introduced into the United Kingdom by the Anglo-American Oil Company in 1928.

C. B. Hunt, Knocking and the Search
for Anti-knock Agents, *School
Science Review*, July, 1979.

Questions

1 Explain in your own words the meaning of 'knock' in internal combustion engines, and why it occurs.

2 How did the periodic table help Midgley in his search for an effective anti-knock agent?

3 TEL is manufactured by reacting chloroethane with a sodium/lead alloy. How is chloroethane manufactured?

4 1,2-Dibromoethane is added to gasoline in conjunction with TEL to act as a 'scavenger' for lead. How is 1,2-dibromoethane manufactured?

5 The manufacture of 1,2-dibromoethane requires large amounts of bromine. How is bromine extracted from sea water?

6 Midgley was granted a UK patent for his discovery in 1923. Why was the economic importance of the discovery 'dramatically re-emphasized fifty years later'?

7 Midgley's initial discovery of iodine's anti-knock properties was made almost accidentally, as the result of the application of a false theory. Discoveries are often made in science by accident, when the discoverer is looking for something else. Suggest another example of such a discovery.

8 The use of TEL as an anti-knock agent is responsible for the discharge into the atmosphere of large quantities of lead, which is known to have toxic effects. In view of this do you think the use of TEL is justified? Discuss.

9 Do you expect TEL will continue to be used in the future? Explain your answer.

12 Accident on the Louisville and Nashville Railroad

This article from the *Daily Telegraph* refers to the derailment in Crestview, Florida, of a goods train carrying a variety of chemicals. The accident, which happened on 8 April 1979, led to the leakage of a number of dangerous chemicals. Read the article and then answer the questions which follow it.

7,000 ready to flee as gas cloud spreads
By IAN BALL in New York

Florida authorities yesterday extended the area of evacuation around the scene of a goods train derailment near Crestview after toxic smoke and fumes began drifting over populous zones.

At least 5 000 people had been moved from their homes to emergency shelters by the morning, but further evacuation was ordered after winds shifted and brought chlorine fumes closer to the town of Crestview. The 7 000 inhabitants of Crestview remained on the alert, ready to leave at any time.

The poisoned-air crisis arose after 28 wagons of a 118-wagon Louisville and Nashville Railroad goods train toppled off the rails. In its long string of tanker wagons, the train was hauling a cargo of liquid chlorine, carbolic acid (phenol), methanol, acetone (propanone), anhydrous ammonia and sulphur.

The derailment occurred as the goods train cleared a trestle bridge spanning the Yellow River in a wooded, swampy area of northern Florida, accessible only from the air and river. One tanker wagon, loaded with deadly carbolic acid was left dangling from the bridge. Several of the wagons burst into flames after the derailment and touched off a fire on the wooden structure of the bridge. Other tankers exploded and soon a deadly mixture of chlorine, acetone, ammonia and other liquid chemicals was billowing in acrid columns from the wrecked train.

The first step was to remove everyone from an 80 square mile area around the derailment. Even with gas masks, firemen were unable to get close to the scene. A decision was taken to let the tanker wagons drain themselves rather than risk trying to plug the leaks.

Astonishingly, no serious injuries were reported. A fisherman who inhaled some of the fumes was taken to hospital for observation.

In February, the Federal Railroad Administration described the Louisville and Nashville Railroad as having 'the worst record of any railroad of the country' in handling hazardous cargoes. Sixteen people died last year when one of its goods trains was derailed at Waverly, Tennessee.

Ian Ball, *Daily Telegraph*, April 10th, 1979.

Questions

1 Draw the structural formula of each of the chemicals in the train's cargo. Use a data book to find the melting and boiling point of each.

2 All the chemicals were transported as liquids. What conditions would be needed in each case to keep the chemical in the liquid phase?

3 The chemicals in the train's cargo are all important industrially. List the chemicals and give one important industrial use of each.

4 'Several of the wagons burst into flames after the derailment . . .'. Which chemicals were probably responsible for this? Explain your answer.

5 The worst danger was from toxic fumes drifting over inhabited regions. Which chemical or chemicals were most hazardous in this respect?

6 What steps might the Louisville and Nashville Railroad have taken to make the carriage of these chemicals safer?

7 Accidents like this, involving the transport of chemicals, are not uncommon in the USA and indeed elsewhere. Do the benefits brought by the use of these chemicals outweigh the risks involved in their transport and handling? Justify your case.

13 Castner's mercury cell

The following passage by Dr D. J. Adam is taken from *Education in Chemistry*. It concerns the development of an early mercury cell, whose modern equivalent is used in the manufacture of sodium hydroxide and chlorine by the electrolysis of brine. The cell was pioneered by an American called Hamilton Young Castner. Read the passage and then answer the questions which follow it.

Hamilton Young Castner

Castner, born in Brooklyn, New York in 1858, was a remarkably inventive chemist. At the age of 28, having already sold out a successful business, he came to Britain with a process he had perfected for making metallic sodium from caustic soda by fusing it with carbon. Lack of home interest caused him to set up a small factory in London to continue his work. The interest of the recently formed Webster Crown Metal Co. Ltd, of Solihull, Birmingham, was aroused since they required metallic sodium for the preparation of metallic aluminium by the reduction of aluminium chloride (Deville process). After investigation and validation of the excellence of Castner's method, it was decided to set up a new company, The Aluminium Company Ltd. based at Oldbury, Birmingham, to exploit the new process, with Castner as managing director. The Aluminium Co. bought Castner's patents for £140 000 and opened a new factory in 1888 with a potential annual output of 50 tonnes. However, just as the business was beginning to develop, the Hall-Héroult electrolytic process for aluminium, patented in 1886, was coming into commercial being, causing the price of aluminium to fall rapidly yet again. So Castner, who was to achieve his greatest triumph in industrial electrolysis, was initially one of its victims.

Castner quickly realised that the company's only real asset was his process for making sodium metal cheaply and so, with characteristic determination, he endeavoured to look for an outlet for this product. He was very successful in developing processes for sodium peroxide (made by controlled oxidation of sodium in aluminium trays) and sodium cyanide (made in two stages by reaction between sodium, charcoal and ammonia).

$$2Na + O_2 \longrightarrow Na_2O_2$$

$$2Na + 2NH_3 \longrightarrow 2NaNH_2 + 2H_2$$

$$2NaNH_2 + 2C \longrightarrow 2NaCN + 2H_2$$

The latter was required by the gold mining industry which was accustomed to working with the potassium salt. To overcome the initial difficulty of getting the sodium salt accepted, it was marketed as 130 per cent potassium cyanide. In fact, so successful was the cyanide process that the Aluminium Co. (which now made no aluminium) was restored to financial health. Furthermore, the demand for sodium was such that Castner set out to improve on his original process for the manufacture of the metal. For this he returned to Sir Humphrey Davy's experiment on the electrolysis of molten sodium hydroxide.

He used an iron pot containing an iron cathode separated from the cylindrical nickel anode by a gauze diaphragm. In theory this process was continuous but impurities in even the best quality caustic soda available to him gave cathodic deposits which caused difficulties. So with typical perseverance he embarked on the task of making the purer caustic soda needed for his electrolytic sodium process. It was well known that on the electrolysis of brine, sodium hydroxide and hydrogen were produced at the cathode and chlorine was evolved at the anode. The difficulty was in separating the caustic soda from the salt. Castner's ultimate solution, the rocking mercury cell (British Patent 16046/1892, 10584/1893) depended on a property again known to Davy, that of the solubility of sodium in mercury to give sodium amalgam. This on reaction with water gave caustic soda solution.

$$2Na/Hg + 2H_2O \longrightarrow 2NaOH + H_2 + Hg \text{ (re-used)}$$

The original cell consisted of a rectangular slate box divided into three compartments by slate partitions almost reaching to the bottom. A sufficient depth of mercury, acting as the cathode, sealed off the compartments. The two end sections, fitted with carbon anodes, contained brine and the centre compartment contained water. Electrolysis gave sodium amalgam in the outer sections. The rocking of the cell caused the mercury to circulate and the amalgam, coming into contact with the water in the centre compartment, formed caustic soda solution and hydrogen, regenerating the mercury. Drawing off the caustic soda and topping up with brine and water made the process continuous. Not only was caustic soda produced continuously but also hydrogen and chlorine, and it obviously had great commercial possibilities.

D. J. Adam, Early Industrial Electrolysis, *Education in Chemistry,* Vol. 17, No. 1, p. 13, 1980

Questions

1 Castner's original process for the production of sodium involved fusing caustic soda (sodium hydroxide) with carbon. Suggest an equation for this reaction.

2 Write an equation for the production of aluminium by the Deville process.

3 Outline, in no more than four lines, the modern electrolytic process for making aluminium (the Hall-Héroult process).

4 Why is this process substantially cheaper than the Deville process?

5 To overcome the initial difficulty of getting sodium cyanide accepted in place of potassium cyanide, Castner's company marketed NaCN as '130 per cent' KCN. Show how the figure 130 per cent is arrived at.

6 Why did Castner decide to look for a method of making purer caustic soda?

7 Draw a labelled diagram of Castner's rocking mercury cell for the electrolysis of brine. Show clearly what happens when the cell is in operation, and write equations for all the chemical reactions which occur.

8 A similar version of Castner's mercury cell was developed by an Austrian called Carl Kellner, and the modern cell is known as the **Castner-Kellner** cell. How does the modern version of the mercury cell differ from Castner's original rocking cell?

9 The Castner-Kellner process for the electrolysis of brine is of great importance to the chemical industry, producing as it does three major industrial chemicals – sodium hydroxide, chlorine and hydrogen – from a cheap and abundant raw material. Give some of the important uses of the products of the process.

10 Can you think of any social or environmental disadvantages of the Castner-Kellner process? Outline them briefly.

14 Chromium plating

The advertisement below originally appeared in *Scientific American*. Read the advertisement, then answer the questions which follow it on the next page.

Man has been plating chromium for over a century – even though puzzled by the electrochemistry involved.

For example, the normal process uses a bath containing Cr^{6+} ions. During plating, the Cr^{6+} ions reduce to Cr^{3+} and then to metallic Cr. Now you might wonder: If the process goes through the Cr^{3+} stage anyway, why not start with those ions? Well, strangely enough, if you try, the process won't work unless considerably modified.

Our scientists have long been intrigued by this enigma here at the General Motors Research Laboratories. And they now believe they've not only explained the Cr^{3+} mystery, but developed a correct theory of the entire chromium plating process as well.

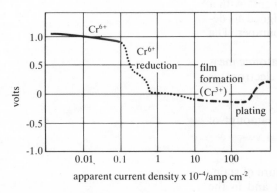

Figure 1 Typical polarisation curve.

By analyzing polarization curves (figure 1) obtained from carefully designed experiments, they concluded that:

● Starting with Cr^{3+} fails because it immediately forms a stable complex with water molecules (figure 2) from which Cr cannot be deposited.

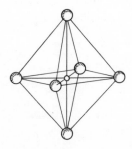

Figure 2 Cr^{+3} bound by water molecules as stable complex ion.

● Starting with Cr^{6+} succeeds because during reduction a chemical film forms around the cathode (the part being plated); and since Cr^{3+} is bound in that film, it does not react with water – but, instead, plates out as chromium metal.

From a murky chromium plating bath, some bright new answers.

General Motors Research Laboratories
Warren, Michigan 48090

Our researchers have, in fact, determined all 10 steps that take place as Cr^{6+} reduces to bright Cr, pinpointed the step at which catalysis begins, and identified the active catalyst. (It's not the sulfate ion, as commonly held, but the hydrogen sulfate ion.)

So have we merely solved a stubborn puzzle? No. More than that, we've gained important new insight to guide us toward a more efficient chromium electroplating process.

General Motors Research Laboratories, *Scientific American*, May, 1979

Questions

1 Why are metals sometimes plated with other metals? What properties make a metal suitable for plating onto others?

2 Name two metals, other than chromium, that are used for plating, and give the reason why they are used.

3 The passage mentions the use of Cr^{6+} ions for electroplating. In fact, simple Cr^{6+} ions do not exist in aqueous solution, but the ion used does contain Cr in the $6+$ oxidation state. Give the probable formula of this ion.

4 Explain briefly in your own words why Cr^{3+} cannot be used successfully in a plating bath, even though it is formed as an intermediate in the reduction of Cr^{6+}.

5 Give the name and formula of the stable complex ion formed by Cr^{3+} with water.

6 Draw a labelled diagram of apparatus that could be used to chromium plate the whole surface of a car door-handle.

7 Write half equations for the reduction of
a the ion containing Cr in the $6+$ state, in acid solution, to Cr^{3+}
b the Cr^{3+} ion to Cr.

8 It is required to plate a car bumper whose total surface area is 3500 cm^2 with a layer of chromium 0.0002 cm thick. The plating bath used is designed to deposit chromium at a uniform rate over the whole surface of the bumper. The plating operation uses a current of 35 A.
   Assume that the density of chromium is 7.2 g cm^{-3}, its relative atomic mass is 52 and that the Faraday constant (the charge on one mole of electrons) is 96 500 coulomb mole^{-1}.
a How many moles of chromium must be deposited on the bumper?
b How much electrical charge would be required to deposit this amount of chromium from the type of plating bath described in the passage?
c How long would the plating operation take?

9 Chromium has many important uses, particularly in the manufacture of stainless steel, and other specialized alloys used in defence, transport and in the home. Supplies of chromium are scarce and uncertain. Do you feel motor manufacturers are justified in using this important metal for plating automobile components? Discuss your answer.

15 The 'direct reduction' process for the production of steel

The following passage, taken from *Endeavour*, describes a new process for the reduction of iron ore to iron. Read the passage, and then answer the questions following it.

The direct reduction route normally involves the reduction of iron ore (usually of high quality) by a reductant gas: usually a hydrogen-carbon monoxide mixture produced by reforming natural gas. The process operates at about 900°C so that the iron ore remains solid, becoming a 'sponge' which contains all the impurities of the original iron ore but with the iron oxide content reduced to iron. This sponge iron is melted in an electric arc furnace where the 'gangue' is removed as a liquid slag and the molten metal refined to a suitable steel composition.

The impressive growth in the capacity installed to produce sponge iron (figure 1) is largely concentrated in those oil producing countries where natural gas must otherwise be flared

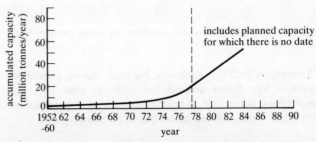

Figure 1 Growth of world direct reduction capacity.

and wasted – this waste occurs on a prodigious scale in the Middle East. The installation of direct reduction plants in these areas enables this gas, an important world resource, to be put to good use. It effectively enables cheap bulk storage and trans-shipment of this energy in the guise of directly reduced pellets, which in suitably dry conditions do not reoxidize. Sponge iron could become a world commodity, bought by many steel producers whose aggregated fluctuating demand would be more smooth than these of an individual company or country. In that way, the direct reduction plants could operate near to constant output.

The augmentation of mini-steelworks by a direct reduction plant, which enables the electric arc furnace to melt not only scrap but sponge iron, is a recent development. Such a works has the form illustrated in figure 2. Already the direct reduction mini-works has advantages over the large integrated steelworks based on blast furnaces and oxygen steelmaking vessels. Its lower capital cost per tonne of steel produced, its simpler manpower and scheduling problems, its greater flexibility, and ability to consume a large proportion of scrap, make it eminently suitable when there is need for, say 0·5 M tonnes of extra steel. In addition, this route has greater scope for continuous as opposed to batch operation. Fragmented scrap together with sponge iron can be fed continuously, and the electric arc furnace itself is amenable to continuous operation with the liquid steel running straight into a continuous casting machine.

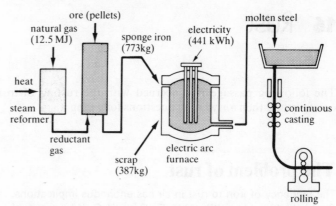

Figure 2 Diagram of a direct reduction electric arc steelworks.

A completely continuous process, which is more readily automated, would permit improved control and consistency at each stage, coupled with reduction in the capital cost of the equipment needed for a given tonnage. In the hesitant times through which the steel industry of the world is now passing, there is likely to be a greater emphasis on developing such a new technology.

Robert S. Barnes, Steel Making and its Future, *Endeavour*, Vol. 2, No. 1, 1978 (By kind permission of Pergamon Press)

Questions

1   In the conventional process for the reduction of iron ore, what reductant is used?

2   What reductant is employed in the direct reduction process? Write equations showing the reactions that occur in the process.

3   Write an equation to show how the reductant gas is made by reforming natural gas (methane).

4   What impurities are likely to be present in the 'sponge iron' produced by direct reduction?

5   Explain what is meant by the statement that the process 'effectively enables cheap bulk storage and trans-shipment of energy in the guise of directly reduced pellets'. Can you see any flaws in this argument?

6   Give three advantages of the direct reduction process compared to the conventional steel-making process.

7   Could direct reduction be used in the extraction of any metals other than iron? If so, which metals might be suitable?

8   In the long-term future, which of the two processes, conventional or direct reduction, is likely to be more important? Explain your answer.

16 Rust

The following passage is concerned with the rusting of iron. Read the passage carefully and then answer the questions following it.

The problem of rust

The tendency of iron to rust in air has enormous implications. Rust is hydrated iron(III) oxide, $Fe_2O_3 . xH_2O$. It has none of the characteristic physical properties possessed by metallic iron, such as high tensile strength. Rusting is therefore an important factor determining the useful life of iron and steel articles. On the other hand, the major ore of iron is virtually identical to rust, and a material with a built-in tendency to revert slowly to the form in which it exists in nature has certain environmental advantages. Nevertheless, rusting is a major problem, costing hundreds of millions of pounds each year in the UK alone, and its control is of the utmost importance.

Most metals tend to corrode in air. Many metals though, produce a protective oxide layer – aluminium is a good example. The dimensions of the unit cell in aluminium oxide are very similar to those in aluminium itself. Aluminium oxide therefore adheres tightly to the metal surface and protects it from further attack by oxygen. Unfortunately, rust and iron differ significantly in packing dimensions. This causes the rust to flake off as it forms, so the rust layer is porous and non-protective.

Corrosion on a car body.

What happens during rusting?

Rusting is an electrochemical process, requiring the presence of both air and water. Anodic (electron-releasing) and cathodic (electron-accepting) regions are set up in the iron (figure 1). At the anodes, iron dissolves:

$$Fe(s) \longrightarrow Fe^{2+}(aq) + 2e^-$$

The electrons produced in this process flow through the iron to a cathodic region, where they are accepted in a reaction involving dissolved oxygen:

$$2H_2O(l) + O_2(aq) + 4e^- \longrightarrow 4OH^-(aq)$$

The Fe^{2+} and OH^- ions produced in these processes migrate outwards, and where they meet, iron(II) hydroxide is precipitated:

$$Fe^{2+}(aq) + 2OH^-(aq) \longrightarrow Fe(OH)_2(s)$$

This precipitate later becomes oxidized to rust, hydrated iron(III) oxide.

Notice that the iron dissolves, that is corrodes, **only in the anodic regions**. At the anodes 'pits' are formed where the iron has dissolved away. There are several factors which determine which regions of the iron become anodic and which cathodic.

1 Oxygen supply

In regions where the oxygen supply is good, cathodes tend to be set up, because the cathodic reaction involves oxygen. It follows that regions of poor oxygen supply, such as crevices in a structure or areas covered with sediment or sand, are more likely to behave as anodes and become corroded.

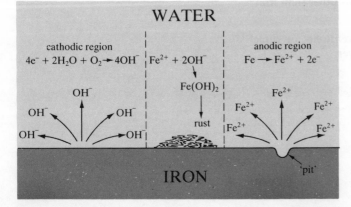

Figure 1 Cathodic and anodic regions in the wet corrosion of iron.

2 The presence of other metals

If iron is in electrical contact with a more reactive metal, such as zinc, the more reactive metal will have a greater tendency than iron to lose electrons and will therefore act anodically and corrode in preference to the iron. The table below lists some metals in order of their anodic reactivity. (This order corresponds roughly to the order of standard electrode potentials of the metals, though it is not exactly the same because conditions are not standard.)

The anodic reactivity series ('Galvanic Series') of some metals in sea water
magnesium
zinc
aluminium
mild steel
cast-iron
lead
tin
brass
copper
nickel
silver
titanium

Zinc blocks bolted to a ships hull to prevent corrosion.

3 The presence of impurities

Certain impurities in the iron may set up cathodic regions, causing other parts of the iron to become anodic.

Preventing rust

There are several methods used to prevent rusting.

1 Use of a protective layer This prevents the surface of the iron coming into contact with air or water. Examples are painting, covering with oil or grease, coating with plastic and plating with another metal (see also **2** below).

2 Cathodic protection The iron is made cathodic by bringing it into electrical contact with a more reactive metal, such as zinc, which corrodes in preference to the iron and is sometimes called a **sacrificial anode**.

3 Alloying By alloying iron with relatively unreactive metals, like nickel or chromium, stainless steels are produced. These alloys are protected from corrosion by thin non-porous films of oxide, such as chromium oxide and nickel oxide.

4 Use of inhibitors Certain chemical substances inhibit rusting by producing insoluble layers in close contact with cathodic or anodic regions. Magnesium sulphate, for example, acts as an inhibitor in cathodic regions by forming insoluble magnesium hydroxide by interaction between magnesium ions and the hydroxide ions produced at the cathodes.

Questions

1 Explain in your own words why iron rusts readily, while aluminium, which is generally more reactive, hardly corrodes at all under normal conditions.

2 When a piece of tin-plated iron is scratched, the exposed iron rusts rapidly. When zinc-plated (galvanized) iron is scratched, hardly any rusting of the exposed iron occurs. Explain this difference.

3 Why should copper rivets never be used to join two pieces of iron? Could aluminium rivets be used?

4 Why is corrosion particularly severe in the 'nooks and crannies' of a car body, such as the door sills, and box sections of the chassis? How could car manufacturers overcome this problem?

5 Rusting in car bodies could be largely overcome if cars were made from zinc-coated steel plate instead of ordinary mild steel plate. Why do you think car manufacturers have not adopted this practice, despite its advantages?

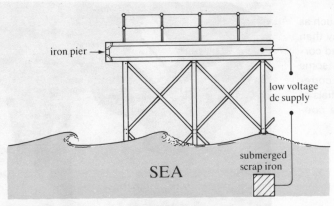

iron pier

low voltage dc supply

submerged scrap iron

SEA

Figure 2 Electrical corrosion protection of an iron pier.

6 Iron structures are sometimes electrically protected from corrosion. An example is illustrated in figure 2. To which terminal of the low-voltage d.c. supply should the iron pier in the figure be connected? Explain your answer.

7 Underwater steel pillars used to support piers are often found to corrode more severely where they are sunk beneath the sea bed than where they are in contact with sea water. Suggest a reason for this.

8 Most of the methods of corrosion protection described above are only temporary. Suppose chemists discovered a cheap way of permanently rust-proofing steel. Such a discovery would obviously have enormous advantages. Are there any disadvantages chemists should consider before marketing their discovery?

17 Fizzy drinks

The following passage refers to fizzy drinks, sometimes called sparkling or carbonated drinks. Read the passage, then answer the questions that follow it.

Equilibria in fizzy drinks

Fizzy drinks contain carbon dioxide dissolved under pressure. When carbon dioxide dissolves in water (and the bulk of any drink is water), several equilibria are established. Carbon dioxide molecules in the gaseous phase (the space above the surface of the drink) are in equilibrium with hydrated molecules in the aqueous phase:

$$CO_2(g) \rightleftharpoons CO_2(aq) \qquad \Delta H \text{ negative} \quad \ldots (1)$$

Carbon dioxide may also react with water to form carbonic acid:

$$CO_2(aq) + H_2O(l) \rightleftharpoons H_2CO_3(aq) \qquad \ldots (2)$$

Carbonic acid is a fairly weak acid, dissociating to give hydrogencarbonate and carbonate ions:

$$H_2CO_3(aq) \rightleftharpoons HCO_3^-(aq) + H^+(aq) \qquad \ldots (3)$$

$$HCO_3^-(aq) \rightleftharpoons CO_3^{2-} + H^+(aq) \qquad \ldots (4)$$

In aqueous solution, carbon dioxide exists mainly as hydrated molecules, $CO_2(aq)$; less than one per cent of the carbon dioxide reacts with water to form carbonic acid.

Most fizzy drinks are made by dissolving carbon dioxide gas in the still drink under pressure, the liquid usually being cooled beforehand. This method is used for soft drinks such as lemonade and Coca-Cola, for keg beer and most bottled beers, and for some cheap sparkling wines.

Some alcoholic drinks are made fizzy by fermentation. After the initial fermentation process, the drink is put in a closed container (usually a bottle) along with a small amount of sugar. Yeast remaining in the drink from the initial fermentation causes a brief second fermentation to occur with the added sugar. This produces a small amount of carbon dioxide, which can only dissolve in the drink because the container is closed.

Champagne is produced in this way, and so is bottled (but not draught) Guinness. It is claimed that drinks made fizzy by this method retain their fizziness for longer than drinks that have merely had carbon dioxide forced in under pressure.

Bubble formation

When a fizzy drink is in a closed container, at a fixed temperature, it is in a state of equilibrium. When the top is taken off the bottle, gaseous carbon dioxide escapes and the equilibrium

$$CO_2(g) \rightleftharpoons CO_2(aq)$$

moves to the left in accordance with Le Chatelier's principle. Gaseous carbon dioxide is released from the aqueous phase and forms bubbles which rise to the surface. The equilibrium position adjusts fairly slowly, however, and gas is evolved from the solution for some time before the new equilibrium position is reached, whereupon the drink is said to be 'flat'. One reason for the slowness of the system to adjust to a new equilibrium position is the difficulty of forming bubbles of gas in the body of a liquid. Bubbles form fastest when there is a nucleus or 'seed' present round which they can start to grow. When the drink enters the mouth, bubbles start to form round nuclei such as taste buds on the tongue, aided by the higher temperature of the mouth. The formation of bubbles in this way is responsible for the slight prickling sensation which is felt when a fizzy drink enters the mouth and which gives these drinks much of their appeal.

A glass of fizzy lemonade. Try to identify some bubble-forming nuclei on the side of the glass.

Questions

1 Use the equilibria listed above to decide how the pH of a fizzy drink differs from that of the same drink without added carbon dioxide.

2 How would the addition of a little sodium hydrogencarbonate (sodium bicarbonate) affect the fizziness of a drink?

3 How would the addition of a slice of lemon affect the fizziness of a drink?

4 Refer to equilibrium (1) to explain why a fizzy drink goes flat more quickly if it is warm.

5 When the top is taken off a fizzy drink, a slight hiss of escaping gas can be heard. If a drink is taken straight from the refrigerator and opened, the hiss is much less noticeable. Explain why.

6 If a fizzy drink is poured into a glass, it is usually noticeable that the bubbles rise to the surface in a stream, and that the stream originates from a particular point on the surface of the glass. Suggest an explanation for this.

7 If a bottle or can of fizzy drink is shaken vigorously before opening, carbon dioxide is released very rapidly when the bottle is opened and the drink often froths right out. A similar effect occurs if sugar is added to fizzy drink in a glass. Suggest an explanation for these observations.

8 Can you see any theoretical justification for the claim that drinks made fizzy by fermentation (e.g. champagne) stay fizzy longer than those produced by forcing in carbon dioxide under pressure?

18 The physical chemistry of ammonia synthesis

The following passage by J. H. J. Peet, taken from the *School Science Review*, discusses some of the factors affecting the yield of ammonia in the Haber Process. Read the passage and then answer the questions which follow.

Figure 1 indicates the variation in the equilibrium constant, K_p, for the reaction

$$\tfrac{1}{2}N_2(g) + \tfrac{3}{2}H_2(g) \rightleftharpoons NH_3(g)$$

with temperature. Clearly, the highest yields are favoured by low temperatures and high pressures, as predicted by Le Chatelier's principle. If an initial reaction mixture containing 3 moles of hydrogen to each mole of nitrogen is used, then the equilibrium yield of ammonia is as shown in figure 2.

The reaction between nitrogen and hydrogen is slow at the temperatures preferred for maximum yield and so a catalyst is required. The rate-controlling step is the chemisorption of nitrogen. There has been a lack of consistency in the value quoted for the activation energy for this step, reports varying between 70 and 220 kJ mol^{-1}.

While the process of selecting a catalyst for a reaction seems to be more of an art than a science, certain clear principles have emerged. Figure 3 illustrates the strength of adsorption, in terms of heat change, of nitrogen on transition metals of different groups. Experimental studies on the catalysts show that the rate of ammonia synthesis reaches a maximum with the iron group metals (d^6 structure), rapidly falling off to either the left or right of iron in the transition series (see the table right). This can be related qualitatively to bonding between the nitrogen and metal being either too strong (and so slowing the rate) or too weak (so preventing activation of the nitrogen). Iron catalysts are poisoned irreversibly by such elements as sulphur,

Metal	Electronic structure	Yields of NH$_3$ (%)
Manganese	$d^5 s^2$	0·8
Iron	$d^6 s^2$	2·0
Cobalt	$d^7 s^2$	0·2
Nickel	$d^8 s^2$	> 0·1
Molybdenum	$d^5 s^1$	1·5
Ruthenium	$d^6 s^2$	~ 1·0
Tungsten	$d^4 s^2$	0·4
Osmium	$d^6 s^2$	2·0

Table 1 Efficiency of metal catalysts in ammonia synthesis. (Equilibrium concentration of NH$_3$ under test conditions was 5%)

selenium, phosphorus, arsenic and also by carbon monoxide. Water, oxygen and carbon dioxide have reversible poisoning effects. Most of the catalysts (except Ni, Mo, W) can be promoted by basic oxides and all show an enhanced effect with certain metals. One commercial combination is usually produced from the fusion of an iron oxide, alumina and potassium oxide, followed by reduction with synthesis gas.

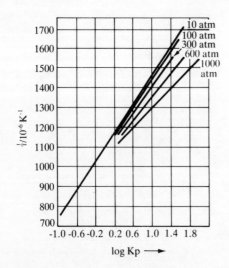

Figure 1 Variation of log Kp with L/T at different pressures.

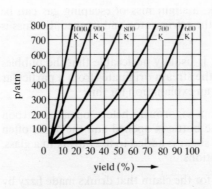

Figure 2 Variation of yield with pressure at different temperatures.

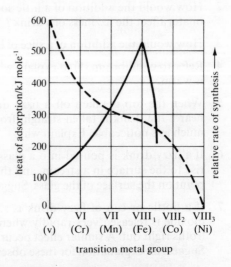

Figure 3 Effectiveness as catalysts of different transition metals. The left-hand axis refers to the broken line, the right-hand axis to the full line.

156

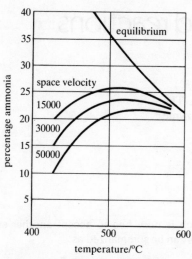

Figure 4 Variation of yield with temperature for different 'space velocities'.

In fact, the mixture is not allowed to react to equilibrium, the reactants being passed over the heated catalyst as a continuous flow. The rate of flow ('space velocity') determines the time of contact and so the yield (figure 4).

J. H. J. Peet, *The Physical Chemistry of Ammonia Synthesis,*
School Science Review, September, 1974

Questions

1   What are the usual sources of the nitrogen and hydrogen required for the Haber process?

2   Write an expression for K_p for the synthesis of ammonia as given in the equation at the beginning of the passage. What are the units of K_p?

3   A typical conversion plant might use a pressure of 300 atm and a temperature of 700 K. Use figures 1 and 2 respectively to obtain values for **a** K_p, and **b** the percentage yield of ammonia using a 3 : 1 nitrogen : hydrogen mixture under these conditions.

4   What would be the new yield in each case if **a** the temperature were reduced from 700 K to 600 K without altering the pressure, and **b** the pressure were increased from 300 atm to 400 atm without altering the temperature?

5   Explain why temperatures below 700 K are seldom used even though this would result in a higher yield.

6   Explain why pressures greater than 300 atm are seldom used even though this would result in a greater yield.

7   What do you understand to be the mechanism of catalysis in this process?

8   Explain **in your own words** why the most effective catalysts for this process are to be found near the centre of the transition series.

9   What is a catalyst poison? What is meant by referring to some poisons as 'reversible'?

10   What is a catalyst promoter?

11   Study figure 4, then try to explain:
   a why the percentage yield of ammonia decreases with increasing rate of flow ('space velocity').
   b why for a given 'space velocity', the percentage yield of ammonia increases to a maximum then decreases with increasing temperature.

12   The Haber process has been described as the most important piece of chemistry on earth. Why is it of such importance to society?

The first passage below is a summary of a simple theory of enzyme action. The second passage describes and gives the results of an experiment involving enzyme kinetics. Read the two passages and then answer the questions which follow.

1 The lock-and-key hypothesis of enzyme action

Enzymes are extremely powerful biological catalysts, far more powerful than inorganic catalysts. They are also far more specific. Most enzymes catalyse only one reaction or one type of reaction. Catalase, for example, an enzyme widely found in living tissue, decomposes hydrogen peroxide extremely fast, yet it has no effect on any other naturally-occurring compound. One molecule of catalase can decompose 6 million molecules of hydrogen peroxide every minute.

Any theory of enzyme action must explain this great efficiency and specificity. The lock-and-key hypothesis provides a relatively simple explanation of enzyme action derived from a study of the kinetics of enzyme-catalysed reactions.

Enzymes are proteins, and like all proteins they have a precise molecular shape which is destroyed by high temperatures and adverse pH conditions. The lock-and-key hypothesis suggests that the enzyme molecule has a particular location on its surface, called the **active site**, into which molecules of the reactant (called the substrate) fit and become attached. It is proposed that the active site is shaped specifically to fit the particular shape of the substrate molecule but no other. Figure 1 represents this diagrammatically. In reality the active site is a cleft in the enzyme molecule formed by folding of the protein chain. The substrate becomes attached to the active site by intermolecular attractions such as hydrogen bonds. The attachment of the substrate to the enzyme facilitates the chemical reaction by weakening bonds or by bringing atoms into the correct configuration for reaction. The substrate is converted to the products at the active site and these products then leave the enzyme which is now free to accept another substrate molecule. All this can be summarized as:

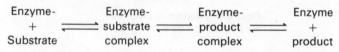

Enzyme- + Substrate ⇌ Enzyme-substrate complex ⇌ Enzyme-product complex ⇌ Enzyme + product

The enzyme and substrate combine for a very brief, but finite period. The substrate concentration is usually in great excess of the enzyme concentration (typically the ratio of substrate molecules to enzyme molecules might be $10^5 : 1$). Above a certain substrate concentration, all the enzyme molecules will be combined with substrate at any given time and the rate of the reaction will then be determined by the concentration of the enzyme and the time taken for the enzyme-substrate complex to form, rearrange and then release the products.

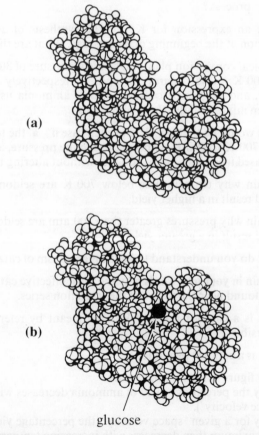

(a)

(b)

glucose

Figure 2 a A molecule of the enzyme hexokinase which catalyses the conversion of glucose to glucose-6-phosphate.
b The same enzyme with a molecule of the substrate, glucose, fitting neatly into the active site.

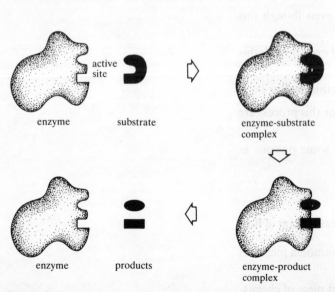

active site
enzyme   substrate   enzyme-substrate complex

enzyme   products   enzyme-product complex

Figure 1 A diagrammatic representation of the lock-and-key model of enzyme action.

2 The kinetics of the enzyme-catalysed hydrolysis of urea

The following account describes two experiments which were performed to investigate the effect of varying enzyme concentration and varying substrate concentration on the rate of the enzyme-catalysed hydrolysis of urea:

$$\underset{\text{urea}}{NH_2CONH_2} + H_2O \xrightarrow{\text{urease}} 2NH_3 + CO_2$$

The reaction was followed by measuring the volume of standard hydrochloric acid needed to react with the ammonia liberated in the reaction.

Experiment 1 In the first experiment, a fixed quantity of the enzyme, urease, was used, and the concentration of the substrate, urea, was varied. For each concentration, the average rate of reaction over the first three minutes was measured. The results are given in the table below.

Experiment 2 In the second experiment, the urea concentration was fixed and the concentration of urease was varied by adding different amounts of urease solution, keeping the total volume of the reaction mixture constant. Once again, the average rate of reaction was measured over the first three minutes. The results of this experiment are given in the table right.

Concentration of urea /mole dm^{-3}	Rate/mole min^{-1}
0	0
0·005	$1·7 \times 10^{-6}$
0·01	$2·3 \times 10^{-6}$
0·02	$3·2 \times 10^{-6}$
0·05	$4·4 \times 10^{-6}$
0·1	$5·9 \times 10^{-6}$
0·2	$7·2 \times 10^{-6}$
0·3	$7·7 \times 10^{-6}$
0·4	$8·0 \times 10^{-6}$
0·5	$8·1 \times 10^{-6}$

Results for experiment 1

Volume of urease solution/cm^3	Rate/mole min^{-1}
0	0
0·005	$0·6 \times 10^{-6}$
0·1	$0·8 \times 10^{-6}$
0·2	$1·8 \times 10^{-6}$
0·3	$3·2 \times 10^{-6}$
0·5	$4·8 \times 10^{-6}$
1·0	$10·4 \times 10^{-6}$
1·5	$14·9 \times 10^{-6}$
2·0	$19·5 \times 10^{-6}$

Results for experiment 2

Questions

1 Use the lock-and-key hypothesis to explain why enzymes are highly specific.

2 Why are enzymes made inactive by small changes in temperature or pH?

3 Enzymes can be rendered inactive by certain chemical substances, called inhibitors. Suggest an explanation for enzyme inhibition using the lock-and-key model.

4 Plot the results of experiment 1 on a graph, with concentration of urea on the horizontal axis. What does the graph tell you about the way in which reaction rate changes as the substrate concentration is increased?

5 Suggest an explanation in terms of the lock-and-key hypothesis for the behaviour you have described in question 4.

6 Plot the results of experiment 2 on a graph, with volume of urease (which is proportional to its concentration) on the horizontal axis. What does the graph tell you about the way in which the reaction rate changes as the enzyme concentration is increased?

7 Explain why the shape of the second graph differs from that of the first.

8 Sketch on your first graph the line you would expect to obtain under the same conditions but with the concentration of urease halved.

9 Summarize the ways in which enzymes differ from inorganic catalysts.

10 The enormous efficiency of enzymes has attracted the attention of industrial chemists, who are interested in the possibility of using them to catalyse reactions which are normally slow or need very high temperatures. Mention one way in which enzymes are used or could be used in an industrial or domestic context.

20 The future of the European plastics industry

The following article is taken from *New Scientist*, 8 March 1979. Read the article,
then answer the questions which follow it.

Among the illusions that have been shattered for polymer producers in Europe during the past few years has been the conviction that the plastics industry had a right to a growth rate about four times the gross national product and a similar one to make a profit from it. Perhaps that is an oversimplification, but this kind of thinking did influence the swashbuckling approach to investment, particularly among the newer European producers. It also caused the massive over-capacity in Western Europe, which at post-OPEC and post-Iran growth rates, could last for over 10 years. The build-up of plastics production capacity in Eastern Europe and in the Far East – plus the inevitability that the Middle Eastern states will eventually move 'downstream' into plastics – is now calling into question the viability of producing plastics at all in Western Europe.

Consequently, most European and US plastics producers see future profitability in the exploitation of their superior technical base. First, there is the development of high-performance polymers (such as polyamides, polycarbonates and so on up to the polyimides) where feedstock (raw materials) costs play a much smaller part. Secondly, there is the continued improvement of the conversion efficiency and operating economics of the processes to produce the existing large-tonnage plastics. This should compensate for possible disadvantages in raw materials costs and keep them in the race.

Pursuing this line of approach has resulted in a minor explosion of process developments in the past few years, centring on the polyolefine group (polyethylene and polypropylene) which, compared with PVC and polystyrene, is still relatively young in its development cycle and thus offers greater potential.

The chief problem has been the cost of feedstock. In the long term, those who can name their own price for feedstock have the advantage. At least they are insulated from the cost increases which bear on Western European producers, and on cost grounds it would seem that the latter must inevitably lose out to the Soviet Union and potential Middle Eastern producers.

Plastics production in Europe, valued recently at around nine billion dollars, totals some 16–17 million tonnes of material, of which five polymers – PVC, polystyrene, low and high density polyethylene (L- & HDPE) and polypropylene – make up about 11–12 million tonnes. It is these large-tonnage materials which are at risk, since they have virtually become commodities, exhibiting considerable price instability, especially over the past twelve months. In particular, the price of polypropylene and of low density polyethylene has been under considerable pressure, partly from surplus capacity but also, in the case of LDPE, from new capacity coming on stream in the USSR, in which part of the original licensing deal was a buy-back agreement by the Western European contractor. What is worrying the West is the realization that there are several other such agreements still to be honoured.

The original technology for polymerizing ethylene, as developed by ICI in the mid-1930's, required high pressures, of the order of 3500 atmospheres, which in the presence of a catalyst almost literally squeezed the molecules together to form a polymer chain. Surprisingly, this superficially crude technology is still widely used, but the percentage of polyethylene produced by the high pressure process is, in the opinion of ICI at least, likely to fall.

The main alternative to high pressure polymerization is, not surprisingly, 'low pressure' technology, in which, following the research of Phillips Petroleum and Professor Karl Zeigler in the 1950's, the catalyst plays the vital part, producing polymer chains with much shorter side-branching. These are capable of much tighter packing, producing a polymer with a dense linear structure. Historically, the high pressure material has been termed low density and the low pressure material, high density.

The important point was that this approach to catalyst technology made way for an advanced method of polymerization, with the possibility of 'tailor-making' olefine polymers and co-polymers (complex molecules) to meet certain specific performance requirements. In particular, the work of Ziegler led directly to the discovery by Professor Giulio Natta of a method for polymerizing propylene.

John Murphy, Plastics Makers Feel the Heat,
New Scientist, March 8th, 1979

Questions

1 Draw the structure of the monomer, and the structure of a section of the polymer chain, for each of the major polymers mentioned in this article – PVC, polystyrene, polyethylene and polypropylene.

2 IUPAC names are not used in this article. Give the IUPAC names for each of the polymers mentioned in **1**.

3 Give examples, one in each case, of articles made from these polymers.

4 How do the physical properties of low density and high density polyethylene compare?

5 What is meant by the 'polyolefine group' of polymers?

6 Why is it inevitable that 'the Middle Eastern states will eventually move down-stream into plastics'?

7 What does the article suggest about the future of the European and US plastics industries?

8 What steps should the industry take to improve its prospects?

21 The Seveso story

The following account of an accident at a chemical factory near Seveso in Northern Italy is taken from an article by W. G. Burton which appeared in the *School Science Review*. Read the passage, then answer the questions following it.

The formation of TCDD

2,4,5-Trichlorophenol (figure 1) has been manufactured for many years. It is used principally as an intermediate in the manufacture of two other chemicals, hexachlorophene (figure 2) – a bactericide, and 2,4,5-T – a herbicide whose structure and systematic name are shown in figure 3.

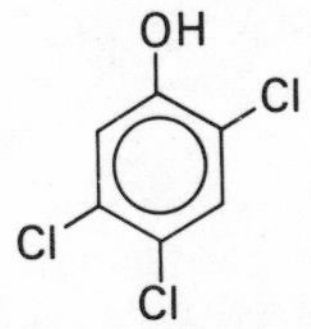

Figure 1 The structure of 2,4,5-trichlorophenol

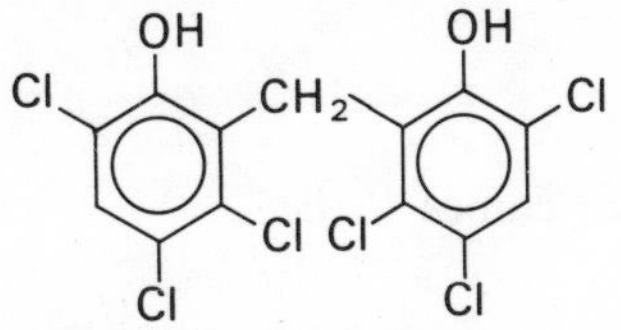

Figure 2 The structure of hexachorophene

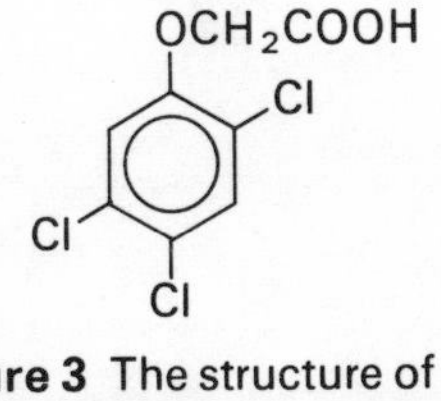

Figure 3 The structure of 2,4,5-trichlorophenoxyethanoic acid (2,4,5-T)

Hexachlorophene was once widely used in toothpaste, soaps, baby lotions and antiseptic washes. However, there is some evidence that hexachlorophene can cause brain damage and its use is being restricted. In 1967, the United States Defence department bought up all the US-manufactured 2,4,5-T for defoliation spraying on Vietnam. Evidence from Vietnam showed that the population exposed to the spraying developed many abnormalities, principally a severe skin rash called chloracne and a high incidence of babies born with severe deformities. There is now evidence that American soldiers who were exposed to the same spraying are suffering from chloracne and may be fathering malformed children.

Although these two chemicals are dangerous in their own right, they contain an even more dangerous contaminant which is produced in a side reaction during the manufacture of

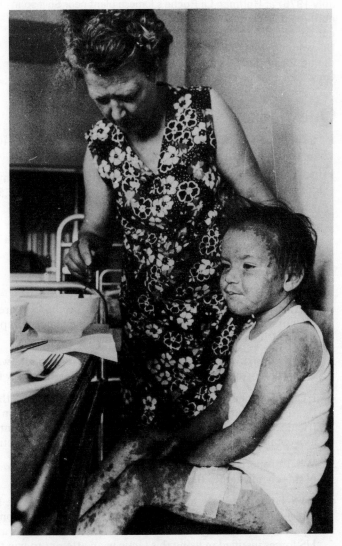

A victim of Seveso. Note the severe skin rash—chloracne.

161

trichlorophenol. This contaminant is known as 2,3,7,8-tetrachlorodibenzo-*p*-dioxin (TCDD or dioxin for short). Its structure is shown in figure 4.

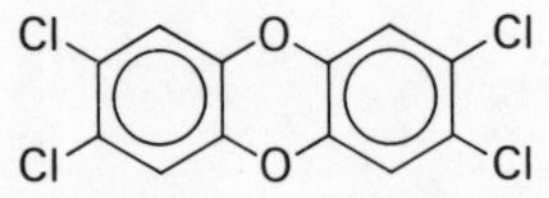

Figure 4 The structure of TCDD (dioxin)

This chemical has been recognized as a major hazard in the chemical industry for almost thirty years. It is extremely stable, being very resistant to attack by heat and other chemicals. Once spilled it is almost impossible to remove from land, buildings or machinery. It has been tentatively linked with liver and kidney damage, cancer and brain damage. It is one of the most powerful teratogens (foetus deformers) known and it is probably the minute traces of TCDD impurity which give rise to this hazard in the use of 2,4,5-T.

The Seveso Incident

Despite this accumulation of knowledge, an accident (on 10 July 1976) at Hoffman La Roche's Icmesa plant near Seveso in northern Italy caused the discharge of 1500 kg of trichlorophenol containing 2·5 kg of TCDD directly into the atmosphere. No warning was given to the local population for five days, during which time pets and poultry died and children fell ill. Only after that time were people warned not to eat contaminated food or plants. Evacuation of the area followed and guarded barbed wire barricades were erected. However, people were still allowed into and out of the contaminated area having paid the guards a small fee. A major road which runs through the most heavily contaminated area is still open.

By summer 1979, over 400 children suffered from the severe skin rash, chloracne, some had skin grafts, others almost permanently wear bandages to hide their disfigurement. The incidence of miscarriages rose from an average 12 per cent to 22 per cent. Presumably many of these foetuses were badly deformed. Of fifty babies born in the area since the accident, five have shown hideous abnormalities. Many women seek abortions but this is not easy in such a strongly Catholic community, and many women are afraid to ask.

The Italian government has set aside £27 000 000 for the relief and rehabilitation of the area and the people, but a satisfactory decontamination plan has not yet been proposed. The incident was seized upon by the local political parties, the governing Christian Democrats denouncing it as a left-wing fabrication and the Communists accusing the local authorities of a 'cover-up'.

Following this incident, a factory in Derbyshire producing the same chemicals which had been very carefully safeguarded by installation of computer-controlled operations, was forced to close by pressure from the local inhabitants.

W. G. Burton, The Seveso Story, *from* Acceptability Equations, *School Science Review*, June, 1979

Questions

1 Explain briefly in your own words how toxic TCDD (dioxin) comes to be formed during the manufacture of the herbicide 2,4,5-T.

2 2,4,5-Trichlorophenol is an intermediate in the manufacture of 2,4,5-T. 2,4,5-trichlorophenol is itself made by reacting 1,2,4,5-tetrachlorobenzene with sodium hydroxide.

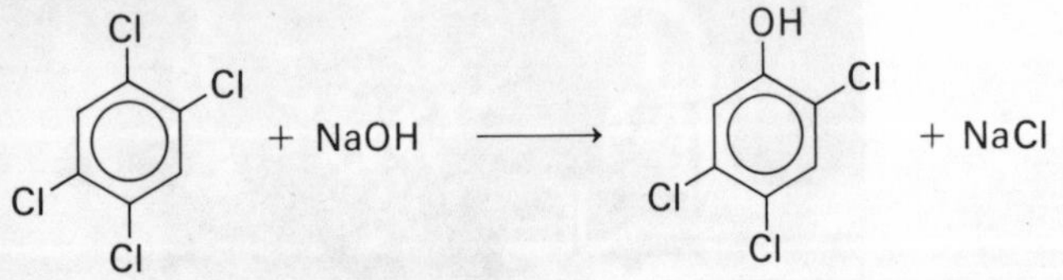

An isomer of 2,4,5-trichlorophenol can be made more simply by reacting phenol with chlorine water – this isomer is the major component of the antiseptic T.C.P. Give the structure of this isomer.

3 TCDD (dioxin) is formed by a condensation reaction between two molecules of 2,4,5-trichlorophenol. Show by means of an equation how this reaction occurs.

4 Hexachlorophene, once widely used as a bactericide, has the structure shown in figure 2. It is made by a condensation reaction between two moles of 2,4,5-trichlorophenol and one mole of methanal, using concentrated sulphuric acid as a catalyst. Show by means of an equation how this reaction occurs.

5 What property of TCDD, apart from its extreme toxicity, makes the discharge of this chemical into the environment a major disaster?

6 2,4,5-T was used as a herbicide during the Vietnam War. It was sprayed on to forests in order to remove the leaf cover in which Viet Cong rebels took refuge. Apart from the danger caused by the TCDD present as an impurity, were the US forces justified in using this tactic in an attempt to win the war? Discuss.

7 The major use of herbicides in peacetime is in agriculture, to kill weeds and thus increase the yield of food crops. In view of the risk involved in the manufacture of at least one herbicide, is the use of these agents justified? Discuss.

22 Aerosols and the ozone layer

The following extract from an article which first appeared in *Chemical Age* relates to the decision of the US government to ban aerosols propelled by chloro-fluorocarbons. The working of a typical aerosol can is summarized in figure 1. The reason for the ban is the suggestion that aerosols, and in particular the chloro-fluorocarbons used to propel them, may interfere with the concentration of ozone, O_3, in the stratosphere. This ozone is thought to have an important influence on the earth's climate. However, little is known about the effect of chloro-fluorocarbons on the ozone layer, and the case against them is not at all clear-cut. Read the passage, then answer the questions which follow on the next page.

US ban on chlorofluorocarbons

The US government has announced its final timetable banning the manufacture and sale of aerosols containing chloro-fluorocarbons which are widely feared to be damaging the earth's upper atmosphere.

Under the rules large-scale production of the gases must cease by 15 October 1978, manufacture of the aerosol cans must stop by 15 December 1978 and the general sale of such aerosols must end by 15 April 1979.

The ban is for 'non-essential uses' of chlorofluorocarbons in spray cans including deodorants, hair sprays, household cleaners and some pesticides. These constitute about 97 to 98 per cent of the annual US market.

Exemptions from the ban will be granted to products for which no substitute propellant exists, including certain respiratory drugs, contraceptive foams, insecticides and electrical and aircraft cleaning fluids.

European reaction

In Europe, the US action could increase pressure on some governments for similar legislation. So far, Sweden is the only country to announce a ban on the production and import of chlorofluorocarbons, to take effect from 1 July 1979.

The most likely outcome of the controversy over the next ten years is that the effect of chlorofluorocarbons on the earth's ozone layer will remain uncertain, and so most West European governments will refrain from imposing any ban.

Prospects for Europe's many chlorofluorocarbon manufacturers seem quite bright as about 60 per cent of the European aerosol propellant market is taken by chlorofluorocarbons. However, the liquefied hydrocarbon propellants, such as mixed butane and propane, are becoming increasingly popular in certain parts of Europe (30 per cent of all UK production).

Objections to hydrocarbons, based on the associated fire risk, have been partly appeased by 'fire-proofing' the gas with additions of dichloromethane.

In addition, carbon dioxide propellant, until now used mainly in Northern European countries in windscreen de-icing cans, now appears to have much wider use. The carbon dioxide gas is dissolved in aqueous solution under pressure. Until now such aerosols have been precluded for many applications where spray pattern is important.

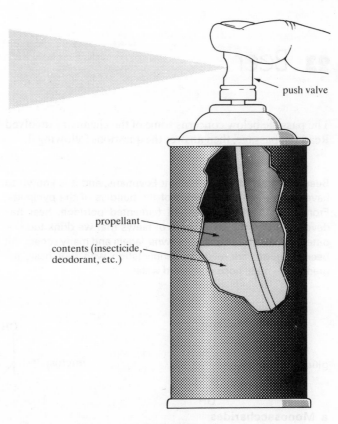

push valve

propellant

contents (insecticide, deodorant, etc.)

A typical aerosal can.

However, this is expected to change in the near future with the introduction of new valve designs. With its inherent low price and environmental acceptance, carbon dioxide could replace 20 per cent of all other propellants.

The last major propellant used is dinitrogen oxide. This is employed almost exclusively in food aerosols (whipped cream and sauces). As the gas is expensive (about £650/tonne), it is not expected to become as popular as other propellants.

Abstracted by Michael Vokins, from *Chemical Age*, March, 1978

Questions

1 What properties of chlorofluorocarbons make them particularly suitable for use as aerosol propellants?

2 Give the formula of one chlorofluorocarbon that could be used as a propellant.

3 What other uses are there for chlorofluorocarbons?

4 List the propellants that have been used as alternatives to chlorofluorocarbons, mentioning the advantages and disadvantages of each.

5 Why is carbon dioxide 'environmentally acceptable?'

6 How important are aerosol packs to society? How readily could we do without them?

7 Was the US government right to ban aerosols containing chlorofluorocarbons when the case against them is not proven?

8 Should these aerosol packs be banned in Britain?

23 Beer

The passage below concerns some of the chemistry involved in the brewing of beer. Read the passage, then answer the questions following it.

Beer was brewed by the ancient Egyptians, and it is known to have been among the rations of the builders of the pyramids. From its origins, as a kind of fermented porridge, beer has developed into the wide range of brews that we drink today – bitters, milds, stouts, pales, browns, lagers and many others. All beers, though, are made by fermentation with four basic ingredients – malt, hops, yeast and water.

Fermentation

Yeast can metabolize a number of carbohydrate substrates, but fermentation occurs fastest with monosaccharides, particularly glucose (figure 1a). In the case of malt, most of the fermentable carbohydrate is in the form of the disaccharide maltose (figure 1b) which is broken down by maltase in the yeast to give glucose. Yeast can ferment glucose aerobically (in the

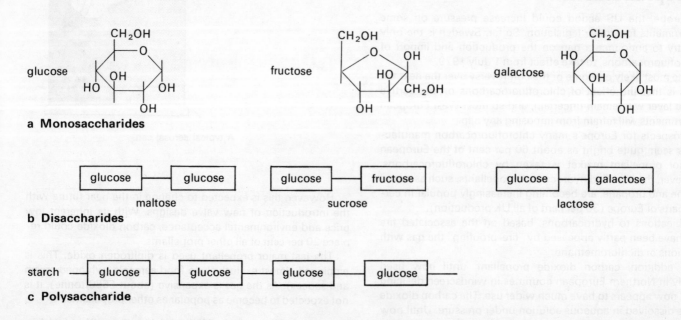

a Monosaccharides

b Disaccharides

c Polysaccharide

Figure 1 Structures of some carbohydrates involved in brewing

Tasting draught beer at the brewery.

presence of air) or anaerobically if there is no air (oxygen) present.

Aerobic fermentation:

$$C_6H_{12}O_6(aq) + 6O_2(aq) \xrightarrow{\text{yeast}} 6CO_2(g) + 6H_2O(l)$$
glucose

Anaerobic fermentation:

$$C_6H_{12}O_6(aq) \xrightarrow{\text{yeast}} 2CH_3CH_2OH(aq) + 2CO_2(g)$$
glucose

Both these reactions are far more complex than the above equations suggest, and involve several enzyme-catalysed stages.

In practice, the unfermented beer (called **wort**) contains much dissolved oxygen, so aerobic fermentation occurs when the yeast is first added. During this stage the yeast multiplies rapidly and fermentation accelerates. When the dissolved oxygen has been used up, anaerobic fermentation takes over and the yeast cells cease rapid division. Fermentation continues until all fermentable carbohydrate is consumed, or until the concentration of ethanol is high enough to inhibit yeast activity.

Ingredients

Malt is the major source of fermentable carbohydrate – it is made by allowing barley to germinate and then killing it. Barley is steeped in water, removed and allowed to stand in warm air for a few days. The barley grains begin to germinate and produce enzymes which start to break down the reserves of starch into maltose (see figure 1, opposite). At the same time proteins in the grain begin to break down, forming amino acids. After a few days, most of the starch has been converted to maltose and the process is stopped by heating the grains, killing the enzymes. Gentle heating produces a light malt, used in light ale, bitter and lagers, while further roasting gives progressively darker malt, used for brown ale and stouts.

Hops

These came relatively late to English beer. The original English brews were flavoured with nettles and other herbs, and called **ale**. The custom of adding hops spread from the continent – originally they were used as much as a preservative as for their flavour, for the pungent oils present in hops have a mild anti-bacterial action. Gradually the flavour of hops became established and the English began to call their ale by the Teutonic name 'bier'.

Yeast

Saccharomyces cerevisiae is a spherical single-celled yeast organism used in both brewing and baking. Different brewers use different yeast strains. The yeast multiplies during fermentation and a small amount is kept back for the next brew, the excess often being sold to yeast extract manufacturers. Some breweries have used the same yeast for hundreds of years.

Water

Fermentation needs an aqueous medium but the quality of the water itself affects the quality of the beer. Dissolved salts are very important. Hard water, containing a lot of calcium and magnesium salts, is good for making pale, fresh bitters. Burton-on-Trent water is particularly suited to making such beers. London water, on the other hand, is only temporarily hard, and becomes softened by boiling during the brewing process. London water is therefore better for making brown beers and stouts than bitters.

Sugar

The best beers contain only malt, hops, yeast and water. Sometimes, though, sugar is added to increase the fermentable material. Generally 'invert sugar' (sucrose which has been hydrolysed to glucose and fructose – figure 1, opposite) is used.

Adjuncts

Starchy additives such as wheat, rice or maize flour are sometimes used to increase body and give extra fermentable material.

Other additives

A number of chemical additives may be added to beer. These can improve the beer's 'head', or make it clearer.

The brewing process

The malt grains are ground into a fine powder called **grist** and soaked in hot water. This extracts soluble materials, mainly maltose and proteins. The malty liquor is then boiled with hops for two hours which extracts the aromatic oils from the hops and kills off bacteria. The liquor is strained and cooled before yeast is added. Fermentation is complete in about five days, after which time the beer is run into conditioning tanks to allow it to clear and to permit further biochemical changes which improve the flavour. Before fermentation the liquor contains many dissolved solids, particularly sugars, so it is denser than water. A typical value for its specific gravity (the density of the beer relative to that of water) might be 1·03. After fermentation, some dissolved solids remain but the specific gravity of the beer is very close to that of water, that is 1·00. From the difference between the initial and final specific gravity values, the amount of sugar fermented and hence the percentage of alcohol in the beer can be calculated. In this example it would be about 4 per cent.

Dispensing beer

There are two ways of serving beer – from barrels (draught beer) or from bottles and cans.

Draught beer

Traditional draught beer is run into metal or wooden casks direct from the conditioning tank and receives no further treatment except for the addition of finings to help it clear. At the pub the cask is usually stored in a cellar and connected via pipes to hand pumps on the bar. Once the cask is opened the beer is in contact with the air. Initially this improves the quality of the beer due to certain oxidation reactions, but after a week or so the beer begins to deteriorate, mainly due to the action of airborne bacteria. Traditional cask beer must therefore be sold quickly to avoid deterioration, and to get over this problem many breweries produce **keg** beer. This beer is pasteurized and filtered before running into metal kegs which are sealed and injected with carbon dioxide. At the pub, the beer is forced out of the keg using carbon dioxide pressure. As the beer is 'killed' before leaving the brewery and because the carbon dioxide prevents it from coming into contact with air, keg beer lasts much longer than traditional beer. Many people, however, feel that it is too fizzy and lacks flavour.

Bottled beer

Most bottled and canned beer is treated in the same way as keg beer, that is pasteurized and pressurized with carbon dioxide. It therefore keeps for a long time.

The food value of beer

The composition of a typical bitter beer is given in the table below.

Total volume	1 dm³ (1 litre)
Ethanol	30 g
Carbohydrate	20 g
Protein	3 g

Malt contains a substantial amount of protein as well as carbohydrate, so the protein content of beer is quite high, particularly in the case of certain stouts and strong bitters which have a high malt content. The protein is responsible for the foamy 'head' on a glass of beer. Protein molecules lower the surface tension of the liquid, thus stabilizing the bubbles that make up the 'head'. Beer always contains some carbohydrate, usually in the form of unfermented maltose, though the quantity varies according to the type of beer. The sweetness of some stouts is increased by adding lactose, a disaccharide whose structure is shown in figure 1b.

Ethanol is a food as well as a drug – it is in fact a more concentrated energy source than carbohydrate, so beer can be very fattening if drunk in quantity. It can also be very intoxicating – the effect of a pint or so can be stimulating, but after two or three pints confusion and loss of co-ordination begin, often with unpleasant and dangerous consequences.

Questions

1 Why must barley be malted before it can be used to make beer?

2 Give two reasons why it is important to exclude air once fermentation has started.

3 Why is 'invert sugar' added to beer when extra sugar is required, rather than ordinary sucrose?

4 Why does the 'specific gravity' of beer decrease during fermentation?

5 Government Customs and Excise men visit breweries regularly and measure the specific gravity of each brew before and after fermentation. Why?

6 Why does keg beer keep longer than traditional draught beer?

7 Suggest a reason why lactose is used to sweeten stout, rather than a more common sugar such as sucrose?

8 Why does a glass of beer often have a large frothy 'head', while a glass of lemonade never has?

9 The heat of combustion of ethanol is about -1370 kJ mole^{-1}, and the heat of combustion of glucose is about -3000 kJ mole^{-1}.
a Calculate the energy value in kJ of the ethanol in 1 dm^3 of bitter of composition given in the composition table above.
b Assuming all the carbohydrate in the beer is present as glucose, calculate the energy value in kJ of the carbohydrate in 1 dm^3 of bitter of composition given in the table.
c Assuming 1 dm^3 (1 litre) = 1·8 pints, and ignoring other components in the beer, calculate the total energy value of 1 pint of this beer.
d Comment on your value. (For comparison the energy value of a slice of bread is about 300 kJ.)

24 Fibres

The following passage concerns some aspects of the chemistry of natural and synthetic fibres. Read the passage and then answer the questions following it on page 170.

Fibres are polymeric materials, either naturally-occurring or synthetically manufactured. They possess certain properties which distinguish them from other polymers such as plastics and rubbers. These include high tensile strength, toughness and low extensibility. The high strength of fibre polymers derives from two properties. First, they have regular molecular structures, so their chains are able to pack closely together in an ordered way. This gives them a high degree of **crystallinity**. Second, they tend to have polar groups in their chain structures, and these give rise to strong interactions between adjoining chains. Synthetic fibres are usually formed by forcing the molten or dissolved polymer through a small hole called a spinnaret, then allowing it to solidify. The fibre formed in this way is then stretched or **drawn**. This process helps align the chains, increasing the fibre's crystallinity and hence its strength.

Until the First World War all fibres in common use were of natural origin. Semi-synthetic fibres (see below) began to make an impression on the textile market in the 1920's. The development of fully synthetic fibres began before the Second World War, but these fibres did not make a serious impact until the 1950's, when the development of the petrochemical industry brought down the price of the raw materials needed for their manufacture.

Natural fibres

Animal fibres, such as wool and silk, are long-chain proteins. The monomers are therefore amino acids, which undergo condensation polymerization under the influence of enzymes, to form peptide links:

MONOMERS
amino acids

POLYMER
protein

(The small side groups R_1, R_2, R_3 etc. may themselves carry functional groups such as NH_2, $COOH$ and OH which can form hydrogen bonds with water and other protein chains.)

The raw materials for the production of animal fibres are plants: sheep feed on grass, silk-worms feed on mulberry leaves. Animal fibres tend to be expensive, but they have many desirable properties – for example they feel pleasant and are comfortable to wear. Plant fibres, such as cotton and linen, are cellulose polymers. The monomer is glucose, and the cellulose is formed by condensation polymerization.

MONOMER
glucose

POLYMER
cellulose

The raw materials for the production of plant fibres are carbon dioxide and water. Cotton comes from the cotton-plant, linen from the flax plant. Cotton is relatively cheap and has a number of attractive features, including being comfortable to wear.

Semi-synthetic fibres

All semi-synthetic fibres are regenerated forms of natural cellulose. Cheap, abundant sources of cellulose such as wood pulp or straw are treated with chemicals to convert the cellulose into a soluble form which can then be extruded through a spinnaret to form fibres.

Viscose rayon is made by treating cellulose with sodium hydroxide and carbon disulphide. This causes partial cross-linking between chains.

Cellulose acetate and **triacetate** are made by treating cellulose with ethanoic acid and ethanoic anhydride so that some of the —OH groups on the cellulose chain are esterified with ethanoate (acetate) groups. 'Tricel' is an example of a cellulose triacetate fibre.

Semi-synthetic fibres are cheap but do not usually have the same quality as natural cellulose fibres. Rayon is often blended with natural fibres, particularly cotton, and it has a number of uses outside the clothing industry, for example in carpets and in tyre reinforcement cord.

Synthetic fibres

There are many different synthetic fibres in use today, but the three major ones are nylon, polyester and acrylic. **Nylons** are polyamides, formed by a condensation reaction between an —NH$_2$ and a —COOH or —COCl group. For example, nylon 66 is made from 1,6-diaminohexane and hexanedioic acid:

$$H_2N-(CH_2)_6-NH_2 \quad HOOC-(CH_2)_4-COOH$$

MONOMERS
1,6-diaminohexane
and
hexanedioic acid

$$-N-(CH_2)_6-N-C-(CH_2)_4-C-$$

$$+ H_2O$$

POLYMER
nylon 66

The linkage is similar to the peptide link in proteins.

Which synthetic fibres might be used for the cardigan, blouse, skirt and tights worn here?

Nylons are named according to the number of carbon atoms in the monomer molecules. Nylon 610, for example, is made from 1,6-diaminohexane and decanedioic acid. Nylon 66 is the dominant polyamide in the UK and the USA. The raw materials for its manufacture are phenol or cyclohexane, both of which can be obtained from coal or oil. Nylon is moderately priced relative to other fibres. It is hard wearing and makes good 'drip-dry' materials and stretch textiles. It is not as comfortable to wear as natural fibre because it lacks moisture absorbency.

Polyesters are condensation polymers formed from dibasic carboxylic acids and diols. The most common polyester is made from benzene-1,4-dicarboxylic acid and ethane-1,2-diol:

$$HOOC-\bigcirc-COOH \quad HOCH_2CH_2OH$$

MONOMERS
benzene-1,4-dicarboxylic
acid and ethane-1,2-diol

$$-CO-\bigcirc-COOCH_2CH_2O-$$

$$+ H_2O$$

POLYMER
polyester

This polyester is marketed under the trade names 'Dacron', 'Terylene' etc.

The raw materials for polyester manufacture are 1,4-dimethylbenzene (xylene) and ethene, both of which are made from oil.

Polyesters are moderately priced relative to other fibres and have the advantage of great tensile strength and resistance to wear. They have good 'wash-and-wear' characteristics and can be permanently creased by ironing. Like nylons, they lack the moisture absorbency of natural fibres, so are often blended with cotton or wool.

Acrylic fibres are addition polymers of propenenitrile, commonly called acrylonitrile.

Monomer
propenenitrile

$$CH_2=CH_2-CN$$

Polymer
acrylic

$$-CH_2-CH-CH_2-CH-CH_2-CH-$$

'Orlon' is a polymer of propenenitrile alone; 'Acrilan' and 'Courtelle' are copolymers, having small amounts of chloroethene (vinyl chloride) copolymerized with the propenenitrile.

Nowadays, the raw material for the manufacture of propenenitrile is propene, obtained from oil, but in the past it was made from ethyne, obtained from coal. Acrylic fibres are fairly low priced. Textiles based on them have a 'woolly' feel and are used as a substitute for wool, having the advantages of being machine-washable and non-shrinking.

Questions

1 For each of the fibres mentioned in the passage give one example of an article of clothing you know to be made from that fibre. In each case, identify the property of the fibre that makes it suitable for the use you mention.

2 For a polymer to show fibre properties it needs to have strong attractive forces between adjacent chains. For each of the fibres mentioned above, identify the groups responsible for the attraction, and name the type of interaction involved.

3 Why does polythene (poly(ethene)) not have fibre properties?

4 One of the main advantages of natural fibres like wool over synthetic fibres like nylon is the greater ability of the natural fibres to absorb moisture. This makes them more comfortable to wear because they do not make you feel 'sweaty' like many synthetic fibres do. Why does wool have greater moisture absorbency than nylon, even though the two polymers are formed by similar peptide condensation linkages?

5 The major polyamide fibre used in France and Germany is nylon 6. Nylon 6 has a structure very similar to nylon 66, but it is derived from only a single monomer molecule, known as caprolactam.

caprolactam

Indicate how caprolactam polymerizes to form nylon 6, and draw a representative section of the polymer chain.

6 Suggest a reason why woollen garments often shrink when washed, but acrylic garments do not.

7 Why has the use of synthetic and semi-synthetic fibres increased so much during the past forty years or so?

8 What is likely to be the future pattern of fibre use? Will synthetic fibres continue to replace natural ones, or will natural fibres make a comeback? Explain your answer.

25 Bread

A recipe for making bread is outlined below. This is followed by notes explaining some of the chemical principles behind breadmaking. Read the two passages and then answer the questions which follow on page 172.

A recipe for bread

1   Mix 25 g fresh yeast with one teaspoonful of sugar. Add 600 cm³ of tepid water and stir.

2   Mix together 1 kg strong flour and two teaspoonfuls of salt. Sift into a bowl.

3   Make a depression in the centre of the flour, pour in the yeasty water and mix well.

4   Knead until the sticky dough becomes elastic and leaves the sides of the bowl clean.

5   Leave the dough in a warm place until it has risen enough to double its original volume.

6   Knead the risen dough well for five minutes.

7   Shape this into an oblong loaf or loaves and put into greased tins.

8   Leave the tins in a warm place until the dough has risen to the top.

9   Cook in a hot oven until the crust is brown and the loaf sounds hollow when tapped.

Kneading dough. Note the cooked loaf alongside, cut open to show the bubbles formed by carbon dioxide gas when the dough rises.

The principles behind breadmaking

Bread can be regarded as a solid foam with a network of gas bubbles trapped in a matrix of solidified starch and protein. The gas bubbles result from the carbon dioxide produced during the fermentation of sugars by yeast:

$$C_6H_{12}O_6 \longrightarrow 2C_2H_5OH + 2CO_2$$

Almost all bread eaten in the United Kingdom is made from wheat flour. Wheat differs from other cereals in that wheat flour contains certain proteins which, when mixed with water, form an elastic proteinaceous material called **gluten**. It is the gluten which makes the dough strong and elastic enough to trap and hold bubbles of carbon dioxide. Without gluten, the starch on its own would be unable to form a foam and the dough would not rise. Consequently it is important that flour for breadmaking contains enough gluten. Flour with a high gluten content is called **strong** flour. Flour made from wheat grown in Britain and north-west Europe is fairly weak, with a total protein con-

tent averaging only 9 per cent. Flour from North American wheat is strong with a total protein content averaging about 14 per cent.

The elasticity of gluten is improved by mechanically stretching the dough. It is thought that this breaks bonds between adjacent protein molecules, allowing them to reform and produce an elastic three-dimensional network. This mechanical stretching occurs when the dough is kneaded, and during the rising caused by the fermentation processes.

When the dough is baked many complex changes occur. The gas bubbles expand and the dough rises further. The starch grains absorb water and burst, forming a rigid network. The gluten loses its elasticity and coagulates. Complex chemical changes occur on the surface of the loaf where the temperature is high, resulting in a crisp brown crust.

Questions

1 Apart from its function as a flavouring agent, why is sugar included in the recipe?

2 Why are bread recipes still successful even when sugar is not used?

3 Why does English bread normally contain a substantial proportion of North American wheat?

4 Suggest a reason why the French, who use little imported wheat, traditionally bake long, low loaves rather than the typical high English loaf.

5 Explain in your own words why it is necessary to knead dough.

6 Suppose a piece of dough initially 1000 cm^3 in volume is allowed to rise until its volume has doubled. It is then kneaded for five minutes, after which time its volume has returned to 1000 cm^3. It is then allowed to rise a second time until its volume has again doubled. What total volume of carbon dioxide must have been evolved during both risings?

7 Use your answer to question 6 to calculate the mass of ethanol formed at the same time as this carbon dioxide. State any assumptions you make.

8 Question 7 suggests that bread contains ethanol, but bread is not normally considered to be intoxicating. Explain.

9 Why does dough stop rising fairly soon after being placed in a hot oven?

10 What structural properties of the protein, gluten, give it elasticity?

11 What changes are likely to be responsible for the coagulation (denaturation) of the gluten during baking? Bear in mind that the temperature of the interior of the loaf never exceeds 100°C.

12 Why do you think bread becomes dry and hard if left in the open, while biscuits, which are also made from flour, become soft and damp?

SECTION THREE Objective questions

Each of the following thirty-two tests is based on the corresponding chapter in the textbook *Chemistry in Context*. Thus, Test 1, entitled 'Atoms, Atomic Masses and Moles' relates to the ideas, concepts and information presented in chapter 1 of *Chemistry in Context* which is also entitled 'Atoms, Atomic Masses and Moles'. Each test contains between fourteen and twenty questions.

TEST 1

Atoms, atomic masses and moles

This test is composed of sixteen questions. For each question, five possible answers are suggested. These answers are labelled **A**, **B**, **C**, **D** and **E**. Select the most appropriate **one** of the answers and write its corresponding letter on a separate answer sheet.

1   The relative atomic mass of neon, which consists of the isotopes $^{20}_{10}Ne$ and $^{22}_{10}Ne$, is 20·2. The percentage of $^{20}_{10}Ne$ atoms in the isotopic mixture is

   A   0·2

   B   2·0

   C   10·0

   D   10·1

   E   90·0

2   A pure liquid is found by experiment to contain 93·7% carbon and 6·3% hydrogen by mass. The empirical formula of the liquid is

   A   CH

   B C_3H_2

   C C_5H_4

   D C_8H_6

   E $C_{15}H$

   (C = 12, H = 1)

3   Which of the following pieces of equipment should be rinsed with the reagent they are to contain rather than with water before beginning a titration?

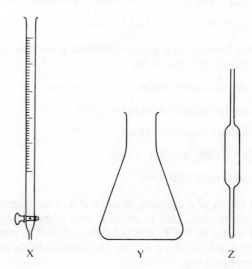

   A   *X* only

   B   *Y* only

   C   *X* and *Z* only

   D   *X* and *Y* only

   E   *X*, *Y* and *Z*

4 One of the compounds in both tea and coffee which acts as a stimulant is caffeine ($M_r = 194$). Analysis shows that caffeine contains 28·9% nitrogen by mass. The number of nitrogen atoms in one molecule of caffeine is

A 1

B 2

C 4

D 7

E 8

(N = 14)

5 100 cm^3 of 0·20 M K$_2$SO$_4$ are added to 100 cm^3 of water and mixed thoroughly. The molarity of K$^+$ ions in the resulting solution is

A 0·05

B 0·10

C 0·15

D 0·20

E 0·40

6 100 g of potassium hydrogencarbonate (KHCO$_3$) were heated to constant mass and the residue weighed 69 g. A possible equation for the decomposition is

A 2KHCO$_3$(s) $\longrightarrow$ K$_2$CO$_3$(s) + CO$_2$(g) + H$_2$O(g)

B KHCO$_3$(s) $\longrightarrow$ KOH(s) + CO$_2$(g)

C 4KHCO$_3$(s) $\longrightarrow$
$\qquad$ K$_2$O(s) + 2KOH(s) + H$_2$O(g) + 4CO$_2$(g)

D 2KHCO$_3$(s) $\longrightarrow$ KOH(s) + KO(s) + 2CO$_2$(g)

E 2KHCO$_3$(s) $\longrightarrow$ K$_2$O(s) + 2CO$_2$(g) + H$_2$O(g)

(K = 39, H = 1, C = 12, O = 16)

7 If the formula of a gaseous element is written as X$_2$(g) we can deduce that it has

A an oxidation number (valency) of two.

B a relative molecular mass of two.

C a mass number of two.

D an atomic number of two.

E an atomicity of two.

8 0·1 moles of a chloride of sulphur was completely oxidized by nitric acid to chloride ions and sulphate ions. 0·1 moles of PbCl$_2$ and 0·2 moles of BaSO$_4$ were precipitated from this solution. The formula of the original compound was

A SCl

B S$_2$Cl

C SCl$_2$

D S$_2$Cl$_2$

E S$_2$Cl$_4$

9 The Avogadro constant is

A the number of electrons required to deposit one mole of atoms of any metal.

B the number of atoms in one mole of atoms of any element.

C the number of grams of any element containing 6×10^{23} atoms.

D the number of atoms in one gram of any element.

E the number of atoms in one mole of molecules of any element.

10 The table below shows the relative molecular masses of three compounds, *I*, *II* and *III*, and the percentage of element X in each of these compounds.

Compound	Relative molecular mass	Percentage of X
I	100	42
II	56	50
III	224	25

What is the probable relative atomic mass of X?

A 7

B 14

C 28

D 42

E 56

11 Careful analysis showed that 2·20 g of a compound containing only phosphorus and sulphur contained 1·24 g of phosphorus. What is the empirical formula of this compound?

(P = 31, S = 32)

A P$_2$S$_3$

B P$_2$S$_5$

C P$_3$S$_2$

D P$_3$S$_4$

E P$_4$S$_3$

12 A molecular formula shows

A the ratio of atoms of the different elements in one molecule.

B the simplest whole number ratio for the atoms of different elements.

C the number of atoms in one molecule of a compound.

D the number of atoms of the different elements in one molecule.

E the number of atoms of the different elements in one mole.

13 Atoms of the same element have the same

    A relative atomic mass.

    B mass number.

    C oxidation number.

    D atomicity.

    E atomic number.

14 60 g of the metal M ($M = 60$) combine with 24 g of oxygen ($O = 16$) to form an oxide. The formula of the oxide is

    A MO

    B M_2O

    C M_2O_3

    D M_3O_2

    E M_5O_2

15 The volume of 0·10 M hydrochloric acid, which is just enough to react completely with 50·0 cm^3 of 0·20 M barium hydroxide is

    A 12·5 cm^3

    B 25·0 cm^3

    C 50·0 cm^3

    D 100·0 cm^3

    E 200·0 cm^3

16 100 cm^3 of 0·5 M H_2SO_4 is added to 400 cm^3 of 0·1 M KOH. The final concentration of hydrogen ions is

    A very low because there is excess of OH^-

    B 10^{-7} since H^+ is neutralized by OH^-

    C 0·01 M

    D 0·06 M

    E 0·12 M

TEST 2

Redox

This test is composed of eighteen questions. For each question, five (or in some cases four) possible answers are suggested. These answers are labelled **A**, **B**, **C**, **D** and **E**. Select the most appropriate **one** of the answers and write its corresponding letter on a separate answer sheet.

1 Which **one** of the following can be described as a 'redox' reaction?

A $Br_2(l) \longrightarrow 2Br(g)$

B $Na^+(g) + Br^-(g) \longrightarrow NaBr(s)$

C $Ag^+(aq) + Br^-(aq) \longrightarrow AgBr(s)$

D $Br_2(l) + H_2O(l) \longrightarrow HBr(aq) + HBrO(aq)$

E $AgBr(s) + 2NH_3(aq) \longrightarrow$
$[Ag(NH_3)_2]^+(aq) + Br^-(aq)$

2 The order of strength as oxidizing agents for the ions Au^{3+}, Li^+, Ni^{2+} and Rb^+ is

strongest $\longrightarrow$ weakest

A $Au^{3+} > Li^+ > Ni^{2+} > Rb^+$

B $Au^{3+} > Ni^{2+} > Li^+ > Rb^+$

C $Ni^{2+} > Au^{3+} > Li^+ > Rb^+$

D $Ni^{2+} > Au^{3+} > Rb^+ > Li^+$

E $Rb^+ > Li^+ > Ni^{2+} > Au^{3+}$

3 The oxidation number of tungsten in $Na_2W_4O_{13} . 10H_2O$ is

A $+4$

B $+6$

C $+8$

D $+11$

E $+12$

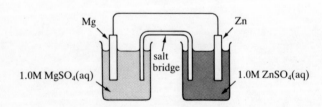

4 In the electrochemical cell shown above,

A the reaction at the zinc electrode is
$Zn \longrightarrow Zn^{2+} + 2e^-$.

B electrons flow through the wire from the zinc to the magnesium electrode.

C a tiny current flows even when the salt bridge is removed.

D magnesium ions migrate into the salt bridge.

E the concentration of Mg^{2+} ions will decrease in the left hand beaker.

5 20 cm³ of an acidified solution of 0·1 M potassium manganate(VII) just reacted with 20 cm³ of a solution of sodium sulphate(IV) (sodium sulphite). What is the concentration, in mole dm⁻³, of the solution of sodium sulphate(IV)?

$$SO_3^{2-} + H_2O \longrightarrow SO_4^{2-} + 2H^+ + 2e^-$$

$$MnO_4^- + 8H^+ + 5e^- \longrightarrow Mn^{2+} + 4H_2O$$

A 0·50

B 0·25

C 0·10

D 0·04

E 0·02

6 Which **one** of the following conversions is an oxidation?

A $Cr_2O_3 \longrightarrow Cr^{3+}$

B $CrO_4^{2-} \longrightarrow Cr_2O_7^{2-}$

C $Cr^{3+} \longrightarrow CrO_4^-$

D $CrO_4^{2-} \longrightarrow CrO_3$

E $CrO_4^{2-} \longrightarrow Cr_2O_3$

7 $VO_4^{3-} + xH^+ + ye^- \longrightarrow V^{3+} + zH_2O$
In the equation above, the values of x, y and z are

	x	y	z
A	8	3	4
B	8	2	4
C	8	2	2
D	4	2	4
E	4	1	2

8 Which of the following species is the strongest oxidizing agent?

A $Li(s)$

B $Li^+(aq)$

C $K(s)$

D $Na^+(aq)$

E $K^+(aq)$

9 In which of the following equations does the **first** stated reagent act as a reducing agent?

A $Cl_2(g) + H_2(g) \longrightarrow 2HCl(g)$

B $H^+(aq) + C_2O_4^{2-}(aq) \longrightarrow HC_2O_4^-(aq)$

C $H^+(aq) + NH_3(g) \longrightarrow NH_4^+(aq)$

D $H_2(g) + 2Na(s) \longrightarrow 2NaH(s)$

E $Mg(s) + 2H^+(aq) \longrightarrow Mg^{2+}(aq) + H_2(g)$

10 In the reaction $CH_4 + Cl_2 \longrightarrow CH_3Cl + HCl$

A C in CH_4 is acting as oxidant and is oxidized.

B C in CH_4 is acting as oxidant and is reduced.

C H in CH_4 is acting as reductant and is reduced.

D C in CH_4 is acting as reductant and is reduced.

E C in CH_4 is acting as reductant and is oxidized.

11 An electron transfer reaction takes place when

A copper is added to $FeSO_4$ solution.

B silver is added to $Zn(NO_3)_2$ solution.

C iron is added to $AgNO_3$ solution.

D zinc is added to $Mg(NO_3)_2$ solution.

E lead is added to $Al_2(SO_4)_3$ solution.

12 Which **one** of the following equations is correctly balanced?

A $H_2O + 2I^- + 2H^+ \longrightarrow 2H_2O + I_2$

B $Cr_2O_7^{2-} + Fe^{2+} + 14H^+ \longrightarrow$
$\qquad\qquad 2Cr^{3+} + Fe^{3+} + 7H_2O$

C $2NO_3^- + 2Cl^- + 4H^+ \longrightarrow$
$\qquad\qquad 2NO + Cl_2 + 2H_2O$

D $2MnO_4^- + 5Sn^{2+} + 16H^+ \longrightarrow$
$\qquad\qquad 2Mn^{2+} + 5Sn^{4+} + 8H_2O$

E $Zn + NO_3^- + 2H^+ \longrightarrow Zn^{2+} + NO_2 + H_2O$

13 0·01 moles of a compound K_2XO_4 will just oxidize 0·04 moles of Fe^{2+} to Fe^{3+}. The final oxidation state of X is therefore

A $+6$

B $+5$

C $+4$

D $+3$

E $+2$

14 In which **one** of the following pairs of substances does the specified element have a different oxidation number in each substance?

A Cr in CrO_3 and $Cr_2O_7^{2-}$

B Cu in $Cu(NH_3)_4^{2+}$ and $CuCl_4^{2-}$

C F in ClF and HF

D Na in $NaCl$ and NaH

E C in CH_4 and C_2H_6

15 A chemist isolated a compound which he described as 'manganese oxide-hydroxide, (MnOOH)'. The oxidation state of manganese in this compound is

A $+1$

B $+2$

C $+3$

D $+4$

E $+5$

Questions 16–18

$$Cr_2O_7^{2-} + 14H^+ + 6e^- \longrightarrow 2Cr^{3+} + 7H_2O$$
$$Fe^{2+} \longrightarrow Fe^{3+} + e^-$$
$$C_2O_4^{2-} \longrightarrow 2CO_2 + 2e^-$$

16 The oxidation number of chromium in $Cr_2O_7^{2-}$ is

A $+2$

B -2

C $+3$

D $+6$

E $+7$

17 The number of moles of Fe^{2+} oxidized by 1 mole of $Cr_2O_7^{2-}$ is

A $\frac{1}{6}$

B 1

C 2

D 3

E 6

18 The number of moles of iron(II) ethanedioate (iron(II) oxalate, FeC_2O_4) oxidized by 1 mole of $Cr_2O_7^{2-}$ is

A $\frac{1}{2}$

B 1

C 2

D 3

E 6

TEST 3
Patterns and periodicity

This test is composed of eighteen questions. For each question, five possible answers are suggested. These answers are labelled **A**, **B**, **C**, **D** and **E**. Select the most appropriate **one** of the answers and write its corresponding letter on a separate answer sheet.

Questions 1 and 5

The statements below relate to five of the elements in the third period of the periodic table, labelled **A**, **B**, **C**, **D** and **E**. In each case, select the letter of the element that has the stated characteristic.

A sodium
B aluminium
C silicon
D phosphorus (white)
E chlorine

1 Which element exists as separate molecules in the solid state at room temperature?

2 Which element forms an amphoteric oxide?

3 Which element forms a gaseous hydride which is slightly basic?

4 Which element has the highest atomic electrical conductance?

5 Which element is the strongest oxidizing agent?

Questions 6–8

Consider the following five elements in the **solid** state in answering the questions below.

A aluminium
B carbon (diamond)
C phosphorus (white)
D potassium
E xenon

6 Which element is composed of monatomic molecules held together by Van der Waals forces?

7 Which element is a solid with a low melting point and high electrical conductivity?

8 Which element is a low melting non-conducting solid composed of symmetrical polyatomic molecules?

9 The elements X, Y and Z are in the same period of the periodic table. The oxide of X is acidic, the oxide of Y is amphoteric and that of Z is basic. The order of these elements in increasing atomic number is likely to be

A XYZ
B XZY
C YZX
D ZXY
E ZYX

10 The elements in the periodic table are listed in order of increasing

A relative atomic mass
B atomic weight
C relative isotopic mass
D nuclear mass
E nuclear charge

11 Element Q forms an ionic compound with sodium. Q is most likely to be a member of groups

A III and IV
B III and V
C IV and V
D IV and VI
E V and VI

12 Elements in the same short period of the periodic table are likely to have

A the same oxidation state.
B similar physical properties.
C similar atomic radii.
D different numbers of electrons in their outer shell.
E similar ionic radii.

13 Which of the following statements is the **best** description of the element of atomic number 14?

   A a reactive metal

   B a poor metal

   C a metalloid

   D a non-metal

   E a transition metal

14 Which **one** of the following elements does not form both ionic and simple molecular binary compounds?

   A hydrogen

   B lithium

   C beryllium

   D nitrogen

   E oxygen

15 Which of the following elements has the highest molar volume at room temperature?

   A lithium

   B carbon

   C sodium

   D aluminium

   E potassium

16 The elements X, Y and Z are in the **same** short period of the periodic table. Element X has a giant molecular structure, element Y is metallic and element Z is composed of simple molecules. The order of these elements in increasing atomic number is

   A XYZ

   B XZY

   C YXZ

   D YZX

   E ZXY

17 0·25 moles of gaseous chlorine produced 0·10 moles of the chloride of element M. The formula of this chloride could be

   A MCl_3

   B $MCl_{2.5}$

   C MCl_5

   D M_2Cl_3

   E M_5Cl_2

18 Which **one** of the following elements forms a soluble basic oxide?

   A caesium

   B nickel

   C phosphorus

   D tin

   E zinc

TEST 4
Atomic structure

This test is composed of fifteen questions. For each question, five possible answers are suggested. The answers are labelled **A**, **B**, **C**, **D** and **E**. Select the most appropriate **one** of the answers and write its corresponding letter on a separate answer sheet.

1 Which **one** of the following instruments could be used to detect a stream of beta-particles?

   A a spectrophotometer

   B a gold-leaf electroscope

   C a mass spectrometer

   D a radiographer

   E a magnetometer

2 X-rays

   A are affected by electric and magnetic fields.

   B are formed by the loss of electrons from atoms.

   C have lower frequencies than visible light.

   D are formed when a solid target is bombarded by cathode rays.

   E consist of smaller particles than electrons.

3 Using a mass spectrometer, it is possible to determine the number of

   A protons in an atom.

   B energy levels in an atom.

   C atoms in one mole of an element.

   D isotopes of an element.

   E neutrons in an atom.

4 When a parallel beam of alpha-particles is directed towards a thin metal foil, most of the particles pass through, but a small fraction appear to rebound from the foil. This is because the alpha particles

   A are very small and have no charge.

   B have widely different velocities.

   C are repelled by electrons in the foil.

   D are repelled by the nuclei of atoms in the foil.

   E obey the law of conservation of momentum.

5 The relative atomic mass of an element is

   A the mass of 6×10^{23} atoms of the element.

   B the weighted mean of the relative isotopic masses on the scale $^{12}_{6}C = 12$.

   C the average of the relative isotopic masses on the scale $^{12}_{6}C = 12$.

   D the mass in atomic mass units of one atom on the scale $^{12}_{6}C = 12$.

   E one twelfth the weighted mean of the isotopic masses.

6 The isotope $^{40}_{19}K$ emits a beta-particle during radioactive decay. The other product is

   A $^{40}_{20}Ca$

   B $^{40}_{18}Ar$

   C $^{39}_{19}K$

   D $^{40}_{20}K$

   E $^{36}_{17}Cl$

7 Which **one** of the following particles has seven protons, eight neutrons and nine electrons?

   A $^{8}_{7}N^{-}$

   B $^{14}_{7}N^{2-}$

   C $^{15}_{7}N^{-}$

   D $^{15}_{8}N^{2-}$

   E $^{15}_{7}N^{2-}$

8 The isotopes W, X, Y and Z form compounds WX, WY and WZ. The compounds WX and WY are both radioactive but WZ is not. Which of the following conclusions is correct?

   A Only one of the isotopes is radioative.

   B Two of the isotopes are radioactive.

   C Three of the isotopes are radioactive.

   D XZ will not be radioactive.

   E W is radioactive in some compounds.

9 Atoms with different atomic numbers must have different

A numbers of electrons.

B numbers of neutrons.

C mass numbers.

D isotopic masses.

E molar masses.

10 The accurate relative isotopic masses of five isotopes are

$$^{1}_{1}H = 1 \cdot 0078, \qquad ^{2}_{1}H = 2 \cdot 0141, \qquad ^{12}_{6}C = 12 \cdot 0000,$$
$$^{14}_{7}N = 14 \cdot 0031, \qquad ^{16}_{8}O = 15 \cdot 9949.$$

Using a high resolution mass spectrometer, a certain gas was found to have a relative molecular mass of 28·0172. The gas could be

A $^{14}_{7}N_2$

B $^{12}_{6}C_2{}^{1}_{1}H_4$

C $^{2}_{1}H^{12}_{6}C^{14}_{7}N$

D $^{12}_{6}C_2{}^{2}_{1}H_2$

E $^{12}_{6}C^{16}_{8}O$

11 The mass spectrometer trace for naturally-occurring magnesium is shown below. Assuming that all three peaks relate to ions with one positive charge, what is the relative atomic mass for magnesium?

A 24·2

B 24·3

C 24·4

D 24·7

E 24·8

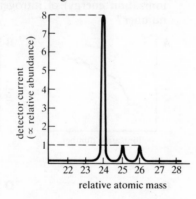

12 The diagram below shows the mass spectrometer trace of the substance X in the region of relative mass 10 to 20 units. Which of the following substances is X most likely to be?

A O_2

B N_2

C NH_3

D H_2O

E CH_4

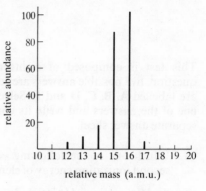

13 The atomic number of an element is the number of

A atoms in one molecule.

B electrons in the neutral atom.

C protons plus neutrons in the atom.

D protons plus electrons in the atom.

E atoms in one mole.

14 Which **one** of the following statements represents the correct relationship between the frequency of X-rays produced by electron bombardment of a metal, v, and the atomic number of the metal, Z? (b is a constant)

A $v \propto Z$

B $v \propto Z^2$

C $v \propto (Z - b)^2$

D $v \propto Z - b$

E $v \propto \sqrt{(Z - b)}$

15 In an attempt to determine the charge carried by **one** electron, a student bombarded tiny oil droplets with a stream of electrons and then measured the total charge on individual oil drops. The charges on 3 different oil drops were found to be $12 \cdot 8 \times 10^{-19}$ coulombs, $16 \cdot 0 \times 10^{-19}$ coulombs and $25 \cdot 6 \times 10^{-19}$ coulombs respectively. These figures suggest that the largest possible charge on the electron is

A $25 \cdot 6 \times 10^{-19}$ C

B $12 \cdot 8 \times 10^{-19}$ C

C $6 \cdot 4 \times 10^{-19}$ C

D $3 \cdot 2 \times 10^{-19}$ C

E $1 \cdot 6 \times 10^{-19}$ C

TEST 5
Electronic structure

This test is composed of eighteen questions. For each question, five possible answers are suggested. These answers are labelled **A**, **B**, **C**, **D** and **E**. Select the most appropriate **one** of the answers and write its corresponding letter on a separate answer sheet.

1 Which **one** of the following equations relates to the second ionization energy of element M?

   A $M(s) \longrightarrow M^{2+}(s) + 2e^-$

   B $M(s) \longrightarrow M^{2+}(g) + 2e^-$

   C $M(g) \longrightarrow M^{2+}(g) + 2e^-$

   D $M^+(s) \longrightarrow M^{2+}(g) + e^-$

   E $M^+(g) \longrightarrow M^{2+}(g) + e^-$

2 Which of the following elements has the highest (most endothermic) second ionization energy?

   A oxygen   **C** neon   **E** magnesium

   B fluorine   **D** sodium

3 The third ionization energies of six consecutive elements in the periodic table are shown below. To which group of the periodic table is element X likely to belong?

   A II

   B III

   C IV

   D V

   E VI

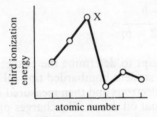

4 The first, second, third and fourth ionization energies of element Y are 740, 1500, 7700 and $10\,500\ \mathrm{kJ\,mole^{-1}}$ respectively. To which group of the periodic table is element Y likely to belong?

   A I   **B** II   **C** III   **D** IV   **E** O

5 In any one group of the periodic table, the first ionization energy decreases with increasing atomic number. Which **one** of the following factors is most responsible for this?

   A the increasing atomic radius

   B the decreasing effective nuclear charge

   C the decreasing bond energy

   D the increasing nuclear charge

   E the increasing atomic mass

6 Which **one** of the graphs below represents a plot of the first ionization energy against atomic number for the first eighteen elements?

7 Which **one** of the following graphs shows the logarithm to base ten of the successive ionization energies (lg ionization energy) of nitrogen against the ionization number?

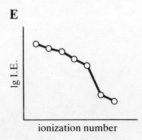

8 Which **one** of the following statements regarding electronic orbitals is correct?

A Each p-orbital can hold a maximum of six electrons.

B The $3p$-orbitals have a higher energy level than the $3s$-orbital.

C The three $3p$-orbitals have slightly different energy levels.

D The charge clouds for p-orbitals are spherical.

E The $1s$-orbital has the same size and shape as the $2s$-orbital.

9 Which **one** of the following diagrams represents the visible region of the atomic hydrogen spectrum?

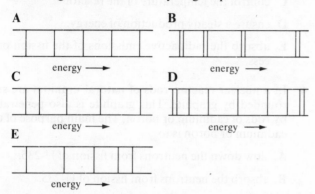

10 When electrons (with a charge of $1\cdot6 \times 10^{-19}$ C) were accelerated through a sample of xenon, ionization first occurred when the accelerating potential was 12 volts. Which **one** of the following values would this suggest for the first ionization energy of xenon in J mole^{-1}?

A $1\cdot6 \times 10^{-19} \times 12$

B $\dfrac{1\cdot6 \times 10^{-19}}{12} \times 6 \times 10^{23}$

C $\dfrac{1\cdot6 \times 10^{-19}}{6 \times 10^{23}} \times 12$

D $\dfrac{12 \times 6 \times 10^{23}}{1\cdot6 \times 10^{-19}}$

E $1\cdot6 \times 10^{-19} \times 12 \times 6 \times 10^{23}$

11 The diagram below shows the logarithms of the first seven ionization energies of element Q. Element Q could be

A aluminium

B beryllium

C boron

D nitrogen

E silicon

ionization number

12 The first ionization energy of oxygen is greater than that of magnesium. Which **one** of the following factors is responsible for this?

A There is a greater nuclear charge on oxygen.

B The number of electrons in the outer shell is greater for oxygen.

C The outermost electrons are further from the nucleus in oxygen.

D The shielding by inner-shell electrons is greater for magnesium.

E Interatomic bonding is stronger in oxygen.

13 The alkali metals are chemically similar because

A their outermost electrons have the same quantum number.

B their outermost electrons have the same energy.

C their outermost shells contain only s electrons.

D their outermost shells contain only one electron.

E they have low first ionization energies.

14 Which **one** of the following notations represents the electronic structure of nitrogen?

A $1s^2 2s^2 3p_x^1 3p_y^1 3p_z^1$ **D** $1s^2 2s^2 2p_x^1 2p_y^1 2p_z^1$

B $1s^2 2s^2 2p_x^2 2p_y^1$ **E** $1s^2 2s^2 3p_x^1 3p_y^1$

C $1s^2 2s^2 2p_x^1 2p_y^2$

15 The electron configuration of the outermost shell of an atom of the element R is $3s^2 3p^5$. The element R is

A aluminium   **D** nitrogen

B chlorine   **E** phosphorus

C fluorine

16 The atomic number of the element with electron configuration (Ne) $3s^2 3p^3$ is

A 5   **D** 13

B 7   **E** 15

C 10

17 The number of electrons required to fill the first seven energy levels ($1s$, $2s$, $2p$, $3s$, $3p$, $3d$, $4s$) is

A 18   **D** 30

B 20   **E** 36

C 24

18 Which **one** of the following represents the electron configuration of the outermost shell of an atom of an element in Group V?

A $2s^2 2p^5$ **D** $4s^2 4p^3$

B $3p^5$ **E** $5s^2 5d^3$

C $3d^3 4s^2$

TEST 6
Nuclear structure and radioactivity

This test is composed of eighteen questions. For each question, five possible answers are suggested. These answers are labelled **A, B, C, D** and **E**. Select the most appropriate **one** of the answers and write its corresponding letter on a separate answer sheet.

1 Gamma-rays are

   A attracted by an electric field.

   B deflected by a magnetic field.

   C composed of particles smaller than electrons.

   D diffracted by a crystal lattice.

   E unable to penetrate thin metal foil.

2 As atomic number increases, the neutron/proton ratio (n/p) for stable isotopes

   A remains constant.

   B slowly increases.

   C slowly decreases.

   D rises and then falls.

   E falls and then rises.

3 Bismuth-212 has a half life of 1 hour. If 16 g of the isotope are allowed to decay, what mass of the isotope remains after 4 hours?

   A 12 g

   B 4 g

   C 2 g

   D 1 g

   E 0 g

4 The emission of gamma-rays by an element results in the loss of

   A electrons

   B charge

   C energy

   D stability

   E atoms

5 In a nuclear reactor, rods of natural uranium are inserted in channels surrounded by blocks of graphite. The main purpose of the graphite is to

   A slow down the neutrons from fission of U-235.

   B absorb the neutrons from fission of U-235.

   C control the temperature of the reactor.

   D ensure a steady production of energy.

   E absorb the radioactive emissions of the fission process.

6 In a nuclear reactor, rods of natural uranium are surrounded by graphite. This graphite is also penetrated by rods of cadmium or boron. The main purpose of the cadmium or boron is to

   A slow down the neutrons from fission of U-235.

   B absorb the neutrons from fission of U-235.

   C conduct away the heat produced from fission of U-235.

   D absorb the radioactive emissions of the fission process.

   E ensure that the uranium never reaches the critical size.

7 The isotope $^{32}_{15}P$ emits a beta-particle during radioactive decay. Which **one** of the following isotopes is the other product of disintegration?

   A $^{32}_{14}Si$

   B $^{32}_{16}S$

   C $^{31}_{15}P$

   D $^{33}_{15}P$

   E $^{33}_{16}S$

8 Rutherford showed that the charge/mass ratio for a certain particle, X, was half the value of the charge/mass ratio for a proton ($^1_1H^+$). Which **one** of the following particles could X be?

   A $^2_1H^{2+}$

   B $^2_2He^{2+}$

   C ^4_2He

   D $^4_2He^+$

   E $^6_3Li^{3+}$

9 Which **one** of the following statements describes the half-life of a radioactive isotope **incorrectly**?

A The time taken for half of the mass of the isotope to disappear.

B The time taken for the concentration of an aqueous isotope to fall to half its initial value.

C The time taken for the number of atoms of the isotope to fall to half their original number.

D The time taken for the rate of decay of the isotope to fall to half its initial value.

E The time taken for the mass of the isotope to fall from half its initial value to zero.

10 Living organisms produce 15.3 ± 0.1 disintegrations of carbon-14 atoms per minute per gram of carbon. A sample, X, produces 1.9 ± 0.1 disintegrations of carbon-14 atoms per minute per gram of carbon. Assuming that the half-life of carbon-14 is 5600 years, which **one** of the following provides the best estimate for the age of X?

A $5600 \times \dfrac{15.3}{1.9}$ years

B $5600 \times \dfrac{1.9}{15.3}$ years

C $5600 \times (15.3 - 1.9)$ years

D 5600×3 years

E $5600 \times 15.3 \times 1.9$ years

11 Certain isotopes, such as $^{14}_{6}C$ have a neutron/proton ratio which is above the stable value. A common process by which these isotopes achieve stability is

A electron capture.

B alpha-decay.

C neutron emission.

D proton capture.

E beta-decay.

12 In a radioactivity experiment the background count was found to be 24 counts per minute, and a radioactive specimen, X, gave a count of 220 counts per minute (including background radiation) at the start. Assuming that X has a half-life of 10 minutes and that it decays directly to a stable isotope, what is the count rate (in counts per minute, including background radiation) 20 minutes after the start of the experiment?

A 0

B 24

C 55

D 73

E 79

13 Which **one** of the following statements provides the most accurate description of atomic fission?
Atomic fission is the disintegration of a nucleus involving

A the formation of alpha-particles.

B the formation of 2 or more neutrons.

C the formation of 2 large nuclei.

D a chain reaction.

E the emission of gamma-radiation.

14 The radioactive isotopes X and Y have half-lives of 2000 years and 6000 years respectively. A sample of rock contains 4 atoms of X for each atom of Y. The rock will contain the same number of atoms of X and Y after

A 2000 years.   **D** 8000 years.

B 4000 years.   **E** 12 000 years.

C 6000 years.

15 Bombardment of $^{7}_{3}Li$ with gamma-rays produces a proton and another nucleus. This other nucleus is

A $^{6}_{3}Li$ **D** $^{4}_{2}He$

B $^{7}_{2}He$ **E** $^{7}_{4}Be$

C $^{6}_{2}He$

16 Naturally-occurring chlorine contains two stable isotopes, $^{35}_{17}Cl$ and $^{37}_{17}Cl$, which occur in the relative proportions $3 : 1$. The relative proportions of the three molecules of chlorine $^{35}_{17}Cl_2$, $^{35}_{17}Cl^{37}_{17}Cl$ and $^{37}_{17}Cl_2$ will be

A $9 : 6 : 3$ **D** $6 : 6 : 1$

B $9 : 6 : 1$ **E** $6 : 3 : 1$

C $9 : 3 : 1$

17 The half-life of strontium-90 ($^{90}_{38}Sr$) is 28 years. This means that

A half the atoms in a sample of strontium-90 decay in 28 years.

B all the atoms in a sample of strontium-90 decay in 56 years.

C all the atoms in a sample of strontium-90 decay in 14 years.

D half a mole of strontium-90 decays in 28 years.

E one mole of strontium-90 decays in 14 years.

18 The nuclear transformation $^{1}_{0}n \longrightarrow {}^{1}_{1}H$ is an example of

A electron capture   **D** beta-decay

B fission   **E** alpha-decay

C fusion

TEST 7
The electronic theory and chemical bonding

This test is composed of eighteen questions. For each question, five possible answers are suggested. These answers are labelled **A, B, C, D** and **E**. Select the most appropriate **one** of the answers and write its corresponding letter on a separate answer sheet.

1 In which **one** of the following substances is ionic bonding present?

   A ice

   B silicon

   C brass

   D lime

   E sugar

2 Which **one** of the following species has a different number of electrons from the K^+ ion?

   A Ar

   B Br^-

   C HS^-

   D PH_4^+

   E SiH_4

3 Element X has atomic number 19 and element Y has atomic number 35. Which **one** of the following binary compounds are X and Y most likely to form?

   A an ionic compound, X^+Y^-

   B an ionic compound, X^-Y^+

   C an ionic compound, $(X^+)_3 Y^{3-}$

   D a molecular compound, XY

   E a molecular compound, XY_3

4 The formation of a co-ordinate (dative) bond between the phosphorus atom in PH_3 and the boron of BF_3 involves

   A increasing the number of electrons in the outer shell of phosphorus.

   B reducing the number of electrons in the outer shell of phosphorus.

   C transferring electrons from phosphorus to boron.

   D transferring electrons from boron to phosphorus.

   E sharing a pair of electrons between boron and phosphorus.

5 The elements X and Y form a compound of formula XY_3. The atomic numbers of X and Y could be

   A 3 and 5

   B 3 and 9

   C 5 and 7

   D 7 and 9

   E 7 and 13

6 Elements Q and R are in the same short period of the periodic table and have 4 and 6 valence electrons respectively. The formula of the compound they form together is most likely to be

   A QR

   B QR_2

   C Q_2R

   D Q_2R_3

   E Q_3R_2

7 Which **one** of the following species has the same electronic structure as Br^-?

   A Cl^-

   B K^+

   C Se^{2-}

   D Ar

   E Xe

8 In 1874, before modern instrumental methods of analysis became available, Van't Hoff and Le Bel predicted quite correctly that the methane molecule was tetrahedral rather than square planar. Which **one** of the following pieces of information would have provided the most convincing evidence for methane's tetrahedral shape?

   A Methane forms only one dichloro-compound.

   B Methane can substitute four chlorine atoms.

   C Methane is a gaseous hydrocarbon.

   D Monochloromethane is a polar substance.

   E Methane is a non-polar substance.

9 Which **one** of the following compounds has the greatest degree of ionic character?

    A beryllium oxide

    B beryllium sulphide

    C calcium oxide

    D calcium sulphate

    E magnesium oxide

10 Which **one** of the following molecules has a bond angle greater than 109° 28'?

    A SCl_2

    B CS_2

    C H_2S

    D CCl_4

    E NH_3

11 In which of the following does the covalent bond show the greatest departure from equal sharing of the electron pair?

    A F_2

    B SiH_4

    C C (graphite)

    D HCl

    E BrCl

Questions **12–18**

The shapes of simple molecules (with respect to their constituent atoms) can be described as

    A linear

    B V-shaped (bent)

    C trigonal planar

    D pyramidal

    E tetrahedral

Select the shape (with respect to atoms) of each of the following species.

12 F_2O

13 NH_4^+

14 NF_3

15 gaseous $BeCl_2$

16 gaseous $AlCl_3$

17 $GeCl_4$

18 C_2H_2

Intermolecular forces

This test is composed of 20 questions. For each question, five possible answers are suggested. These answers are labelled **A, B, C, D** and **E**. Select the most appropriate **one** of the answers and write its corresponding letter on a separate answer sheet.

1 Which **one** of the following compounds has no permanent dipole?

  A CH_2Cl_2

  B C_2Cl_6

  C $CH_3.O.CH_3$

  D NCl_3

  E $CH_3.CCl_3$

2 In which **one** of the following compounds would hydrogen bonding not occur?

  A $CH_3 . O . CH_3$

  B NH_2Cl

  C $(CH_3)_2NOH$

  D C_6H_5OH

  E H_2O_2

3 Compound X, empirical formula CHBr, was found to have zero dipole moment. Which one of the following structural formulae could be that of X?

4 Molecules of the compound ZCl_3 have zero dipole moment. What is their shape, with respect to constituent atoms?

  A linear

  B trigonal planar

  C tetrahedral

  D pyramidal

  E square planar

5 Which **one** of the following values would you use in estimating an approximate value for the strength of the hydrogen bond between ammonia molecules?

  A the heat of fusion of $NH_3(s)$

  B the heat of vaporization of $NH_3(l)$

  C the N—H bond energy in $NH_3(l)$

  D the heat of formation of $NH_3(l)$

  E the heat of solution of $NH_3(g)$

6 In which **one** of the following molecules does the covalent bond show the greatest departure from equal sharing of the electron pair?

  A Cl_2

  B HI

  C F_2

  D HCl

  E ClF

7 As atomic number increases from sodium to chlorine across period 3, the electronegativity of the elements

  A increases steadily.

  B decreases steadily.

  C increases to a maximum and then decreases.

  D decreases to a minimum and then increases.

  E stays almost constant.

8 According to the kinetic theory of gases, the molecules of an ideal gas

  A all have the same velocity.

  B are tiny charged particles.

  C attract each other strongly.

  D occupy a negligible volume.

  E lose energy on collision.

9 In the Van der Waals' Real Gas Equation,

$$\left(p + \frac{a}{V^2}\right)(V - b) = RT,$$

the term $\frac{a}{V^2}$ is inserted to allow for

A the volume of the molecules.

B the non-elastic molecular collisions.

C the intermolecular attractions.

D the different molecular velocities.

E collisions with the walls of the vessel.

10 Which **one** of the following types of bond is responsible for the surface tension of liquid tetrachloromethane (CCl_4)?

A covalent bonding

B hydrogen bonding

C permanent dipole attractions

D induced dipole attractions

E co-ordinate bonding

11

	melting point/K	boiling point/K
XBr	823	1583
ZBr	315	389

From the data above concerning the two bromides, XBr and ZBr, it is reasonable to conclude that

A ZBr has an ionic structure.

B X and Z are in the same group of the periodic table.

C XBr has a giant molecular structure.

D ZBr is very soluble in water.

E X is a metal and Z is a non-metal.

12 The boiling point of SiH_4 is higher than that of CH_4 because

A molecules of SiH_4 are polar unlike those of CH_4.

B molecules of SiH_4 are hydrogen-bonded but those of CH_4 are not.

C SiH_4 has a giant structure but CH_4 is composed of simple molecules.

D molecules of SiH_4 have permanent dipole-dipole attractions.

E intermolecular forces are stronger in SiH_4 than in CH_4.

13 The atoms P and Q have the electron configurations $1s^2 2s^2 2p^6 3s^2$ and $1s^2 2s^2 2p^3$ respectively. The formula of the compound they form together is most likely to be

A PQ_2 D P_3Q

B P_2Q_3 E P_3Q_2

C P_2Q_5

14 Which **one** of the following represents the order of increasing dipole moment for the trichlorobenzenes X, Y and Z shown below?

X Y Z

A XYZ B YXZ C YZX D ZXY E ZYX

15 The heat of sublimation of solid chlorine is 25 kJ mole^{-1} and the Cl—Cl bond energy is 245 kJ mole^{-1}. The energy required to tear one chlorine molecule away from its neighbours in solid chlorine is therefore

A $\dfrac{25}{6 \times 10^{23}}$ kJ   D $\dfrac{220}{6 \times 10^{23}}$ kJ

B $\dfrac{245}{6 \times 10^{23}}$ kJ   E $\dfrac{2 \times 270}{6 \times 10^{23}}$

C $\dfrac{270}{6 \times 10^{23}}$ kJ

Questions **16–20** concern the following classes of organic compound:

A carbohydrate
B nucleic acid
C protein
D fat
E amino acid

Select from **A** to **E** the class into which you would place a compound which

16 catalyses the decomposition of urea in mammals.

17 is more soluble in tetrachloromethane than in water.

18 relies entirely on —OH groups for hydrogen bonding.

19 acts as a structural polymer in plants.

20 is responsible for the similarity of a child to its parents.

TEST 9
Structure, bonding and properties: the solid state

This test is composed of 18 questions. For each question, four or five possible answers are suggested. These answers are labelled **A, B, C, D** and sometimes **E**. Select the most appropriate **one** of the answers and write its corresponding letter on a separate answer sheet.

1 Crystalline solids have good cleavage planes because the particles in the crystal are

   A weakly bonded together.

   B separated by large distances.

   C arranged in a regular fashion.

   D spherically symmetrical.

   E sometimes separated by dislocations.

2 The number of nearest neighbours to each metal atom in a body-centred cubic structure is

   A 4

   B 6

   C 8

   D 10

   E 12

3 Which **one** of the following would favour the formation of an ionic compound with the largest degree of covalent character?

   A a small anion with a multiple charge

   B a small anion with a single charge

   C a large cation with a multiple charge

   D a small cation with a single charge

   E a small cation with a multiple charge

4 The co-ordination number of carbon in graphite is

   A 3

   B 4

   C 5

   D 6

   E 7

5 The co-ordination number of carbon in diamond is four. This means that each carbon atom has

   A four covalent bonds.

   B four outer electrons.

   C an oxidation number of four.

   D four nearest neighbours.

   E four hybrid sp^3 electrons.

6 Hydrogen atoms are difficult to detect by X-ray diffraction because

   A the mass of the hydrogen atom is too small.

   B the electron density of hydrogen atoms is too low.

   C the radius of the hydrogen atom is too small.

   D the size of the hydrogen nucleus is too small.

   E the bonds to hydrogen atoms are too short.

7 Graphite and diamond are allotropes of carbon because

   A they are different physical states of the same element.

   B their atoms contain different numbers of neutrons.

   C they are different crystalline structures of the same element.

   D their molecules contain different numbers of atoms.

   E they give equal volumes of carbon dioxide on heating.

Questions 8–11

The following diagram shows a unit cell of the compound formed between the metal X and the non-metal Y.

   The ion of X is represented by an open circle at the centre of the unit cell and ions of Y are represented by black circles at the corners of the cell.

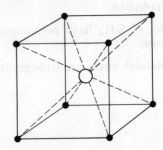

8 The formula of the compound formed between X and Y is

    A XY

    B XY_2

    C XY_4

    D XY_6

    E XY_8

9 The number of ions of Y in the unit cell is

    A $\frac{1}{2}$

    B 1

    C 2

    D 4

    E 8

10 The co-ordination numbers of ions of X and Y are respectively

    A $1:1$

    B $1:4$

    C $1:8$

    D $4:4$

    E $8:8$

11 If the volume of 1 mole of the compound is V dm^3 and the volume of a unit cell is v dm^3, the Avogadro constant, L, is given by

    A $\dfrac{V}{v} \times \dfrac{1}{8}$ **D** $\dfrac{V}{v} \times 4$

    B $\dfrac{V}{v} \times \dfrac{1}{4}$ **E** $\dfrac{V}{v} \times 8$

    C $\dfrac{V}{v}$

Questions 12–18

Crystalline solids may be divided into four distinct structures labelled **A**, **B**, **C** and **D** below:

A   giant metallic

B   giant ionic

C   giant molecular

D   simple molecular

In each of the following questions select the most likely structure for the substance with the stated property.

12 A hard, brittle substance which conducts electricity when molten.

13 A clear substance which gradually softens between 400 and 800°C.

14 An element whose relative molecular mass is four times its relative atomic mass.

15 An element which boils at 1100°C to give a monatomic vapour.

16 A solid mixture of two elements which conducts electricity.

17 A monatomic substance held together by Van der Waals' forces.

18 A substance boiling at 190 K which reacts with water to form a solution which conducts electricity.

TEST 10
The gaseous state

This test is composed of nineteen questions. For each question five possible answers are suggested. These answers are labelled **A**, **B**, **C**, **D** and **E**. Select the most appropriate **one** of the answers and write its corresponding letter on a separate answer sheet.

1 The temperature of 4 dm³ of an ideal gas rises from 200 K to 400 K and, at the same time, the pressure on the gas is halved. What is the final volume of the gas?

   A 1 dm³

   B 2 dm³

   C 4 dm³

   D 8 dm³

   E 16 dm³

2 What volume of oxygen is required to react completely with a mixture of 10 cm³ of hydrogen and 20 cm³ of carbon monoxide?
(All volumes are measured at the same temperature and pressure.)

   A 10 cm³

   B 15 cm³

   C 20 cm³

   D 25 cm³

   E 30 cm³

3 10 cm³ of a hydrocarbon react completely with exactly 40 cm³ of oxygen to produce 30 cm³ of carbon dioxide. What is the formula of the hydrocarbon?
(All volumes are measured at the same temperature and pressure.)

   A CH_4 **D** C_3H_6

   B C_2H_6 **E** C_3H_8

   C C_3H_4

Questions **4** to **6** concern the analysis of a volatile hydrocarbon, X. (C = 12, H = 1)

4 X was found to contain 85·7% by mass of carbon. What is the empirical formula of X?

   A $C_{\frac{1}{2}}H$

   B CH

   C CH_2

   D C_6H

   E C_7H_4

5 When 0·14 g of X was vaporized at 100°C, it occupied 62 cm³. Assuming that one mole of X occupies 22 400 cm³ at s.t.p., the relative molecular mass of X is

   A $\dfrac{0·14 \times 22\,400}{373 \times 62}$

   B $\dfrac{0·14 \times 22\,400 \times 373 \times 62}{273}$

   C $\dfrac{0·14 \times 62 \times 373}{22\,400}$

   D $\dfrac{0·14 \times 22\,400 \times 373}{62 \times 273}$

   E $\dfrac{0·14 \times 22\,400 \times 273}{62 \times 373}$

6 The accurate relative molecular mass of X (to the nearest integer) is

   A 37

   B 65

   C 69

   D 70

   E 73

Questions **7** to **10**

Two identical flasks at the same temperature contain 4 g of helium and 4 g of hydrogen respectively.

(He = 4, H = 1)

7 The ratio, number of molecules of He : number of molecules of hydrogen is

   A 1 : 1

   B 1 : 2

   C 1 : 4

   D 2 : 1

   E 4 : 1

8 The ratio, pressure of helium : pressure of hydrogen is

A 1 : 1

B 1 : 2

C 1 : 4

D 2 : 1

E 4 : 1

9 The ratio, average kinetic energy per helium molecule : average kinetic energy per hydrogen molecule, is

A 1 : 1

B 1 : 2

C 1 : 4

D 2 : 1

E 4 : 1

10 The ratio, mean square velocity of helium molecules : mean square velocity of hydrogen molecules is

A 1 : 1

B .1 : 2

C 1 : 4

D 2 : 1

E 4 : 1

11 10 cm³ of a gaseous element, Y reacts with excess hydrogen forming 40 cm³ of a gaseous compound containing Y and hydrogen, all volumes being measured under the same conditions of temperature and pressure. From this information, we can deduce that one molecule of the reactant, Y, could contain

A one atom of Y.

B two atoms of Y.

C one or two atoms of Y.

D two or four atoms of Y.

E four atoms of Y.

12 The molecules of a **real** gas

A occupy zero volume.

B exert no forces on one another.

C all have the same speed.

D collide with equal momentum.

E collide inelastically.

13 The mass of air inside a telephone kiosk (1 m × 1 m × 2 m) is about

A 0·25 kg

B 1·3 kg

C 2·5 kg

D 5 kg

E 25 kg

14 Flask A contains 1 dm³ of nitrogen at 1 atm and flask B contains 2 dm³ of oxygen at 3 atm. If the flasks are connected, at constant temperature, what is the final pressure in atmospheres?

A $1\frac{1}{3}$

B 2

C $2\frac{1}{3}$

D $2\frac{1}{2}$

E 4

Questions **15–19** refer to the five graphs of a quantity Y plotted against a quantity X labelled **A, B, C, D** and **E** in the figure below.

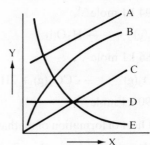

Choose the graph which describes the relationship between X and Y most accurately in each of the following questions. Each graph may be used once, more than once, or not at all.

15 Y is the pressure of nitrogen (p) and X is the volume of the nitrogen (V) at constant temperature.

16 Y is the pressure of nitrogen (p) and X is the reciprocal of the volume of nitrogen ($1/V$) at constant temperature.

17 Y is the product, pressure × volume (i.e. pV) and X is the volume (V) for a constant number of moles of a perfect gas at constant temperature.

18 Y is the volume of gas (V) and X is the temperature in degrees centigrade ($T/°C$) for a constant number of moles of a perfect gas at constant pressure.

19 Y is the product, pressure × volume (i.e. pV) and X is the absolute temperature (T/K) for a constant number of moles of ideal gas.

TEST 11
Energy changes and bonding

This test is composed of twenty questions. For each question, five possible answers are suggested. These answers are labelled **A**, **B**, **C**, **D** and **E**. Select the most appropriate **one** of the answers and write its corresponding letter on a separate answer sheet.

1 Which **one** of the following equations relates to the heat of atomization of bromine?

A $Br_2(g) \longrightarrow 2Br(g)$

B $\frac{1}{2}Br_2(g) \longrightarrow Br(g)$

C $Br_2(l) \longrightarrow 2Br(g)$

D $\frac{1}{2}Br_2(l) \longrightarrow Br(g)$

E $\frac{1}{2}Br_2(s) \longrightarrow Br(g)$

2 Using the data:

$C \text{ (graphite)} + O_2(g) \longrightarrow CO_2(g)$

$\Delta H^{\ominus} = -394 \text{ kJ mole}^{-1}$,

$H_2(g) + \frac{1}{2}O_2(g) \longrightarrow H_2O(l)$

$\Delta H^{\ominus} = -286 \text{ kJ mole}^{-1}$,

$CH_4(g) + 2O_2(g) \longrightarrow CO_2(g) + 2H_2O(l)$

$\Delta H^{\ominus} = -890 \text{ kJ mole}^{-1}$,

the standard heat of formation of methane is

A $-1856 \text{ kJ mole}^{-1}$.   **D** $+76 \text{ kJ mole}^{-1}$.

B $-210 \text{ kJ mole}^{-1}$.   **E** $+210 \text{ kJ mole}^{-1}$.

C -76 kJ mole^{-1}.

3 When 100 cm^3 of 0.1 M NaOH is neutralized by excess hydrochloric acid, the final volume of the solution is 130 cm^3 and the temperature of the mixture rises by $1.1°C$. Ignoring the thermal capacity of the container and assuming the specific heat capacity of all the solutions to be $4.0 \text{ J g}^{-1} \text{ K}^{-1}$, what is the heat of neutralization of NaOH by HCl in J mole^{-1}?

A $130 \times 1.1 \times 4$ **D** $100 \times 1.1 \times 4$

B $130 \times 1.1 \times 4 \times 10$ **E** $100 \times 1.1 \times 4 \times 100$

C $130 \times 1.1 \times 4 \times 100$

4 The heats of formation of $P_4(s)$, $P_4(g)$ and $P(g)$ at 298 K are 0, 52 and 316 kJ mole^{-1} respectively. What is the heat of atomization of phosphorus in J mole^{-1}?

A 52   **D** $316 + \frac{52}{4}$

B $\frac{316}{4}$ **E** $316 + 52$

C 316

5 The energy released when **one** electron is added to a gaseous atom would be expected to be highest for elements in group

A I   **B** II   **C** VI   **D** VII   **E** 0

6 Which **one** of the following spontaneous changes is endothermic?

A the mixing of two ideal liquids at room temperature.

B the condensation of water vapour at room temperature.

C the decomposition of ozone into oxygen at room temperature.

D the conversion of cyclopropane to propene at room temperature.

E the melting of sulphur at its melting point.

7 The standard heats of formation for CO(g) and $COCl_2(g)$ are $-110 \text{ kJ mole}^{-1}$ and $-223 \text{ kJ mole}^{-1}$ respectively. The standard enthalpy change in kJ mole^{-1} for the reaction

$$CO(g) + Cl_2(g) \longrightarrow COCl_2(g)$$

is therefore

A -333 **C** $+113$ **E** not obtainable without further data.

B -113 **D** $+333$

8 Which **one** of the following substances supports the fact that the thermal stability of a compound **cannot** always be related to its enthalpy of formation?

	Substance	$\Delta H_f^{\ominus}$/kJ mole^{-1}	Stability
A	V	-10	stable
B	W	-1000	stable
C	X	-100	unstable
D	Y	$+10$	unstable
E	Z	$+1000$	unstable

9 Which of the following pairs of substances will have the closest values for their molar heats of combustion?

A propane and butane

B cyclohexane and hexane

C cyclopropane and propene

D propan-1-ol and methoxyethane

E *cis*-but-2-ene and *trans*-but-2-ene

10 In order to calculate the average C—H bond energy in methane it is necessary to know the standard heat of formation of methane and also

A the molar heat of vaporization of methane.

B the standard heats of atomization of carbon and hydrogen.

C the standard heats of combustion of methane, hydrogen and carbon.

D the electron affinity of hydrogen and the first four ionization energies of carbon.

E the standard heats of formation of carbon and hydrogen.

11 Some average bond energies at 298 K in kJ mole^{-1} are:
C≡C 835; C=C 610; C—C 346; C—H 413; H—H 436.
What is the value of $\Delta H^{\ominus}_{298}$, in kJ mole^{-1}, for the reaction $2H_2(g) + HC≡CH(g) \longrightarrow CH_3CH_3(g)$?

A −665   **D** +319

B −291   **E** +665

C +291

12 Which **one** of the following equations summarizes the process relating to the lattice energy of calcium oxide?

A $Ca^{2+}(g) + O^{2-}(g) \longrightarrow CaO(s)$

B $Ca^{2+}(s) + O^{2-}(g) \longrightarrow CaO(s)$

C $Ca^{2+}(aq) + O^{2-}(aq) \longrightarrow CaO(s)$

D $Ca^{2+}(g) + O^{2-}(g) \longrightarrow CaO(g)$

E $Ca(g) + \frac{1}{2}O_2(g) \longrightarrow CaO(s)$

13 $\Delta H^{\ominus}_f$ for the hypothetical compound Ar^+Br^- would have a high positive value. The main reason for this is that

A bromine has a large enthalpy of atomization.

B bromine has a small electron affinity.

C bromine has a large first ionization energy.

D argon has a large enthalpy of atomization.

E argon has a large first ionization energy.

14 Assuming that both NaCl(s) and MgS(s) have simple cubic structures, that the lattice energy of sodium chloride is −800 kJ mole^{-1} and that the distance between the centres of Na$^+$ and Cl$^-$ ions in NaCl is very similar to that between Mg^{2+} and S^{2-} ions in MgS, which of the following provides the best estimate for the lattice energy of magnesium sulphide?

A −3200 kJ mole^{-1} **D** −400 kJ mole^{-1}

B −1600 kJ mole^{-1} **E** −200 kJ mole^{-1}

C −800 kJ mole^{-1}

15 The first, second and third ionization energies of element X are 500, 1000 and 4000 kJ mole^{-1} respectively. The hydration energies of ions X^+, X^{2+} and X^{3+} are −400, −1800 and −5800 kJ mole^{-1} respectively. A consideration of enthalpy changes suggests that in aqueous solution, X is likely to form

A X^+(aq) ions only.   **D** X^+(aq) and X^{2+}(aq) ions.

B X^{2+}(aq) ions only.   **E** X^{2+}(aq) and X^{3+}(aq) ions.

C X^{3+}(aq) ions only.

Questions **16** to **20** refer to the diagram below.

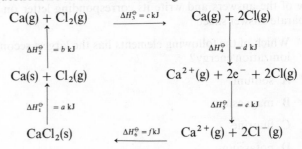

16 Which **one** of the following changes represents the standard heat of formation of CaCl$_2$(s)?

A $\Delta H^{\ominus}_1$ **D** $-\Delta H^{\ominus}_2 - \Delta H^{\ominus}_1$

B $-\Delta H^{\ominus}_1$ **E** $-\Delta H^{\ominus}_3 - \Delta H^{\ominus}_2 - \Delta H^{\ominus}_1$

C $\Delta H^{\ominus}_6$

17 What is the standard heat of atomization of chlorine in kJ mole^{-1}?

A $\dfrac{c}{2}$ **D** $c + e$

B $-c$ **E** $\dfrac{c + e}{2}$

C c

18 What is the standard enthalpy change in kJ for the process,
$$Ca(s) \longrightarrow Ca^{2+}(g) + 2e^- ?$$

A $b + c$ **D** $b + c + d$

B $b + d$ **E** $b + c + d + e$

C $b + e$

19 Which of the six changes shown in the diagram is/are exothermic in the direction indicated?

A $\Delta H^{\ominus}_1$ only   **D** $\Delta H^{\ominus}_1$, $\Delta H^{\ominus}_5$ and $\Delta H^{\ominus}_6$

B $\Delta H^{\ominus}_5$ only   **E** $\Delta H^{\ominus}_5$ and $\Delta H^{\ominus}_6$

C $\Delta H^{\ominus}_6$ only

20 The enthalpy change denoted by $\Delta H^{\ominus}_5$ represents

A the sum of the first two electron affinities of Cl(g).

B the sum of the first two ionization energies of Cl(g).

C the sum of the electron affinities of Cl(g) and Cl$^-$(g).

D twice the electron affinity of Cl(g).

E twice the first ionization energy of Cl(g).

TEST 12
Patterns across the periodic table

This test is composed of twenty questions. For each question five possible answers are suggested. These answers are labelled **A**, **B**, **C**, **D** and **E**. Select the most appropriate **one** of the answers and write its corresponding letter on a separate answer sheet.

1 Which of the following elements has the largest second ionization energy?

   A sodium

   B magnesium

   C fluorine

   D potassium

   E neon

2 In which of the following is the radius ratio the smallest?

   A $\dfrac{Li^+}{Li}$ **D** $\dfrac{Na^+}{Na}$

   B $\dfrac{B^{3+}}{B}$ **E** $\dfrac{F^-}{F}$

   C $\dfrac{N^{3-}}{N}$

3 Which **one** of the following sets of data represents the first five successive ionization energies of magnesium in kJ mole^{-1}?

   A 500   4600   6900   9500   13400

   B 2080   4000   6100   9400   12200

   C 580   1800   2700   11600   14800

   D 740   1500   7700   10500   13600

   E 790   1600   3200   4400   16100

4 The first ionization energy of phosphorus is greater than the first ionization energy of potassium. One of the factors responsible for this is that

   A the outermost electron is further from the nucleus in potassium atoms than in phosphorus atoms.

   B there are more electrons in the outer quantum shell of phosphorus than potassium.

   C the inner quantum shells provide greater shielding in phosphorus than in potassium.

   D the nuclear charge is greater for phosphorus than potassium.

   E the covalent bonding in P_4 molecules is stronger than the metallic bonding in potassium.

5 The covalent radius for chlorine is half the distance between the centres of

   A two chlorine atoms in gaseous chlorine.

   B two chlorine atoms in solid chlorine.

   C two chlorine atoms in a chlorine molecule.

   D two neighbouring, but unbonded chlorine atoms.

   E two chlorine atoms in a dichloro-compound.

6 The series of elements Na, Mg, Al, P and S were heated in a stream of chlorine to obtain their chlorides. The heat evolved per mole of chlorine atoms will

   A increase to a maximum and then decrease along the series.

   B decrease to a minimum and then increase along the series.

   C rise progressively along the series.

   D fall progressively along the series.

   E remain approximately constant.

7 Separate samples of NaCl, $MgCl_2$, Al_2Cl_6, PCl_3 and S_2Cl_2 were added to water and the resulting mixtures were tested with full-range indicator paper. The pH's of the solutions obtained will

   A increase from 7 to a maximum and then decrease.

   B increase from 7 to a maximum and remain high.

   C decrease from 11 to a minimum and then increase.

   D decrease from 11 to a minimum and remain low.

   E decrease from 7 to a minimum and remain low.

8 Elements X, Y and Z are in the same short period of the periodic table. The hydrides of X and Y are gaseous, but that of Z is a solid at room temperature. The hydride of X is insoluble in water, that of Y gives an acidic solution with water and that of Z gives an alkaline solution. The elements arranged in order of atomic number are

   A XZY

   B YXZ

   C YZX

   D ZXY

   E ZYX

196

9 The hydride of element Q reacts with water producing hydrogen and an alkaline solution. Q is probably

A an alkali metal.

B a transition metal.

C a metal low in the activity series.

D an element in group V.

E a metal in group IV.

Questions **10** to **14** refer to the five graphs below showing various physical quantities, labelled Y, against atomic number.

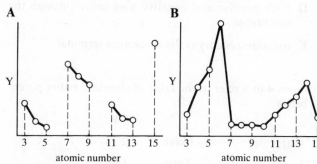

A

B

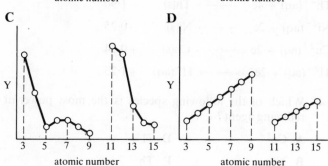

C

D

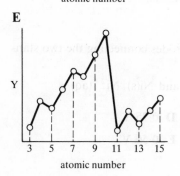

E

Which of the graphs is appropriate when Y is the physical quantity indicated below? Each graph may be used once, more than one, or not at all.

10 Electronegativity

11 The ionic radius of the most stable ion

12 The boiling point of the chloride

13 The first ionization energy

14 The melting point of the element

Questions **15–20**

Five classes into which oxides may be divided are

A acidic
B amphoteric
C basic
D neutral
E peroxide

Assign each of the following oxides to its appropriate class.

15 Cl_2O

16 CO

17 Cs_2O

18 SiO_2

19 BaO_2

20 Al_2O_3

TEST 13

Competition processes

This test is composed of sixteen questions. For each question five possible answers are suggested. These answers are labelled **A, B, C, D** and **E**. Select the most appropriate **one** of the answers and write its corresponding letter on a separate answer sheet.

Questions **1** to **3** concern the arrangement shown in the diagram below which is composed of the following half-cells:

$$Cr^{3+}(aq)/Cr(s), \quad E^{\ominus} = -0.74 \text{ V} \quad \text{and} \quad Ag^+(aq)/Ag(s), \quad E^{\ominus} = +0.80 \text{ V}.$$

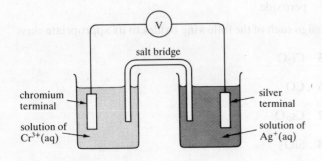

1 What is the standard e.m.f. of the cell

$Cr(s)|Cr^{3+}(aq)\vdots Ag^+(aq)|Ag(s)$?

A $(0.80 + 0.74)$ V

B $(0.80 - 0.74)$ V

C $\left(0.80 + \dfrac{0.74}{3}\right)$ V

D $(-0.80 + 0.74)$ V

E $\left(\dfrac{0.80 + 0.74}{3}\right)$ V

2 The e.m.f. of the cell can be reduced by reducing

A the area of the silver terminal.

B the cross-sectional area of the salt bridge.

C the thickness of the connecting wire.

D the concentration of $Cr^{3+}(aq)$.

E the concentration of $Ag^+(aq)$.

3 When the cell operates normally

A electrons flow through the voltmeter from silver to chromium.

B the silver terminal is more negative than chromium.

C the concentration of $Cr^{3+}(aq)$ will fall.

D both positive and negative ions move through the salt bridge.

E reduction occurs at the chromium terminal.

Questions **4** to **6** refer to the table of standard redox potentials below.

	$E^{\ominus}/V$
$Ce^{3+}(aq) + 3e^- \longrightarrow Ce(s)$	-2.33
$Th^{4+}(aq) + 4e^- \longrightarrow Th(s)$	-1.90
$Ni^{2+}(aq) + 2e^- \longrightarrow Ni(s)$	-0.25
$Cu^{2+}(aq) + 2e^- \longrightarrow Cu(s)$	$+0.34$
$Tl^{3+}(aq) + 2e^- \longrightarrow Tl^+(aq)$	$+1.25$

4 Which of the following species is the most powerful reducing agent?

A Ce^{3+} **D** Tl^+

B Ni   **E** Th

C Tl^{3+}

5 The e.m.f. between electrodes connecting the two standard half-cells

$Pt(s)|Tl^+(aq), Tl^{3+}(aq)$ and $Ni(s)|Ni^{2+}(aq)$ is

A 2.00 V   **D** 0.75 V

B 1.50 V   **E** 0.50 V

C 1.00 V

6 Excess copper was added to 100 cm³ of 0.1 M $Tl(NO_3)_3$ solution. What mass of copper would react? (Cu = 64)

A 1.28 g   **D** 0.42 g

B 0.96 g   **E** 0.32 g

C 0.64 g

Questions **7** and **8** refer to the following equations involving reactions of ammonia.

A $NH_3 + NH_3 \longrightarrow NH_4^+ + NH_2^-$
B $2Na + 2NH_3 \longrightarrow 2NaNH_2 + H_2$
C $3CuO + 2NH_3 \longrightarrow 3Cu + N_2 + 3H_2O$
D $NH_3 + NH_2Cl \longrightarrow H_2NNH_2 + HCl$
E $NH_3 + H_2S \longrightarrow NH_4HS$

7 In which reaction does ammonia behave as an oxidizing agent?

8 In which reaction does ammonia behave as both an acid and a reducing agent?

9 Which **one** of the following is a redox reaction?

A $Cu^+ + 2NH_3 \longrightarrow [Cu(NH_3)_2]^+$

B $Cu^+ + 4Cl^- \longrightarrow [CuCl_4]^{3-}$

C $Cu^+ + Cl^- \longrightarrow CuCl$

D $2Cu^+ \longrightarrow Cu + Cu^{2+}$

E $[Cu(NH_3)_2]^+ + 2H^+ \longrightarrow Cu^+ + 2NH_4^+$

10 In the equation, $Cu + Cu^{2+} + 4Cl^- \longrightarrow 2[CuCl_2]^-$, Cl^- is behaving as

A an oxidizing agent.

B a reducing agent.

C a ligand.

D a base.

E a complex ion.

11 When hydrogen reacts with sodium to form sodium hydride, it behaves as

A an acid.

B a base.

C a nucleophile.

D an oxidizing agent.

E a reducing agent.

12 30 cm^3 of an 0.1 M solution of a stable cation of an element Q react exactly with 12 cm^3 of 0.1 M acidified potassium manganate(VII).

$$MnO_4^- + 8H^+ + 5e^- \longrightarrow Mn^{2+} + 4H_2O$$

Which **one** of the following equations could represent the change in oxidation of Q correctly?

A $Q^+ \longrightarrow Q^{2+}$

B $Q^+ \longrightarrow Q^{3+}$

C $Q^+ \longrightarrow Q^{4+}$

D $Q^{2+} \longrightarrow Q^{3+}$

E $Q^{4+} \longrightarrow Q^{2+}$

13 The size and sign of an electrode potential provides an indication of

A the energetic favourability of a reaction.

B the kinetic feasibility of a reaction.

C the overall stoichiometry of a reaction.

D the activation energy of a reaction.

E the quantity of materials reacting.

14 Which **one** of the following species is the conjugate base to HSO_4^-?

A H_2SO_4

B SO_4^{2-}

C SO_3

D OH^-

E HSO_3^-

15 Lead(II) chloride is precipitated from solutions of lead salts by adding dilute hydrochloric acid. However, this precipitate dissolves when sufficient concentrated hydrochloric acid is added because

A lead chloride is soluble in acid.

B lead chloride complexes with Cl^- ions.

C chloride ions combine readily with H^+ ions.

D chloride ions complex with HCl.

E excess Cl^- ions reverse the equilibrium.

16 When the acid, H_3PO_2 is treated with excess sodium hydroxide, the only sodium salt obtained has the formula NaH_2PO_2. Which **one** of the following structures for H_3PO_2 best fits this data?

199

TEST 14

Groups I and II – the alkali metals and the alkaline-earth metals

This test is composed of fifteen questions. For each question, five possible answers are suggested. These answers are labelled **A**, **B**, **C**, **D** and **E**. Select the most appropriate **one** of the answers and write its corresponding letter on a separate answer sheet.

1 On descending group I from lithium to francium there is a steady increase in

   A electronegativity.

   B ionization energy.

   C boiling point.

   D standard electrode potential.

   E molar volume.

2 The sodium ion, Na^+, is isoelectronic with

   A P^{3-}

   B Ar

   C K^+

   D F^-

   E Na

3 Given below are the first four ionization energies, in kJ $mole^{-1}$, of aluminium, calcium, magnesium, rubidium and strontium (not in order). Which is the series for calcium?

   A 590   1100   4900   6500

   B 400   2700   3800   5100

   C 740   1500   7700   10500

   D 580   1800   2700   11600

   E 550   1050   4200   5500

4 Alkali metals

   A are reduced by hydrogen.

   B form reactive cations.

   C are stored under water.

   D readily form complex ions.

   E are oxidized by water.

5 Metals in group I of the periodic table have **no**

   A coloured compounds.

   B insoluble compounds.

   C amphoteric oxides.

   D reaction with hydrogen.

   E stable nitrides.

6 Which **one** of the following equations represents correctly the thermal decomposition of potassium nitrate?

   A $2KNO_3 \longrightarrow K_2O + 2NO_2 + \frac{1}{2}O_2$

   B $2KNO_3 \longrightarrow K_2O + 2NO + \frac{3}{2}O_2$

   C $2KNO_3 \longrightarrow K_2O + NO + NO_2 + O_2$

   D $2KNO_3 \longrightarrow K_2O_2 + 2NO_2$

   E $2KNO_3 \longrightarrow 2KNO_2 + O_2$

7 Lithium differs from sodium and potassium in that it forms

   A an ionic hydride which reacts with water.

   B a carbonate which decomposes readily to the oxide.

   C a peroxide when it reacts with oxygen.

   D a nitrate which decomposes on heating.

   E a hydroxide which does not decompose on heating.

8 Which **one** of the following metal ions in aqueous solution would give a white precipitate with aqueous sodium ethanedioate (oxalate), but no precipitate with aqueous potassium chromate(VI)?

   A Mg^{2+}

   B Ca^{2+}

   C Sr^{2+}

   D Ba^{2+}

   E Ra^{2+}

9   A solid compound of calcium was warmed with an equal volume of water evolving a gas which neither burned nor supported combustion. The formula of this compound could be

    A CaC_2

    B $CaCl_2$

    C CaH_2

    D Ca_3N_2

    E CaO_2

10 On descending group II from magnesium to barium, there is a steady increase in the solubility in water of the

    A   carbonates.

    B   hydroxides.

    C   phosphates(V).

    D   chromates(VI).

    E   sulphates(VI).

11 Magnesium does **not** form Mg^{3+} ions because, relatively speaking, it has

    A   a low second ionization energy.

    B   a low third ionization energy.

    C   a low fourth ionization energy.

    D   a high second ionization energy.

    E   a high third ionization energy.

12 Which **one** of the following compounds would be expected to have the highest degree of ionic character?

    A   caesium sulphide

    B   caesium selenide

    C   francium sulphide

    D   rubidium sulphide

    E   rubidium selenide

13 The alkali metals and the alkaline-earth metals are usually extracted by

    A   reduction of the carbonate with coke.

    B   electrolysis of the molten chloride.

    C   electrolysis of the oxide in molten cryolite.

    D   thermite reduction of the oxide with aluminium.

    E   roasting the carbonate in air.

14 Which **one** of the following sodium salts could decolorize an aqueous solution of iodine?

    A   sodium chloride

    B   sodium sulphate(VI)

    C   sodium thiosulphate

    D   sodium nitrate(V)

    E   sodium chromate(VI)

15 Which **one** of the following ions might be responsible for the hardness of a sample of tap water?

    A Cl^-

    B CO_3^{2-}

    C HCO_3^-

    D Mg^{2+}

    E SO_4^{2-}

TEST 15

Aluminium

This test is composed of fifteen questions. For each question, five possible answers are suggested. These answers are labelled **A**, **B**, **C**, **D** and **E**. Select the most appropriate **one** of the answers and write its corresponding letter on a separate answer sheet.

1 Which **one** of the following represents the ground-state electronic configuration of gallium, the element immediately below aluminium in group III?

A $1s^22s^22p^63s^23p^64s^14p^2$

B $1s^22s^22p^63s^23p^64s^24p^1$

C $1s^22s^22p^63s^23p^64d^{10}4s^24p^1$

D $1s^22s^22p^63s^23p^63d^{10}4s^24p^1$

E $1s^22s^22p^63s^23p^63d^{10}4s^24p^3$

2 Which **one** of the following minerals does **not** contain aluminium?

A emery

B mica

C anhydrite

D emerald

E beryl

3 The elements calcium and aluminium are similar in having

A amphoteric hydroxides.

B soluble hydroxides.

C insoluble sulphates.

D volatile chlorides.

E ionic fluorides.

4 The chemistry of Be^{2+}(aq) resembles that of Al^{3+}(aq). The principal reason for this is that Be^{2+} and Al^{3+} have similar

A ionic radii.

B charge densities.

C hydration energies.

D electron structures.

E solvation numbers.

5 What are the relative positions of the chlorine atoms around one of the aluminium atoms in a molecule of Al_2Cl_6?

A tetrahedral

B trigonal planar

C square planar

D octahedral

E pyramidal

6 Which of the following properties could refer to aluminium chloride?

	State at 20°C	b.pt.	pH of aqueous solution
A	solid	1400	7
B	solid	800	6
C	solid	200	3
D	solid	220	9
E	liquid	200	4

7 The elements in group III, considered in order of increasing atomic number, show a **decrease** in

A molar volume.

B electronegativity.

C ionic radius.

D nuclear charge.

E strength as reducing agents.

8 With which **one** of the following reagents does aluminium react most readily?

A concentrated H_2SO_4

B dilute HNO_3

C hot dilute NaOH

D concentrated NH_3

E concentrated $KMnO_4$

9 Aluminium is obtained from bauxite – hydrated aluminium oxide containing impurities such as iron(III) oxide and silicon(IV) oxide. On which **one** of the following essential features of aluminium oxide is the purification of bauxite based?

A its amphoteric nature

B its stability on heating

C its insolubility in water

D its conductivity when molten

E its insolubility in weak acids

10 Aluminium is obtained industrially by

A reducing aluminium oxide with coke.

B electrolysing molten aluminium chloride.

C electrolysing molten aluminium oxide.

D electrolysing aluminium oxide in molten cryolite.

E reducing aluminium oxide with iron.

11 Aluminium articles form only a thin coating of oxide on exposure to the atmosphere, but iron objects eventually corrode away completely. Which **one** of the following statements helps to explain this difference?

A Aluminium oxide is covalently bonded.

B Aluminium oxide forms a giant structure.

C Aluminium oxide is insoluble in water.

D Rust is readily penetrated by water.

E Rust is soluble in rain water.

Questions **12** to **15**

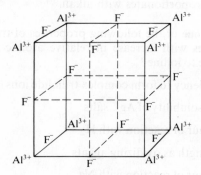

The diagram above represents a unit cell of aluminium fluoride.

12 What is the number of F^- ions in **one** unit cell?

A 12

B 6

C 4

D 3

E 1

13 If the volume of 1 mole of aluminium fluoride is V dm^3 and the volume of a unit cell is v dm^3, what is the value of the Avogadro constant, L?

A $\dfrac{V}{v} \times \dfrac{1}{12}$

B $\dfrac{V}{v} \times \dfrac{1}{8}$

C $\dfrac{V}{v} \times \dfrac{1}{3}$

D $\dfrac{V}{v}$

E $\dfrac{V}{v} \times 3$

14 What are the co-ordination numbers of the Al^{3+} ion and the F^- ion respectively?

A 1 and 3

B 3 and 1

C 6 and 2

D 8 and 12

E 12 and 8

15 In the crystalline structure of aluminium fluoride, the arrangement of Al^{3+} ions is

A tetrahedral.

B body-centred cubic.

C face-centred cubic.

D cubic close-packed.

E simple cubic.

TEST 16

Group VII – the halogens

This test is composed of seventeen questions. For each question, five possible answers are suggested. These answers are labelled **A, B, C, D** and **E**. Select the most appropriate **one** of the answers and write its corresponding letter on a separate answer sheet.

1   The halogens differ from the alkali metals in having some

   A insoluble compounds.

   B coloured compounds.

   C covalent compounds.

   D reactive compounds.

   E abundant compounds.

2   When bromine is prepared by warming a mixture of solid manganese(IV) oxide, solid sodium bromide and concentrated sulphuric acid, the liquid bromine which collects is contaminated with hydrogen bromide. Which **one** of the following techniques would purify the bromine most effectively?

   A Washing with water

   B Washing with conc. H_2SO_4

   C Ether extraction

   D Fractional distillation

   E Standing over NaOH(s)

3   Fluorine is manufactured by electrolytic rather than chemical oxidation because chemical oxidants

   A will not oxidize fluorides.

   B produce impure fluorine.

   C are difficult to control.

   D react with fluorine.

   E attack the reaction vessel.

4   Which **one** of the following statements helps to explain why sodium and fluorine react more vigorously than sodium and chlorine?

   A Fluorine is more volatile than chlorine.

   B Fluorine has a smaller nuclear charge than chlorine.

   C Fluorine has a larger first ionization energy than chlorine.

   D Fluorine has a smaller bond energy than chlorine.

   E Fluorine has a more positive electron affinity than chlorine.

5   When $Cl_2(g)$ is bubbled into cold dilute NaOH(aq) the principal products of the reaction are

   A NaCl and NaClO

   B NaClO and $NaClO_2$

   C NaCl and $NaClO_3$

   D NaClO and $NaClO_3$

   E NaCl and $NaClO_4$

6   Br^- is produced in the reaction between Br_2 and

   A Cl_2

   B Cl^-

   C HCl

   D I_2

   E I^-

7   Fluorine

   A readily forms fluorates(V).

   B cannot form a stable oxide.

   C forms an insoluble silver salt.

   D forms a compound with krypton.

   E disproportionates with alkali.

8   Which **one** of the following properties of the halogens **increases** with increase in relative atomic mass from fluorine to iodine?

   A tendency to form complex trihalide ions (Hal_3^-)

   B the solubility of Ag^+ salts

   C vigour of reaction with H_2

   D strength as oxidizing agents

   E vigour of reaction with Na

9   Which **one** of the following properties of the halogens **decreases** as relative atomic mass increases from fluorine to iodine?

   A absorption of visible light

   B melting point

   C metallic character

   D atomic radius

   E first ionization energy

10 Which **one** of the following reactions could be used to prepare gaseous hydrogen iodide?

   A Pass $H_2(g)$ over heated $I_2(s)$.

   B Warm concentrated HNO_3 with $I_2(s)$.

   C Warm concentrated H_3PO_4 with $NaI(s)$.

   D Warm concentrated H_2SO_4 with $NaI(s)$ and $MnO_2(s)$.

   E Heat $H_2O(g)$ and $I_2(g)$ to red heat.

11 The hydrogen halides

   A are coloured compounds.

   B form soluble lead salts.

   C increase in stability from HF to HI.

   D donate protons to water.

   E are liquids at 20°C.

12 When a solution of potassium bromide is treated with silver nitrate(V) solution it forms

   A a white precipitate soluble in dilute $NH_3(aq)$.

   B a white precipitate insoluble in dilute $NH_3(aq)$.

   C a cream precipitate soluble in dilute $NH_3(aq)$.

   D a cream precipitate insoluble in dilute $NH_3(aq)$.

   E a yellow precipitate insoluble in dilute $NH_3(aq)$.

13 Iodine exists in different oxidation states from -1 to $+7$. Which **one** of the following iodine-containing ions could **not** undergo disproportionation?

   A I^+

   B I^-

   C IO^-

   D ICl_2^+

   E IO_3^-

14 The measured electrode potential for the reaction $2Cl^-(aq) \rightleftharpoons Cl_2(g) + 2e^-$ is independent of

   A the temperature.

   B the size of the electrode.

   C the nature of the electrode.

   D the pressure of $Cl_2(g)$.

   E the purity of $Cl^-(aq)$.

15 Naturally-occurring chlorine is composed of two isotopes, $^{35}_{17}Cl$ and $^{37}_{17}Cl$, in the relative proportions $3:1$. What are the relative proportions of $^{35}_{17}Cl_2 : ^{35}_{17}Cl^{37}_{17}Cl : ^{37}_{17}Cl_2$ in gaseous chlorine?

   A $3:2:1$ **C** $6:3:1$ **E** $9:6:1$

   B $6:2:1$ **D** $9:3:1$

16 The mass of sodium thiosulphate-5-water ($Na_2S_2O_3 . 5H_2O$) required to make up $250\ cm^3$ of a $0.5\ M$ solution is ($Na = 23$, $S = 32$, $O = 16$, $H = 1$)

   A $158 \times 0.5 \times \dfrac{250}{1000}$ **D** $248 \times 0.5 \times 250$

   B $158 \times 0.5 \times \dfrac{1000}{250}$ **E** $248 \times \dfrac{1}{0.5} \times \dfrac{250}{1000}$

   C $248 \times 0.5 \times \dfrac{250}{1000}$

17 Which **one** of the following pairs of compounds contains halogen atoms whose combined oxidation numbers total $+8$?

   A $KClO_4$, $NaBrO_4$

   B $KClO_3$, $NaBrO_3$

   C ClO_2, I_2O_5

   D $NaClO_2$, HIO_3

   E $NaCl$, $KClO_4$

Group IV – carbon to lead, non-metal to metal

This test is composed of fourteen questions. For each question, five possible answers are suggested. These answers are labelled **A, B, C, D** and **E**. Select the most appropriate **one** of the answers and write its corresponding letter on a separate answer sheet.

1   Diamond and graphite are classed as allotropes because

   A they are different crystalline forms of the same element.

   B they have different melting points and boiling points.

   C their atoms contain different numbers of neutrons.

   D the bonding in their structures is different.

   E they form different oxides on burning in oxygen.

2   Graphite conducts electricity because

   A it is crystalline.

   B it contains an excess of electrons.

   C its atoms are close together.

   D it contains carbon ions.

   E its electrons are de-localized.

3   The half-life of radioactive $^{14}_{6}C$ is 5000 years. 12 g of this isotope will

   A contain 6×10^{23} atoms.

   B decay to 3 g in 10 000 years.

   C decay to 9 g in 2500 years.

   D disappear in 10 000 years.

   E eventually decay to $^{12}_{6}C$.

4   All the elements in group IV form

   A simple molecular chlorides.

   B amphoteric oxides.

   C more than one hydride.

   D stable hydrides.

   E ionic sulphides.

5   Which **one** of the following oxides is most stable on heating in air at 750°C?

   A CO

   B SiO

   C GeO

   D SnO

   E PbO

6   CO_2 and SiO_2 are both

   A gaseous at room temperature.

   B linear molecules.

   C soluble in dilute HCl(aq).

   D soluble in concentrated NaOH(aq).

   E amphoteric oxides.

7   In which **one** of the following equations is silicon(IV) oxide acting as a base?

   A $SiO_2 + 4HF \longrightarrow SiF_4 + 2H_2O$

   B $SiO_2 + 2KOH \longrightarrow K_2SiO_3 + H_2O$

   C $SiO_2 + CaO \longrightarrow CaSiO_3$

   D $SiO_2 + 4Mg \longrightarrow Mg_2Si + 2MgO$

   E $SiO_2 + 4F_2 \longrightarrow SiF_4 + 2F_2O$

8   As the atomic number of the group IV elements increases, their dioxides become

   A more covalent.

   B stronger oxidizing agents.

   C more acidic.

   D more stable to heat.

   E more soluble in water.

9   Which **one** of the following chlorides decomposes most readily on heating?

   A $SiCl_4$

   B $SnCl_2$

   C $SnCl_4$

   D $PbCl_2$

   E $PbCl_4$

10　Tin(IV) chloride

    A　is formed when hydrogen chloride reacts with molten tin.

    B　is a strong reducing agent at room temperature.

    C　conducts electricity in the liquid state.

    D　has a lower boiling point than tin(II) chloride.

    E　has more ionic character than tin(II) chloride.

11　Which **one** of the following equations correctly shows the reaction of a group IV tetrachloride with water?

    A　$CCl_4 + 2H_2O \longrightarrow CO_2 + 4HCl$

    B　$SiCl_4 + H_2O \longrightarrow SiOCl_2 + 2HCl$

    C　$GeCl_4 + H_2O \longrightarrow GeO + Cl_2 + 2HCl$

    D　$SnCl_4 + 4H_2O \longrightarrow Sn(OH)_4 + 4HCl$

    E　$3PbCl_4 + 4H_2O \longrightarrow Pb_3O_4 + 8HCl + 2Cl_2$

12　Which **one** of the following reagents will give a precipitate when added in excess to aqueous lead(II) nitrate?

    A　conc. HNO_3

    B　conc. HCl

    C　dil. HCl

    D　conc. $NaOH$

    E　dil. $NaOH$

13　What is the shape of the methyl carbonium ion, CH_3^+, with respect to atoms?

    A　pyramidal

    B　square planar

    C　tetrahedral

    D　trigonal planar

    E　V-shaped

14　As the atomic number of elements in group IV increases

    A　the hydride XH_4 becomes more stable.

    B　the oxidation state $+4$ becomes more stable.

    C　the element becomes more electronegative.

    D　the oxide XO becomes more acidic.

    E　the chloride XCl_2 has increasing ionic character.

The transition metals

This test is composed of eighteen questions. For each question, five possible answers are suggested. These answers are labelled **A**, **B**, **C**, **D** and **E**. Select the most appropriate **one** of the answers and write its corresponding letter on a separate answer sheet.

1   Which **one** of the following represents the correct electronic structure of an isolated Cr^{3+} ion?
   ((Ar) $\equiv$ electron structure of an argon atom)

   A $(Ar)3d^3 4s^1$

   B $(Ar)3d^2 4s^1$

   C $(Ar)3d^1 4s^2$

   D $(Ar)3d^4$

   E $(Ar)3d^3$

2   Which **one** of the following provides the most satisfactory definition of a transition metal?
   A transition metal is an element which

   A occurs in the 'd-block' of the periodic table.

   B has variable oxidation state in its compounds.

   C forms an ion with a partially filled 'd sub-shell'.

   D has similar properties to its neighbours in the periodic table.

   E lies between scandium and zinc in the periodic table.

3   Transition metals are the only elements which

   A form complex ions with hydroxide ions.

   B form coloured ions in aqueous solution.

   C form amphoteric oxides and hydroxides.

   D form covalently-bonded anhydrous chlorides.

   E show an oxidation number higher than 4.

4   Compared to the 's-block' metals, transition metals generally form

   A oxides which are more basic.

   B oxides which are more soluble.

   C salts which are more stable.

   D salts which are more ionic.

   E ions which are reduced more easily.

5   Which **one** of the following ions has the smallest radius?

   A Cr^{3+}

   B Ni^{2+}

   C Ni^{3+}

   D Ti^{2+}

   E Ti^{3+}

6   Which **one** of the following ions will catalyse the reaction of $I^-(aq)$ with $S_2O_8^{2-}(aq)$ least effectively?

   A Cr^{3+}

   B Fe^{3+}

   C Sc^{3+}

   D Ti^{3+}

   E V^{3+}

7   Which **one** of the following metals forms the complex ions $[M(NH_3)_4]^{2+}$ and $[M(NH_3)_2]^+$?

   A Cu

   B Ag

   C Sc

   D Pb

   E Zn

8   Which **one** of the following pairs of ions can form isomorphous salts?

   A Fe^{2+}, Fe^{3+}

   B $Cr_2O_7^{2-}$, CrO_4^{2-}

   C MnO_4^{2-}, MnO_4^-

   D CrO_4^{2-}, MnO_4^{2-}

   E Fe^{3+}, Mn^{2+}

9   The three commonest oxidation states of **manganese** are

   A $+2, +3, +4$

   B $+2, +3, +6$

   C $+2, +4, +7$

   D $+3, +4, +7$

   E $+3, +6, +7$

10 One of the isomers of $CrCl_3 . 6H_2O$ dissolves in water forming a solution from which only one third of the total chloride present is precipitated by $AgNO_3(aq)$. Which **one** of the following represents the correct structure for the complexed chromium ion in the aqueous solution?

A $[Cr(H_2O)_6]^{3+}$

B $[CrCl_2]^+$

C $[CrCl]^{2+}$

D $[Cr(H_2O)_5Cl]^{2+}$

E $[Cr(H_2O)_4Cl_2]^+$

Questions **11** to **18** refer to the five transition elements labelled **A** to **E** below.

A Cr
B Co
C Ni
D Ag
E V

Which **one** of these transition elements

11 has the greatest number of oxidation states?

12 most resembles the element with electronic structure 2, 8, 18, 15, 2?

13 is used as a catalyst in the hardening of vegetable oils?

14 might be left in the sludge at the bottom of a cell after the electrolytic refining of copper?

15 forms a complex anion of formula XO_3^-?

16 forms complex ions in which the cation normally has a co-ordination number of two?

17 has the smallest density?

18 forms compounds with similar properties to analogous sulphur compounds?

TEST 19

Metals and the activity series

This test is composed of twenty questions. For each question five possible answers are suggested. These answers are labelled **A, B, C, D** and **E**. Select the most appropriate **one** of the answers and write its corresponding letter on a separate answer sheet.

1 Which **one** of the following **never** occurs as the uncombined element in the earth's crust?

A carbon

B copper

C lead

D silver

E sulphur

2 Which **one** of the following statements helps to explain why limestone ($CaCO_3$) is added to the blast furnace during the manufacture of iron?

A $CaCO_3$ forms an acidic gas on decomposition.

B $CaCO_3$ forms a basic oxide on decomposition.

C $CaCO_3$ provides a refractory lining for the furnace.

D $CaCO_3$ floats on the hot molten iron.

E $CaCO_3$ increases the furnace temperatures.

3 Which **one** of the following metals is used as a protective layer and as a sacrificial metal for iron and steel objects?

A copper

B magnesium

C zinc

D tin

E nickel

4 Copper is extracted from copper pyrites ($CuFeS_2$) by roasting the concentrated ore in air to form copper(I) sulphide. The impure copper(I) sulphide is then heated in air forming copper(I) oxide. The copper(I) oxide is finally reduced to copper by

A heating strongly in the absence of air.

B heating with unchanged Cu_2S in the absence of air.

C heating with silica in a closed furnace.

D heating with coke in the absence of air.

E heating with coke in a blast furnace.

5 In which **one** of the following formulae does copper show the highest oxidation state?

A $CuFeS_2$

B $CuCl.CO.2H_2O$

C H_2CuCl_3

D $Cu(NH_3)Cl . H_2O$

E $K_3Cu(CN)_4$

6 Unlike the compounds and complex ions of most transition metals, those of the copper(I) ion are often white or colourless. The lack of colour in these compounds may be explained by the fact that the Cu^+ ion has

A no unpaired electrons.

B no electrons in the fourth quantum shell.

C no unfilled d-orbitals.

D no delocalized electrons.

E d-orbitals of equal energy.

Questions **7** to **13** refer to the five methods (**A, B, C, D** and **E**) listed below which are used to extract metals from their ores.

A electrolysis of the molten chloride
B electrolysis of a molten mixture containing the oxide
C reduction of the oxide with coke
D displacement from aqueous solution
E reduction of the oxide or chloride with a more reactive metal

Which of these methods is used industrially to obtain

7 aluminium

8 barium

9 magnesium

10 silver

11 tin

12 titanium

13 zinc

Questions **14** to **20** refer to the five ions labelled **A, B, C, D** and **E** below.

A $Ca^{2+}(aq)$
B $Cu^{2+}(aq)$
C $Fe^{2+}(aq)$
D $Fe^{3+}(aq)$
E $Pb^{2+}(aq)$

Which **one** of these aqueous ions

14 gives a precipitate with NaOH(aq) which dissolves in excess?

15 gives a precipitate with NH_3(aq) which dissolves in excess?

16 gives a precipitate with both dilute H_2SO_4(aq) and dilute HCl(aq)?

17 is the weakest oxidizing agent?

18 gives a blood red solution with NCS^-(aq)?

19 gives a dark blue precipitate with $[Fe(CN)_6]^{3-}$(aq)?

20 gives a cream precipitate and a brown solution with I^-(aq)?

This test is composed of eighteen questions. For each question five possible answers are suggested. These answers are labelled **A, B, C, D** and **E**. Select the most appropriate **one** of the answers and write its corresponding letter on a separate answer sheet.

1 The value of K, the partition coefficient for the distribution of a solute between two immiscible solvents changes when

  A the volumes of the solvents change.

  B the concentration of solute in one solvent approaches zero.

  C the volumes of the solvents differ greatly.

  D the mass of solute changes.

  E the temperature of the system changes.

2 The partition coefficient of a solute W between solvents X and Y is a. This statement can be written as

  A $\dfrac{[W \text{ in } X]}{[W \text{ in } Y]} = a$ **D** $\dfrac{[X \text{ in } W]}{[Y \text{ in } W]} = a$

  B $\dfrac{[W \text{ in } X]}{[W \text{ in } Y]} = \dfrac{1}{a}$ **E** $\dfrac{[X \text{ in } W]}{[Y \text{ in } W]} = \dfrac{1}{a}$

  C $\dfrac{[W \text{ in } Y]}{[W \text{ in } X]} = a$

3 The partition coefficient of solute X between tetrachloromethane and water is 4. If a solution of 20 g of X in 100 cm^3 of water is extracted with two 100 cm^3 portions of tetrachloromethane in succession, what mass of X is removed by the tetrachloromethane?

  A 4·8 g   **D** 18·75 g

  B 8·75 g   **E** 19·2 g

  C 16·0 g

4 Which **one** of the following systems is **not** in dynamic equilibrium at room temperature?

  A iodine in a closed container

  B a strip of copper in $CuSO_4$(aq)

  C solid NaCl in saturated NaCl(aq)

  D a cell through which a constant current passes

  E gaseous NO_2 in a closed container

5 Which **one** of the following gases will most closely obey Henry's Law when the solvent is water?

  A NH_3 **D** H_2S

  B CH_4 **E** CO_2

  C SO_2

6 Which **one** of the following will change the value of K_c?

  A change in concentration

  B change in pressure

  C change in temperature

  D employing a catalyst

  E adding more reactant

7 At a particular temperature, the equilibrium constant for the reaction $P + Q \rightleftharpoons R + S$ is 10^{10}. From this information we can deduce that P and Q react

  A rapidly to form a high proportion of R and S at equilibrium.

  B slowly to form a high proportion of R and S at equilibrium.

  C at an unknown rate to form a high proportion of R and S at equilibrium.

  D rapidly to form a low proportion of R and S at equilibrium.

  E slowly to form a low proportion of R and S at equilibrium.

8 What are the units of K_p for the reaction represented by the equation $NO(g) + \frac{1}{2}O_2(g) \rightleftharpoons NO_2(g)$?

  A atm$^{-1/2}$ **D** atm^{-1}

  B atm$^{1/2}$ **E** atm$^{1\frac{1}{2}}$

  C atm

9 The equilibrium constant, K_c, for the reaction

$$N_2(g) + 3H_2(g) \rightleftharpoons 2NH_3(g)$$

is 2 mole^{-2} dm^6 at 620 K. What is the concentration of NH_3 at equilibrium at 620 K, when the equilibrium concentrations of both N_2 and H_2 are 2 mole dm^{-3}?

  A 2 mole dm^{-3} **D** $\sqrt{32}$ mole dm^{-3}

  B $\sqrt{8}$ mole dm^{-3} **E** 32 mole dm^{-3}

  C 4 mole dm^{-3}

10 The equilibrium constant for the reaction represented by the following equation is 4.

$$W(g) + X(g) \rightleftharpoons Y(g) + Z(g)$$

How many moles of X are present in the equilibrium mixture formed when one mole of W is mixed with one mole of X?

A 2

B $1\frac{1}{2}$

C $\frac{3}{4}$

D $\frac{2}{3}$

E $\frac{1}{3}$

11 The numerical value of K_P for the reaction

$$X_2(g) + 3Y_2(g) \rightleftharpoons 2XY_3(g)$$

is 10^2 at a particular temperature.
What is the numerical value of K_P for the reaction

$$XY_3(g) \rightleftharpoons \tfrac{1}{2}X_2(g) + \tfrac{3}{2}Y_2(g)$$

at the same temperature?

A $\dfrac{1}{10^2} \times \dfrac{1}{2}$

B $\dfrac{1}{10}$

C $\dfrac{1}{10^2}$

D $\dfrac{-10^2}{2}$

E 10^2

12 The equilibrium constants for the following reactions were measured at a certain temperature.

$$CO(g) + H_2O(g) \rightleftharpoons CO_2(g) + H_2(g) \quad K_P = \tfrac{4}{10}$$

$$CO(g) + \tfrac{1}{2}O_2(g) \rightleftharpoons CO_2(g) \quad K_P = 10^6 \text{ atm}^{-1/2}$$

What is the numerical value of K_P for the reaction below at the same temperature?

$$H_2O(g) \rightleftharpoons H_2(g) + \tfrac{1}{2}O_2(g)$$

A 4×10^5

B $\dfrac{1}{4 \times 10^5}$

C $\dfrac{4}{10^7}$

D $\dfrac{10^7}{4}$

E It is not calculable from the data given

13 The partial pressures of SO_2, O_2 and SO_3 in equilibrium at a particular temperature are

$$P_{SO_2} = 4\tfrac{1}{2} \text{ atm}, \quad P_{O_2} = P_{SO_3} = 2\tfrac{1}{4} \text{ atm}.$$

What is K_P, for the reaction,

$$2SO_2(g) + O_2(g) \rightleftharpoons 2SO_3(g),$$

at this temperature?

A $\frac{1}{9}$ atm^{-1}

B $\frac{1}{9}$ atm

C $\frac{2}{9}$ atm^{-1}

D $\frac{2}{9}$ atm

E $\frac{9}{2}$ atm^{-1}

14 When nickel is heated with carbon monoxide at 70°C, the following equilibrium is established.

$$Ni(s) + 4CO(g) \rightleftharpoons Ni(CO)_4(g)$$

What is the equilibrium constant, K_c, for this reaction?

A $\dfrac{[Ni(s)][CO(g)]^4}{[Ni(CO)_4(g)]}$

B $\dfrac{[Ni(CO)_4(g)]}{[Ni(s)][CO(g)]^4}$

C $\dfrac{[CO(g)]^4}{[Ni(CO)_4(g)]}$

D $\dfrac{[Ni(CO)_4(g)]}{[CO(g)]^4}$

E $\dfrac{[Ni(CO)_4(g)]}{[CO(g)]}$

Questions **15** and **16** refer to the following equilibrium.

$$[Ag(NH_3)_2]^+(aq) + I^-(aq) \rightleftharpoons AgI(s) + 2NH_3(aq)$$

moles a b c d
at equil.

15 The equilibrium constant, K_c, for this reaction is

A $\dfrac{ab}{cd^2}$

B $\dfrac{d}{ab}$

C $\dfrac{ab}{d^2}$

D $\dfrac{cd^2}{ab}$

E $\dfrac{d^2}{ab}$

16 When concentrated $NH_3(aq)$ is added to the equilibrium mixture, which **one** of the following will be unchanged once equilibrium is restored?

A the concentration of $[Ag(NH_3)_2]^+(aq)$

B the concentration of $I^-(aq)$

C the concentration of $NH_3(aq)$

D the mass of AgI

E the value of K_c

Questions **17** and **18**. A mixture, initially containing 2 moles of $CO(g)$ and 1 mole of $Cl_2(g)$, reached equilibrium with $COCl_2(g)$ when 75% of the Cl_2 had reacted.

17 If the total final pressure was P, what is the partial pressure of $COCl_2(g)$ in the equilibrium mixture?

A $\frac{1}{4}P$

B $\frac{1}{3}P$

C $\frac{3}{4}P$

D $\frac{3}{2}P$

E $\frac{9}{4}P$

18 What is the equilibrium constant, K_P, for this reaction?

A $\dfrac{3}{4} P$ atm^{-1}

B $\dfrac{12}{5P}$ atm^{-1}

C $\dfrac{27}{5P}$ atm^{-1}

D $\dfrac{3}{5P}$ atm^{-1}

E $\dfrac{12}{5}$ atm^{-1}

TEST 21
Factors affecting equilibria

This test is composed of twenty questions. For each question, five possible answers are suggested. These answers are labelled **A**, **B**, **C**, **D** and **E**. Select the most appropriate **one** of the answers and write its corresponding letter on a separate answer sheet.

1 The concentration of $PCl_5(g)$ in the equilibrium,

$$PCl_3(g) + Cl_2(g) \rightleftharpoons PCl_5(g)$$

is increased by

A adding phosphorus to the equilibrium mixture.

B decreasing the total pressure.

C adding a noble gas to the equilibrium mixture.

D decreasing the volume of the container.

E using an appropriate catalyst.

2 In which of the following equilibria will an increase in pressure at constant temperature increase the yield of products on the right-hand side of the equation?

A $NO_2(g) \rightleftharpoons NO(g) + \frac{1}{2}O_2(g)$

B $BaCO_3(s) \rightleftharpoons BaO(s) + CO_2(g)$

C $H_2(g) + Br_2(g) \rightleftharpoons 2HBr(g)$

D $3Fe(s) + 4H_2O(g) \rightleftharpoons Fe_3O_4(s) + 4H_2(g)$

E $4NO(g) + 6H_2O(g) \rightleftharpoons 4NH_3(g) + 5O_2(g)$

3 In the endothermic homogeneous gas-phase reaction

$$P + Q \rightleftharpoons R + S,$$

the yield of R at equilibrium is increased by

A raising the total pressure.

B using a suitable catalyst.

C reducing the volume of the container.

D removing S as it is produced.

E lowering the temperature.

4 For a certain reaction, K_P at 300 K is 1·0 and K_P at 600 K is 2·0. From this, we can deduce that

A K_P at 450 K is 1·5.

B increase in pressure favours the formation of products.

C the reaction is endothermic.

D K_P increases with increase in pressure.

E there is a decrease in volume on reaction.

5 In an equilibrium reaction, a positive catalyst increases

A the rate of the reverse reaction.

B the kinetic energy of the reacting particles.

C the equilibrium constant of the reaction.

D the activation energy of the reaction.

E the exothermicity of the reaction.

6 Positive catalysts

A are unchanged physically at the end of a reaction.

B are unchanged chemically throughout a reaction.

C affect the position of equilibrium of a reaction.

D are involved in the mechanism of a reaction.

E require the presence of a promoter.

7 Exothermic reactions always have

A fast reaction rates.

B small activation energies.

C large equilibrium constants.

D negative enthalpy changes.

E unstable reactants.

8 For the reaction, $N_2(g) + 3H_2(g) \rightleftharpoons 2NH_3(g)$; $\Delta H = -92$ kJ. Increasing the temperature at constant pressure will increase

A the yield of ammonia at equilibrium.

B the time to reach equilibrium.

C the rates of forward and backward reactions.

D the value of the equilibrium constant.

E the partial pressure of ammonia at equilibrium.

9 Which **one** of the following sets of conditions would give the highest yield of $CO_2(g)$ from $CaCO_3(s)$ in the equilibrium below?

$$CaCO_3(s) \rightleftharpoons CaO(s) + CO_2(g); \quad \Delta H = +178 \text{ kJ}$$

A 1000°C and 10 atm.   **D** 500°C and 1 atm.

B 1000°C and 1 atm.   **E** 100°C and 1 atm.

C 500°C and 10 atm.

10 The synthesis of ammonia from nitrogen and hydrogen is exothermic. During the manufacture of ammonia, this reaction is carried out at only moderate temperatures because high temperatures

A reduce the efficiency of the catalyst.

B reduce the yield of ammonia at equilibrium.

C are very difficult to maintain.

D result in a reaction rate which is too fast.

E are attained once the reaction is underway.

11 $2SO_2(g) + O_2(g) \rightleftharpoons 2SO_3(g); \quad \Delta H = -197 \text{ kJ}$

Which **one** of the following can be deduced from the information above?

A the qualitative effect of temperature on the equilibrium mixture

B the composition of the equilibrium mixture

C the ideal reaction conditions

D the fact that a catalyst is necessary

E the activation energy of the reaction

12 The value of the equilibrium constant for the reaction

$$H_2(g) + I_2(g) \rightleftharpoons 2HI(g)$$

can be altered by

A adding iodine to the system.

B adding oxygen to the system.

C increasing the total pressure.

D increasing the temperature.

E adding a suitable catalyst.

13 Assuming that K_1 and K_2 are the respective equilibrium constants for the reactions

$$H_2(g) + Cl_2(g) \rightleftharpoons 2HCl(g)$$

and

$$2H_2(g) + O_2(g) \rightleftharpoons 2H_2O(g)$$

under given conditions, what is the equilibrium constant for the reaction

$$4HCl(g) + O_2(g) \rightleftharpoons 2H_2O(g) + 2Cl_2(g)$$

under the same conditions?

A $\dfrac{2K_1}{K_2}$ C $\dfrac{K_2}{2K_1}$ E $\dfrac{K_2}{K_1^2}$

B $\dfrac{K_2}{K_1}$ D $\dfrac{K_1^2}{K_2}$

14 Sulphuric acid is used in cleaning steel because it

A dissolves grease.   D renders iron passive.

B reacts with iron.   E prevents rusting.

C removes rust.

15 Which **one** of the following substances can be used to dry moist ammonia?

A conc. sulphuric acid

B anhydrous copper(II) sulphate

C calcium oxide

D phosphorus pentoxide

E glacial acetic acid

16 All explosives

A contain oxygen.

B decompose spontaneously.

C liberate gases on explosion.

D contain carbon compounds.

E are made from concentrated HNO_3.

Questions **17** to **20**

17 Ethene may be obtained by the catalytic cracking of ethane.

$$C_2H_6(g) \rightleftharpoons C_2H_4(g) + H_2(g); \quad \Delta H = +137 \text{ kJ}$$

$K_c = 10^{-5} \text{ mole dm}^{-3}$

The proportion of ethane converted to ethene at equilibrium could be increased by

A raising the initial pressure of ethane.

B using a more efficient catalyst.

C raising the temperature.

D reducing the volume of the container.

E mixing hydrogen with ethane before reaction.

18 What is the approximate equilibrium concentration of ethene in mole dm^{-3}, if the initial concentration of ethane is 10^{-1} mole dm^{-3} and the initial concentrations of hydrogen and ethene are both zero?

A 10^{-6} B $\frac{1}{2} \times 10^{-6}$ C 10^{-4} D 10^{-3} E 10^{-2}

19 In order to obtain a correct result in the last question, it is necessary to assume that

A only a small proportion of ethane has reacted.

B almost all the ethane has formed ethene.

C the equilibrium constant is effectively zero.

D one mole of gas occupies 22.4 dm^3 at s.t.p.

E the catalyst exerts zero pressure.

20 If the initial concentration of ethane were increased from 0.1 mole dm^{-3} to 0.2 mole dm^{-3}, the concentration of ethene at equilibrium would change by a factor of about

A $\dfrac{1}{2}$ B $\dfrac{1}{\sqrt{2}}$ C $\sqrt{2}$ D 2   E 4

Ionic equilibria in aqueous solution

This test is composed of twenty questions. For each question, five possible answers are suggested. These answers are labelled **A**, **B**, **C**, **D** and **E**. Select the most appropriate **one** of the answers and write its corresponding letter on a separate answer sheet.

1 What is the expression for the solubility product of silver chromate(VI)?

A $2[Ag^+][CrO_4^{2-}]$

B $4[Ag^+]^2[CrO_4^{2-}]$

C $[Ag^+]^2[CrO_4^{2-}]$

D $\dfrac{2[Ag^+][CrO_4^{2-}]}{[Ag_2CrO_4]}$

E $\dfrac{[Ag^+]^2[CrO_4^{2-}]}{[Ag_2CrO_4]}$

2 If the solubility of calcium carbonate, $CaCO_3$, is s mole dm^{-3}, what is the numerical value of its solubility product at the same temperature?

A $\sqrt{s}$ D s^2

B s E $4s^2$

C $\dfrac{s^2}{4}$

3 What is the solubility of lead chloride ($PbCl_2$) in water at 298 K, if $K_{sp} = 2 \times 10^{-5}$ $mole^3$ dm^{-9}?

A $\dfrac{2 \times 10^{-5}}{3}$ mole dm^{-3}

B $\sqrt{2 \times 10^{-5}}$ mole dm^{-3}

C $\sqrt[3]{2 \times 10^{-5}}$ mole dm^{-3}

D $\sqrt[3]{\dfrac{2 \times 10^{-5}}{2}}$ mole dm^{-3}

E $\sqrt[3]{\dfrac{2 \times 10^{-5}}{4}}$ mole dm^{-3}

4 If the solubility of bismuth sulphide (Bi_2S_3) in water at a given temperature is x mole dm^{-3}, what is the solubility product of bismuth sulphide at the same temperature?

A x^5 D $(2x)^2(3x)^3$

B $5x^5$ E $(5x)^5$

C $(2x)^5$

5 The solubility product of aluminium hydroxide is 1×10^{-33} $mole^4$ dm^{-12}. What is the maximum $[OH^-]$ in mole dm^{-3}, in a solution in which $[Al^{3+}]$ is 10^{-9} mole dm^{-3}?

A $\dfrac{10^{-24}}{3}$ D $\dfrac{10^{-8}}{3}$

B 10^{-24} E 10^{-8}

C 10^{-12}

6 The solubility of lead sulphate in pure water at 298 K is $1 \cdot 20 \times 10^{-4}$ mole dm^{-3}. What is the solubility of lead sulphate in $1 \cdot 00$ M H_2SO_4 at this temperature?

A $1 \cdot 44 \times 10^{-8}$ D $1 \cdot 00 - 1 \cdot 2 \times 10^{-4}$

B $1 \cdot 20 \times 10^{-4}$ E $1 \cdot 00$

C $\sqrt{1 \cdot 2 \times 10^{-4}}$

7 The solubility products of four metal sulphates in $mole^2$ dm^{-6} are $K_{sp}(BaSO_4) = 1 \times 10^{-10}$, $K_{sp}(PbSO_4) = 2 \times 10^{-8}$, $K_{sp}(SrSO_4) = 3 \times 10^{-7}$ and $K_{sp}(CaSO_4) = 9 \times 10^{-6}$. Which of these metal sulphates will be precipitated if a $0 \cdot 0001$ M solution of sulphate ions, SO_4^{2-}, is mixed with an equal volume of a $0 \cdot 0001$ M solution of the appropriate metal ions?

A $BaSO_4$ only

B $BaSO_4$ and $PbSO_4$ only

C $BaSO_4$, $PbSO_4$ and $SrSO_4$ only

D $CaSO_4$ and $SrSO_4$ only

E $CaSO_4$, $SrSO_4$ and $PbSO_4$ only

8 The solubility of AgI will be greater in

A $KNO_3(aq)$ than in water.

B $KI(aq)$ than in water.

C $AgNO_3(aq)$ than in water.

D water than in $KNO_3(aq)$.

E water than in $AgNO_3(aq)$.

9 If the pH of a strong monobasic acid is $2 \cdot 17$, what is the concentration of the acid in mole dm^{-3}?

A between $0 \cdot 0001$ and $0 \cdot 001$

B between $0 \cdot 001$ and $0 \cdot 01$

C between $0 \cdot 01$ and $0 \cdot 10$

D between $0 \cdot 10$ and $1 \cdot 00$

E greater than $1 \cdot 00$

10 What is the pH of a solution of sulphuric acid containing 0.2 mole dm^{-3}? (Assume the acid is completely dissociated and take $\lg 2 = 0.3$)

A 0.03

B 0.33

C 0.40

D 0.60

E 1.60

11 The pH of a 0.10 M solution of a weak alkali could be

A 4

B 6

C 9

D 12

E 14

12 The pH of an alkaline solution is 8. Which **one** of the following expressions is correct?

A $[OH^-] = 10^{-8}$

B $\lg[OH^-] = 8$

C $-\lg[OH^-] = 8$

D $-\lg[H^+] = 6$

E $\lg[H^+] = -8$

13 At 298 K, the numerical value of the ionic product of water, K_w, is 10^{-14}. The corresponding value of the equilibrium constant for
$H_2O + H_2O \rightleftharpoons H_3O^+ + OH^-$ is

A $10^{-14} \times \dfrac{18}{1000} \times \dfrac{18}{1000}$

B $10^{-14} \times \dfrac{18}{1000}$

C $10^{-14} \times \dfrac{18}{1000} \times 2$

D $10^{-14} \times \dfrac{1000}{18}$

E $10^{-14} \times \dfrac{1000}{18} \times \dfrac{1000}{18}$

14 The dissociation constant of butanoic acid ($CH_3CH_2CH_2COOH$) at 298 K is 1.5×10^{-5} mole dm^{-3}. What is the approximate hydrogen ion concentration, in mole dm^{-3}, of a 0.1 M solution of butanoic acid at 298 K?

A $\sqrt{1.5 \times 10^{-4}}$

B $\sqrt{1.5 \times 10^{-6}}$

C 1.5×10^{-4}

D 1.5×10^{-5}

E 1.5×10^{-6}

15 1.0 M NaOH is added to $100 \ cm^3$ of 1.0 M CH_3COOH. How many cm^3 of 1.0 M NaOH must be added before the maximum buffering effect occurs?

A $33\frac{1}{3}$

B 50

C $66\frac{2}{3}$

D 75

E 150

16 The pH ranges and colour changes for two acid–base indicators are given below.

Indicator	pH range
Congo red	violet 3.0–5.0 red
Bromothymol blue	yellow 6.0–7.6 blue

A solution in which congo red is violet and bromothymol blue is yellow is

A strongly acidic

B weakly acidic

C neutral

D weakly alkaline

E strongly alkaline

17 The indicator, HX, is a weak acid.

$$HX(aq) \rightleftharpoons H^+(aq) + X^-(aq)$$
$$K_a(HX) = 10^{-10} \ \text{mole} \ dm^{-3}$$

When HX is used in the titration of a strong acid with a strong base, the indicator will change colour when

A $[H^+] = 10^{-10}$ mole dm^{-3}

B $[H^+] = 10^{-7}$ mole dm^{-3}

C $[H^+] = 10^{-5}$ mole dm^{-3}

D $[H^+] = [X^-]$

E $[H^+] = [HX]$

Questions **18** to **20** refer to the graph below which shows how the pH of a solution changes during a certain titration.

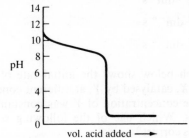

18 At which pH does the maximum buffering effect occur?

A 2

B 5

C 8

D 9

E 11

19 To which **one** of the following titrations does the curve relate?

A strong base with strong acid

B strong base with weak acid

C weak base with strong acid

D weak base with weak acid

E weak base with concentrated acid

20 If the initial concentration of the base is 0.1 M, what is the approximate value of its dissociation constant? (Assume the base is monoacidic.)

A 10^{-2}

B 10^{-4}

C 10^{-5}

D 10^{-6}

E 10^{-7}

TEST 23
Reaction rates

This test is composed of fifteen questions. For each question, five possible answers are suggested. These answers are labelled **A**, **B**, **C**, **D** and **E**. Select the most appropriate **one** of the answers and write its corresponding letter on a separate answer sheet.

1 The velocity constant of a reaction is

   A independent of reactant concentrations.

   B independent of the activation energy.

   C proportional to the absolute temperature.

   D proportional to the reaction rate.

   E unaffected by the presence of a catalyst.

2 The rate of a reaction between aqueous solutions of M and N is given by   Rate $= k[M][N]$. What are the units of k?

   A mole $dm^{-3}\ s^{-1}$

   B $mole^{-1}\ dm^3\ s$

   C $mole^{-1}\ dm^3\ s^{-1}$

   D $mole^2\ dm^{-6}\ s$

   E $mole^2\ dm^{-6}\ s^{-1}$

3 The graph below shows the initial rate of decomposition of X, catalysed by Y, at different concentrations of X. The concentration of Y was constant in all experiments. Which **one** of the following sets of conclusions is correct?

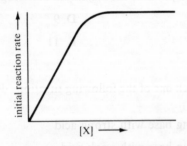

	Order with respect to X at low concentration of X	Order with respect to X at high concentration of X
A	0	1
B	1	0
C	1	2
D	2	1
E	2	0

4 On which **one** of the following values will the initial rate of a reaction depend?

   A ΔH

   B ΔS

   C E_A

   D $E^{\ominus}$

   E K_c

5 The half-life of bismuth-212 is 1 hour. If 12 g are allowed to decay, what mass of bismuth-212 is left after 3 hours?

   A 9 g

   B 4 g

   C 3 g

   D $1\frac{1}{2}$ g

   E 0 g

6 The activation energy of a reaction is

   A the minimum energy change for the forward reaction.

   B the minimum potential energy of the activated complex.

   C the minimum kinetic energy of colliding particles.

   D the minimum energy required to start a reaction.

   E the minimum energy needed by colliding molecules for a reaction.

7 In a reaction between substances V and W, some collisions between molecules of V and W do not result in reaction. Which **one** of the following suggestions may explain this?

   A The system has already reached equilibrium.

   B The activation energy of the back reaction is very low.

   C The molecules of V and W may collide inappropriately.

   D The activated complex is very unstable.

   E The molecules of V and W react endothermically.

8 The reaction between L and M,

$$L(g) + M(g) \longrightarrow LM(g),$$

is zero order with respect to L. It may be inferred from this that

A the reaction rate is constant.

B the reaction is first order with respect to M.

C the reaction takes place in one step.

D the rate determining step does not involve L.

E L is present in large excess.

9 P and Q react according to the equation

$$P + 2Q \longrightarrow PQ_2$$

The reaction is zero order with respect to P and second order with respect to Q. Which **one** of the following suggestions would be a possible mechanism for the reaction?

A $P + Q \xrightarrow{\text{fast}} PQ$ $PQ + Q \xrightarrow{\text{slow}} PQ_2$

B $P + Q \xrightarrow{\text{slow}} PQ$ $PQ + Q \xrightarrow{\text{fast}} PQ_2$

C $Q + Q \xrightarrow{\text{fast}} Q_2$ $Q_2 + P \xrightarrow{\text{slow}} PQ_2$

D $Q + Q \xrightarrow{\text{slow}} Q_2$ $Q_2 + P \xrightarrow{\text{fast}} PQ_2$

E $P + Q + Q \xrightarrow{\text{slow}} PQ_2$

Questions **10** to **12**

Two gases, X and Y, react according to the stoichiometric equation

$$X(g) + 3Y(g) \longrightarrow XY_3(g)$$

The following table gives the results of five experiments carried out at 350 K in order to determine the order of the reaction.

Expt. Number	Initial concentration of X/mole dm^{-3}	Initial concentration of Y/mole dm^{-3}	Initial rate of formation of XY_3/mole dm^{-3} s^{-1}
1	0·10	0·10	0·01
2	0·10	0·20	0·04
3	0·10	0·30	0·09
4	0·20	0·10	0·02
5	0·30	0·10	0·03

10 What is the order of the reaction between X and Y with respect to Y?

A 0   **D** 3

B 1   **E** 4

C 2

11 What is the initial rate of formation of XY_3, in mole dm^{-3} s^{-1}, when the initial concentration of X is 0·05 M and that of Y is 0·2 M?

A 0·005

B 0·01

C 0·02

D 0·04

E 0·08

12 By what factor does the reaction rate increase if the concentrations of both X and Y are trebled?

A 6   **D** 18

B 9   **E** 27

C 12

Questions **13** to **15**

A substance R undergoes a first order reaction forming the product S. The equation for the reaction is

$$R \underset{k_{-1}}{\overset{k_1}{\rightleftharpoons}} S$$

(k_1 and k_{-1} are the velocity constants for the forward and backward reactions respectively.)

13 Which **one** of the following expressions represents the rate of the **forward** reaction at any instant?

A $k_1([R] - [S])$

B $k_1 - k_{-1}$

C $k_1[R]$

D $(k_1 - k_{-1})[R]$

E $k_1[R] - k_{-1}[S]$

14 Which **one** of the following expressions represents the **overall** rate of the reaction at any instant?

A $k_1([R] - [S])$

B $k_1 - k_{-1}$

C $k_1[R]$

D $(k_1 - k_{-1})[R]$

E $k_1[R] - k_{-1}[S]$

15 Which **one** of the following conditions will apply when the reaction is at equilibrium?

A $k_1 = k_{-1}$

B $[R] = [S]$

C $\dfrac{[R]}{[S]} = 1$

D $k_1[R] = 0$

E $k_1[R] = k_{-1}[S]$

Introduction to carbon chemistry

This test is composed of sixteen questions. For each question, five possible answers are suggested. These answers are labelled **A**, **B**, **C**, **D** and **E**. Select the most appropriate **one** of the answers and write its corresponding letter on a separate answer sheet.

1   Silicon cannot form a stable series of hydrides (SiH_4, Si_2H_6, etc.) similar to carbon, because compounds containing silicon and hydrogen, unlike those of carbon and hydrogen

  A are energetically unstable in the presence of oxygen.

  B are kinetically unstable in the presence of oxygen.

  C react exothermically with oxygen.

  D form a solid product on reaction with oxygen.

  E decompose easily at room temperature and pressure.

2   To which of the following homologous series does the compound of molecular formula CH_2O belong?

  A alcohols

  B ethers

  C ketones

  D aldehydes

  E carboxylic acids

3   Which **one** of the following compounds contains the typical functional groups of both alkenes and ethers?

  A $HC{\equiv}C{-}CH_2{-}O{-}CH_3$

  B $H_2C{=}CH{-}C\begin{smallmatrix}{\diagdown} H \\ {\diagup}{\parallel} O\end{smallmatrix}$

  C $HC{\equiv}C{-}\underset{\underset{O}{\parallel}}{C}{-}CH_3$

  D $H_2C{=}CH{-}\underset{\underset{O}{\parallel}}{C}{-}CH_2{-}CH_3$

  E $CH_3{-}CH{=}CH{-}CH_2{-}O{-}CH_3$

4   Which **one** of the following groups of compounds contains three members of the same homologous series?

  A $CH_3CH_2CH_2OH$, $CH_3CH_2OCH_3$, $CH_3OCH_2CH_3$

  B C_3H_8, C_4H_{11}, C_5H_{14}

  C C_2H_2, C_4H_4, C_6H_6

  D CH_3OH, CH_3OCH_3, $CH_3CH_2OCH_3$

  E $HCHO$, CH_3CHO, CH_3CH_2CHO

5   Organic compounds are sometimes identified by first preparing their derivatives. If this technique is used it is important that the derivative should

  A be a solid at room temperature.

  B have a melting point below 100°C.

  C have a low boiling point.

  D be prepared and purified easily.

  E reform the original compound easily.

6   Which **one** of the following methods would be most appropriate in identifying the position of hydrogen atoms in an organic compound?

  A mass spectrometry

  B nuclear magnetic resonance

  C combustion analysis

  D X-ray diffraction

  E polarimetry

7   The hydrocarbon, Q, was subjected to combustion analysis. On complete combustion, 0·2 g of Q gave 0·66 g of carbon dioxide and 0·18 g of water. The empirical formula of Q is

  A CH_3

  B C_2H_3

  C C_3H_2

  D C_3H_4

  E C_3H_8

8 The structural formula of 3,4,4,5-tetramethylheptane can be written as

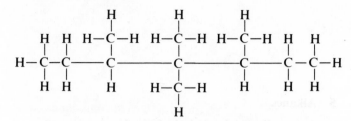

Which **one** of the following represents a correct abbreviation for this structural formula?

A $CH_3.CH(CH_3).C(CH_3)_2.CH(CH_3).CH_2CH_3$

B $CH_3.CH_2.CH(CH_3).CH(CH_3).CH(CH_3).CH_2CH_3$

C $CH_3.CH_2.CH(CH_3).C(CH_3)_2.CH.CH_3.CH_2.CH_3$

D $CH_3.CH_2.CH(CH_3).C(CH_3)_2.CH(CH_3).CH_3$

E $CH_3.CH_2.CH(CH_3).C(CH_3)_2.CH(CH_3).CH_2.CH_3$

9 Isomers do **not** necessarily have the same

A empirical formula

B relative molecular mass

C number and type of bonds

D number of atoms per molecule

E kind of atoms per molecule

10 Isomers have the same

A structural formula

B functional groups

C crystal structure

D optical activity

E percentage composition

11 The number of isomers of formula $C_2H_3Cl_3$ is

A 1

B 2

C 3

D 4

E 6

12 How many isomers, including enantiomers (optical isomers), are there of formula $C_3H_6Cl_2$?

A 3

B 4

C 5

D 6

E 7

13 The number of asymmetric carbon atoms in one molecule of the compound $CHO.CH(OH).CH(OH).CH(OH).CH_2OH$ is

A 1

B 2

C 3

D 4

E 5

14 An organic compound contains 45·86% carbon, 8·92% hydrogen and 45·22% chlorine. (H = 1, C = 12, Cl = 35·5.) Its empirical formula is

A CH_2Cl

B CH_6Cl

C C_3H_7Cl

D $C_6H_7Cl_2$

E $C_6H_{14}Cl$

15 A compound contains 24% carbon and 76% fluorine by mass (C = 12, F = 19). Its molecular formula could be

A CF

B CF_2

C C_2F_2

D C_3F_6

E $2CF_2$

16 Which **one** of the following compounds exists in optically active forms?

A $CH_3.CH_2.O.CH_3$

B $CH_3.CH=CH_2$

C $CH_3.CH(OH).CH_2.CH_3$

D $CH_3.CH_2.CHO$

E $CH_3.CH(OH).CH_3$

TEST 25
Petroleum and alkanes

This test is composed of twenty questions. For each question, five possible answers are suggested. These answers are labelled **A, B, C, D** and **E**. Select the most appropriate **one** of the answers and write its corresponding letter on a separate answer sheet.

1 Which **one** of the following fractions from crude oil contains the highest percentage of carbon?

A diesel oil

B gasoline

C kerosine

D naphtha

E paraffin

2 If the cylinder of a motor-car engine was filled with nothing but petrol vapour and then sparked in the usual way, the petrol would

A explode violently.

B burn quietly.

C remain unreacted.

D decompose to soot.

E ignite prematurely.

3 The simplest alkane to possess at least one structural isomer is

A ethane

B propane

C butane

D pentane

E hexane

4 How many isomers are there of formula C_5H_{12}?

A 2

B 3

C 4

D 5

E 6

5 Alkanes

A are all straight-chain hydrocarbons.

B have the general formula C_nH_{n+2}.

C are unsaturated hydrocarbons.

D include aliphatic and aromatic compounds.

E undergo substitution reactions.

6 Alkanes will react with

A molten Na.

B concentrated H_2SO_4.

C concentrated KOH.

D gaseous Cl_2.

E concentrated $KMnO_4$.

7 Three hydrocarbons were completely burnt in oxygen. In each case, the volume of carbon dioxide produced was found to be equal to the volume of water vapour produced, all measurements being made at the same temperature and pressure. What is the general formula of the hydrocarbons?

A C_nH_n

B C_nH_{n+2}

C C_nH_{2n+1}

D C_nH_{2n}

E C_nH_{2n+2}

8 Which **one** of the following molecules is most likely to undergo homolytic fission?

A I_2

B HCl

C BrF

D CH_4

E CH_3Cl

9 Cracking

A is the opposite of reforming.

B cannot occur without a catalyst.

C must involve decomposition.

D does not occur with branched-alkanes.

E never produces branched-alkanes.

Questions **10** to **13**

The chlorination of an alkane involves several stages, some of which are shown below.

I $Cl_2 \longrightarrow Cl\cdot + Cl\cdot$

II $RCH_3 + Cl\cdot \longrightarrow RCH_2\cdot + HCl$

III $RCH_2\cdot + Cl_2 \longrightarrow RCH_2Cl + Cl\cdot$

IV $RCH_2\cdot + Cl\cdot \longrightarrow RCH_2Cl$

V $RCH_2\cdot + RCH_2\cdot \longrightarrow RCH_2CH_2R$

10 The stages which involve propagation are

   A I and II

   B II and III

   C III and IV

   D I, II and III

   E II, III and IV

11 Stage V of the process is an example of

   A an initiation reaction.

   B homolytic fission.

   C heterolytic fission.

   D catalytic reforming.

   E free radical addition.

12 Termination of the reaction takes place when

   A two free radicals collide and combine.

   B all the chlorine radicals have reacted.

   C all the chlorine is converted to $Cl\cdot$.

   D the initiation process has finished.

   E the reaction has just finished.

13 Given the following bond energies (bond energy terms) in kJ mole^{-1}, $E(Cl-Cl) = 242$; $E(C-H) = 413$; $E(C-Cl) = 339$; $E(H-Cl) = 431$; what is the enthalpy change for the reaction below?

      $RCH_3 + Cl_2 \longrightarrow RCH_2\cdot + HCl + Cl\cdot$

   A $+1086$ kJ mole^{-1}

   B $+655$ kJ mole^{-1}

   C $+563$ kJ mole^{-1}

   D $+339$ kJ mole^{-1}

   E $+224$ kJ mole^{-1}

Questions **14–20** refer to the five structures below.

A

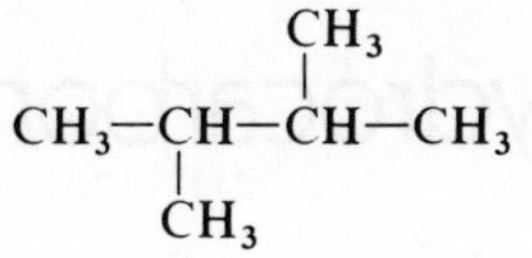

               CH_3
               |
$CH_3-CH-CH-CH_3$
           |
          CH_3

D

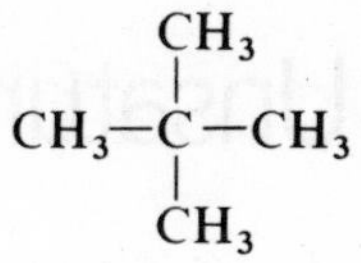

             CH_3
             |
CH_3-C-CH_3
             |
            CH_3

B

            CH_3 CH_3
            | |
$CH_3-CH_2-CH-CH-CH_3$

E

          CH_3
          |
         CH
       /   \\
   H_2C CH_2
    | |
   H_2C-CH_2

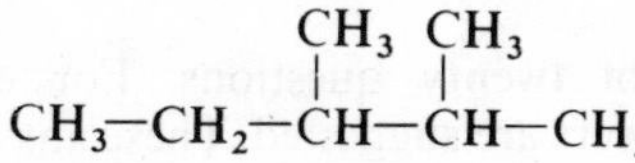

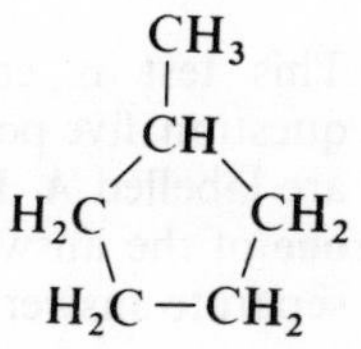

C

              H
              |
   H_2C-C
     | \\CH_3
   H_2C-CH_2

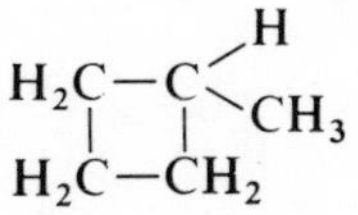

14 Which structure is under the greatest strain?

15 Which structure will exist as optical isomers?

16 Which is isomeric with pentane?

17 Which has a systematic name ending in -propane?

18 Which structure could form **two** and only two different monochloro-compounds when hydrogen atoms are replaced by chlorine?

19 Which structure has the most carbon atoms in **one** plane?

20 Which structure requires the least amount of oxygen per mole for complete combustion?

TEST 26
Unsaturated hydrocarbons

This test is composed of twenty questions. For each question, five possible answers are suggested. These answers are labelled **A**, **B**, **C**, **D** and **E**. Select the most appropriate **one** of the answers and write its corresponding letter on a separate answer sheet.

1 What is the systematic name for $CH_3CH=CHCH=CH_2$?

A penta-1, 2-diene

B penta-1, 3-diene

C penta-1, 4-diene

D penta-2, 3-diene

E penta-2, 4-diene

2 Which **one** of the following compounds is formed as the major product when but-1-ene reacts with iodine monochloride, ICl?

A $CH_3CH_2CHClCH_2I$

B $CH_3CH_2CHICH_2Cl$

C $CH_3CHICHClCH_3$

D $CH_3CHICH_2CH_2Cl$

E $CH_2ClCHICH_2CH_3$

3 Using the bond energies given below, what is the energy change in kJ for the following reaction?

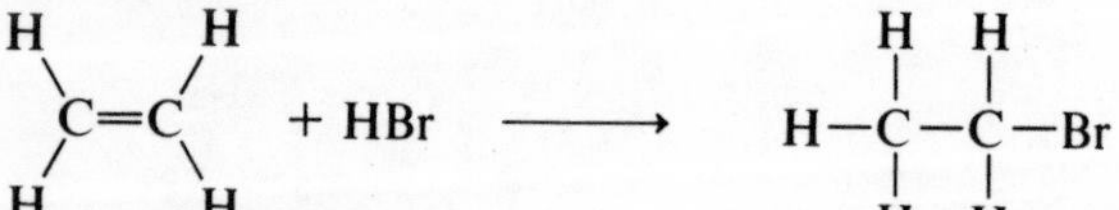

$$\underset{H}{\overset{H}{>}}C=C\underset{H}{\overset{H}{<}} + HBr \longrightarrow H-\underset{\underset{H}{|}}{\overset{\overset{H}{|}}{C}}-\underset{\underset{H}{|}}{\overset{\overset{H}{|}}{C}}-Br$$

$E(C-C) = 346 \text{ kJ mole}^{-1}$,
$E(C=C) = 610 \text{ kJ mole}^{-1}$,
$E(C-H) = 413 \text{ kJ mole}^{-1}$,
$E(C-Br) = 280 \text{ kJ mole}^{-1}$,
$E(H-Br) = 365 \text{ kJ mole}^{-1}$

A -2014

B -282

C -64

D $+64$

E $+282$

4 Which **one** of the following compounds is formed as the major organic product when propan-1-ol is heated with excess concentrated H_2SO_4 at 150°C?

A $CH_3CH=CH_2$

B $\underset{CH_3}{\overset{CH_3}{>}}C=CH_2$

C $CH_2=C=CH_2$

D $CH_3CH_2CH_2OCH_2CH_2CH_3$

E $CH_3CH_2CH=CHCH_2CH_3$

5 A simple reaction scheme is shown below.

$$CH_3CH=CH_2 \xrightarrow{\text{HBr(g)}} X \xrightarrow[\text{NaOH(aq)}]{\text{Reflux with}} Y$$

What is the formula of Y?

A $CH_3CH_2CH_2OH$

B $CH_3CHOHCH_3$

C $CH_3CHBrCH_2OH$

D $CH_3CHOHCH_2Br$

E $CH_3CHBrCH_3$

6 Which **one** of the following reagents can be used in a simple chemical test to distinguish between propene and propyne?

A HBr(g)

B $Br_2(aq)$

C $[Ag(NH_3)_2]^+(aq)$

D $AgNO_3(aq)$

E $KMnO_4(aq)$

7 A plastic with highly-branched polymer chains is likely to have a

A low tensile strength.

B high melting point.

C rigid structure.

D high density.

E high crystallinity.

Questions **8** to **14** refer to the five compounds labelled **A, B, C, D** and **E** below.

A   cyclohexane
B   cyclohexene
C   hexane
D   hex-1-yne
E   hex-2-yne

Which **one** of the above compounds

8   decolorizes bromine water most rapidly?

9   gives a pale yellow precipitate with ammoniacal silver nitrate solution?

10   has the empirical formula CH_2?

11   has only **one** monochloro-derivative?

12   has four carbon atoms arranged linearly?

13   undergoes the largest volume change when the vapour burns in excess oxygen to form gaseous products?

14   has one σ-bond and one π-bond between adjacent carbon atoms?

For each of questions **15** to **20**, one or more of the numbered alternatives (**1, 2** and **3**) listed below may be correct. Decide whether each of the alternatives is or is not correct and then choose

A   if **1, 2** and **3** are all correct.
B   if **1** and **2** only are correct.
C   if **2** and **3** only are correct.
D   if **1** only is correct.
E   if **3** only is correct.

(No other combination is used as a correct answer.)

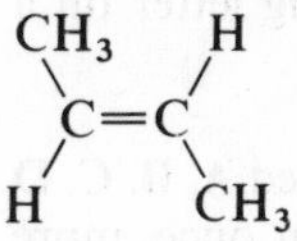

15   Which of these compounds will have a geometrical isomer?

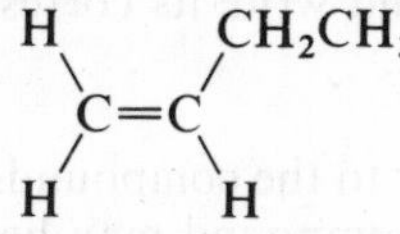

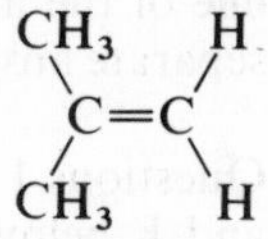

16   Which of these compounds will not produce butane on catalytic hydrogenation?

17   Which of these compounds could produce optical isomers on treatment with hydrogen bromide?

18   Which of these compounds will have zero dipole moment?

19   Which of these compounds will decolorize a very dilute acidified solution of manganate(VII) ions?

20   Which of these compounds could be obtained by partial hydrogenation of an alkyne?

TEST 27

Aromatic hydrocarbons

This test is composed of nineteen questions. For each question five possible answers are suggested. These answers are labelled **A**, **B**, **C**, **D** and **E**. Select the most appropriate **one** of the answers and write its corresponding letter on a separate answer sheet.

Questions **1** to **6** refer to the compounds labelled **A, B, C, D** and **E** below. Each compound may be chosen once, more than once or not at all.

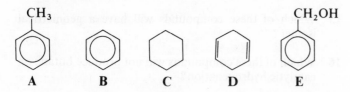

A   **B**   **C**   **D**   **E**

1   Which compound has no reaction with bromine plus iron filings in the dark?

2   Which compound absorbs two moles of hydrogen per mole of the compound at 150°C in the presence of a powdered nickel catalyst?

3   Which compound reacts with sodium at room temperature?

4   Which compound will decolorize dilute acidified potassium manganate(VII) at room temperature?

5   Which compound has the highest degree of unsaturation?

6   Which compound has a puckered ring of carbon atoms?

Questions **7** to **11**

Choose from the list **A–E** the reaction conditions most suitable for carrying out the conversions in questions **7** to **11**. Each letter may be chosen once, more than once, or not at all.

A   Boil the reactant with dilute hydrochloric acid.
B   Warm the reactant with chloromethane in the presence of aluminium chloride.
C   Treat the reactant with phosphorus(III) chloride at room temperature.
D   Pass chlorine into the boiling reactant in sunlight.
E   Pass chlorine through the reactant in the presence of aluminium chloride in the dark.

7

8

9

10

11

12   What is the total number of trichlorobenzenes of formula $C_6H_3Cl_3$?

   A   2

   B   3

   C   4

   D   5

   E   6

13   X-ray diffraction studies show that the molecule of benzene has

   A   a distorted hexagonal shape.

   B   alternate double and single bonds.

   C   twelve atoms in the same plane.

   D   longer C—C bonds than in ethane.

   E   shorter C—C bonds than in ethene.

14 Which **one** of the following conditions and reagents is most suitable for preparing nitrobenzene from benzene?

   A Mix benzene with conc. HNO_3 at room temperature.

   B Mix benzene with conc. HNO_3 plus dilute H_2SO_4 at 15°C.

   C Warm benzene with conc. HNO_3 at 50°C.

   D Warm benzene with conc. HNO_3 plus conc. H_2SO_4 at 50°C.

   E Reflux benzene with conc. HNO_3.

15 Which **one** of the following could **not** be obtained as a main product when methylbenzene is heated with a mixture of conc. HNO_3 and conc. H_2SO_4?

   A 2-nitromethylbenzene

   B 3-nitromethylbenzene

   C 4-nitromethylbenzene

   D 2,4-dinitromethylbenzene

   E 2,4,6-trinitromethylbenzene

16 When benzene, methylbenzene and nitrobenzene are nitrated, the order of **increasing** rate of nitration is

   A benzene, methylbenzene, nitrobenzene.

   B benzene, nitrobenzene, methylbenzene.

   C nitrobenzene, benzene, methylbenzene.

   D nitrobenzene, methylbenzene, benzene.

   E methylbenzene, benzene, nitrobenzene.

17 Iodobenzene can be obtained by reacting benzene with iodine(I) chloride. What is the electrophile attacking the benzene in this reaction?

   A Cl^+

   B I

   C I_2

   D I^-

   E I^+

Questions 18 and 19

Study the experimentally determined enthalpy changes in the figure below.

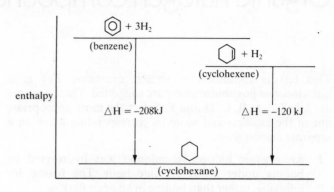

18 If benzene had the hypothetical cyclohexa-1,3,5-triene structure,   , as suggested by Kekulé, what would be the likely value of its heat of hydrogenation per mole?

   A -120 kJ

   B -208 kJ

   C -328 kJ

   D -360 kJ

   E -448 kJ

19 By how much is benzene more stable than the hypothetical cyclohexa-1,3,5-triene structure suggested by Kekulé, judging from the information in the figure above?

   A 0 kJ per mole

   B 88 kJ per mole

   C 120 kJ per mole

   D 152 kJ per mole

   E 240 kJ per mole

TEST 28
Organic halogen compounds

This test is composed of sixteen questions. For each question, five possible answers are suggested. These answers are labelled **A, B, C, D** and **E**. Select the most appropriate **one** of the answers and write its corresponding letter on a separate answer sheet.

1 An organic halogen compound was hydrolysed by boiling under reflux for one hour. The reason for refluxing, rather than boiling in an open flask, is

 A to boil the mixture at a higher temperature.

 B to prevent escape of volatile reagents.

 C to increase the area of contact between reagents.

 D to avoid drops of liquid spitting out of the flask.

 E to alternately heat and cool the reacting substances.

2 Which **one** of the following reagents is most suitable for converting

$$\bigcirc\!\!-CH_2OH \quad into \quad \bigcirc\!\!-CH_2Br \;\; ?$$

 A $Br_2(l)$

 B $NaBr(aq)$

 C $NaBrO_3(aq)$

 D $CH_3Br(l)$

 E $PBr_3(l)$

3 For a given alkyl group, R, the iodo-compound reacts most readily, the bromo-compound less so and the chloro-compound reacts least readily. This is because

 A chlorine is more reactive than iodine.

 B chloro-compounds are more polar than iodo-compounds.

 C the C—Cl bond is shorter than the C—I bond.

 D the C—Cl bond is stronger than the C—I bond.

 E the chlorine atom is smaller than the iodine atom.

4 Which **one** of the following reagents is most suitable for converting $CH_3CH_2CH_2Br$ into $CH_3CH_2CH_2OCH_2CH_3$?

 A NaOH(s) in ethanol

 B CH_3CH_2COCl

 C dry $CH_3CH_2COO^-Ag^+$

 D $CH_3CH_2O^-Na^+$ in ethanol

 E CH_3CH_2OH

5 Which **one** of the following reagents is most suitable for converting.

$$H-\overset{\overset{\displaystyle H}{|}}{C}-\overset{\displaystyle O}{\overset{\|}{C}}\!\!\underset{OH}{} \quad into \quad Cl-\overset{\overset{\displaystyle H}{|}}{\underset{H}{C}}-\overset{\displaystyle O}{\overset{\|}{C}}\!\!\underset{OH}{} \;\; ?$$

 A PCl_5 at 50°C

 B conc. HCl at 80°C

 C $SOCl_2$ at room temperature

 D NaCl in hot conc. sulphuric acid

 E Cl_2 in bright sunlight at 100°C

6 Which **one** of the following is the main organic product when $BrCH_2COBr$ is treated with cold water?

 A $HOCH_2COBr$

 B $BrCH_2COOH$

 C $HOCH_2COOH$

 D CH_3COOH

 E $HOCH_2CHO$

7 Which **one** of the following is the major organic product when $ClCH_2CH_2COCl$ is refluxed with excess aqueous sodium hydroxide?

 A $HOCH_2CH_2COCl$

 B $ClCH_2CH_2COOH$

 C $HOCH_2CH_2COOH$

 D $HOCH_2CH_2COO^-Na^+$

 E $Na^+{}^-OCH_2CH_2COO^-Na^+$

8 Which **one** of the following sets of bromo-compounds is arranged in order of increasing reactivity of the bromine atom?

A CH_3CH_2Br, C_6H_5Br, CH_3COBr

B CH_3CH_2Br, CH_3COBr, C_6H_5Br

C C_6H_5Br, CH_3CH_2Br, CH_3COBr

D C_6H_5Br, CH_3COBr, CH_3CH_2Br

E CH_3COBr, C_6H_5Br, CH_3CH_2Br

Questions **9–16** refer to the five organic halogen compounds labelled **A** to **E** below.

A CH_3Cl

B

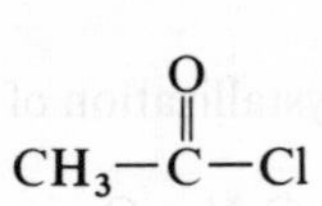

$$CH_3-\overset{\overset{\displaystyle O}{\|}}{C}-Cl$$

C

Br
(cyclohexane ring with Br)

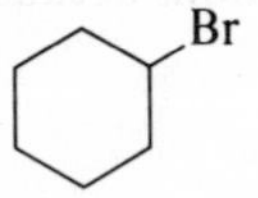

D CCl_3F

E

Cl
(benzene ring with Cl)

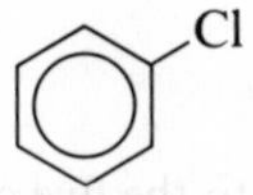

Which **one** of the above compounds

9 is the most volatile?

10 is the most suitable for use as an aerosol propellant?

11 is an acyl halide?

12 reacts least readily with aqueous sodium hydroxide?

13 reacts most readily with cold water?

14 reacts with alcoholic sodium hydroxide to form an alkene?

15 has a planar molecule?

16 reacts with alcohols to form esters?

TEST 29

Alcohols, phenols and ethers

This test is composed of twenty questions. For each question, five possible answers are suggested. These answers are labelled **A, B, C, D** and **E**. Select the most appropriate **one** of the answers and write its corresponding letter on a separate answer sheet.

1 Equal volumes of ethanol and water are completely miscible whereas equal volumes of ethoxyethane (ether) and water form separate layers on mixing. The main reason for this is that

   A molecules of ether are larger than those of ethanol.

   B the densities of water and ethanol are closer than those of water and ethoxyethane.

   C molecules of ethanol are polar, but those of ethoxyethane are non-polar.

   D the boiling points of water and ethanol are closer than those of water and ethoxyethane.

   E ethanol can form hydrogen bonds with water, but ethoxyethane cannot do so.

2 With which **one** of the following reagents do ethanol and phenol react in a similar fashion?

   A $FeCl_3(aq)$

   B NaOH(aq)

   C $HNO_3(aq)$

   D Na(s)

   E $CH_3COOH(l)$

3 The best conditions for preparing ethoxyethane (ether) by the dehydration of ethanol using concentrated sulphuric acid are

   A excess conc. H_2SO_4 at 140°C.

   B excess ethanol at 140°C.

   C excess conc. H_2SO_4 at 170°C.

   D excess ethanol at 170°C.

   E excess conc. H_2SO_4 at 250°C.

4 Phenol could be separated from a mixture of phenol and benzene by

   A extracting the phenol with trichloromethane.

   B dissolving the benzene in conc. sulphuric acid.

   C dissolving the phenol in sodium hydroxide solution.

   D extracting the benzene with methylbenzene (toluene).

   E fractional crystallization of the phenol.

5 Ethoxyethane, $(C_2H_5)_2O$, is useful as a solvent for many organic reactions because it is

   A unreactive.

   B volatile.

   C immiscible with water.

   D less dense than water.

   E non-polar.

Questions **6** to **13** refer to the five compounds labelled **A, B, C, D** and **E** below.

A methanol
B ethanol
C propan-2-ol
D phenylmethanol (benzyl alcohol)
E phenol

6 Which is most frequently used as a solvent in the laboratory and in industry?

7 Which has the highest vapour pressure at 20°C?

8 Which, in aqueous solution, reacts most vigorously with sodium?

9 Which is oxidized to a ketone?

10 Which, when oxidized strongly, forms a solid mono-carboxylic acid?

11 Which reacts with ethanoyl chloride to form an ester, but does not react with ethanoic acid?

12 Which reacts with methanoic acid to form an ester which is isomeric with propanoic acid?

13 Which reacts readily with bromine water to form a brominated derivative?

For each of questions **14–20** one or more of the numbered alternatives (**1, 2** and **3**) listed below may be correct. Decide whether each of the alternatives is or is not correct and then choose

A if **1, 2** and **3** are all correct.
B if **1** and **2** only are correct.
C if **2** and **3** only are correct.
D if **1** only is correct.
E if **3** only is correct.

(No other combination is used as a correct answer.)

$CH_3.CH_2.CH(OH).CH_3$ $\qquad$ $CH_3.CH_2.CH_2.CH_2OH$
1 $\qquad\qquad\qquad\qquad$ **2**

$(CH_3)_3COH$
3

14 Which will decolorize dilute acidified potassium manganate(VII) (potassium permanganate) on warming?

15 Which will give gaseous hydrogen chloride on treatment with phosphorus pentachloride?

16 Which will give yellow crystals of CHI_3 on warming with iodine in alkaline solution?

17 Which could **not** be oxidized to a ketone?

18 Which could form an ester with ethanoyl chloride (acetyl chloride)?

19 Which could form methylpropene on dehydration?

20 Which will have optically active forms?

TEST 30

Carbonyl compounds

This test is composed of twenty questions. For each question, five possible answers are suggested. These answers are labelled **A, B, C, D** and **E**. Select the most appropriate **one** of the answers and write its corresponding letter on a separate answer sheet.

1   The order of **increasing** rate of reaction of ethanal, methanal and propanone with cyanide ion, CN^-, is

   A ethanal, methanal, propanone.

   B ethanal, propanone, methanal.

   C methanal, ethanal, propanone.

   D methanal, propanone, ethanal.

   E propanone, ethanal, methanal.

2   The reaction of compounds of the type HX with carbonyl compounds is best described as

   A addition-elimination.

   B electrophilic addition.

   C electrophilic substitution.

   D nucleophilic addition.

   E nucleophilic substitution.

3   The reaction of ethanal with HCN(aq) is slow, but a rapid reaction results if ethanal is treated first with NaCN(aq) and then with dilute HCl(aq). The most likely explanation for this is that the first step of the reaction involves attack of the

   A carbonyl oxygen atom by H^+.

   B carbonyl oxygen atom by Na^+.

   C carbonyl carbon atom by HCN.

   D carbonyl carbon atom by Cl^-.

   E carbonyl carbon atom by CN^-.

4   Which **one** of the following compounds would undergo the iodoform reaction?

   A CH_3CH_2CHO

   B HCHO

   C ⬡—CHO

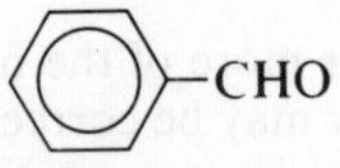

   D ⬡—$COCH_3$

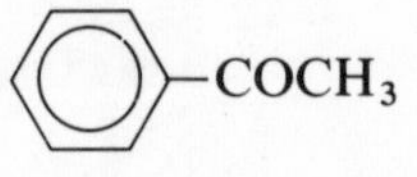

   E $CH_3CH_2COCH_2CH_3$

5   Which **one** of the following sets contains three ions, all of which react with ethanal?

   A CN^-, Cl^-, H^+

   B CN^-, H^+, HSO_3^-

   C CN^-, HSO_3^-, HSO_4^-

   D Cl^-, H^+, HSO_3^-

   E H^+, HSO_3^-, HSO_4^-

6   Butanone could be distinguished from butanal by observing their separate reactions with

   A hydroxylamine

   B iron(III) chloride

   C sodium hydrogensulphate(IV)

   D iodine in alkaline solution

   E hydrogen in the presence of nickel

Questions 7 to 16

For each of questions **7–16** one or more of the numbered alternatives (**1, 2** and **3**) listed below may be correct. Decide whether each of the alternatives is or is not correct and then choose

A if **1, 2** and **3** are all correct.
B if **1** and **2** only are correct.
C if **2** and **3** only are correct.
D if **1** only is correct.
E if **3** only is correct.

(No other combination is used as a correct answer.)

H.CHO	CH₃.CHO	CH₃.CO.CH₃
1	**2**	**3**

7 Which could be prepared by oxidation of a secondary alcohol?

8 Which will polymerize on evaporation of an aqueous solution?

9 Which will reduce diamminosilver(I) ions, $[Ag(NH_3)_2]^+$, to silver?

10 Which will undergo a condensation reaction with 2,4-dinitrophenylhydrazine?

11 Which will be oxidized to carbon dioxide and water on warming with dilute acidified $KMnO_4$ solution?

12 Which will convert Fehling's solution to either a red precipitate of copper(I) oxide or a deposit of copper?

13 Which will give a yellow precipitate on warming with a mixture of iodine and alkali?

14 Which will be gaseous at $0°C$?

15 Which will undergo an addition reaction on treatment with HCN?

16 Which will be reduced to a primary alcohol with hydrogen in the presence of a platinum catalyst?

Questions **17–20** refer to the structures of five compounds **A, B, C, D** and **E** shown below. (The open-chain forms of compounds **A, B, D** and **E** shown here normally exist in equilibrium with the ring-form.)

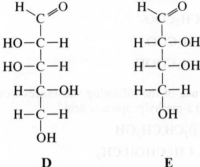

A   **B**   **C**

D   **E**

17 Which compound is a ketose?

18 Which compound is **not** a carbohydrate?

19 Which structure has four asymmetric carbon atoms?

20 Which compound is a component of RNA and vitamin B-12?

TEST 31
Carboxylic acids

This test is composed of eighteen questions. For each question, five possible answers are suggested. These answers are labelled **A, B, C, D** and **E**. Select the most appropriate **one** of the answers and write its corresponding letter on a separate answer sheet.

1   Which **one** of the following substances is the main organic product when $ClCH_2COOH$ is refluxed with aqueous $NaOH$?

   A $ClCH_2COO^-$

   B $HOCH_2COOH$

   C $HOCH_2COO^-$

   D $^-OCH_2COO^-$

   E $^-OCH_2COOH$

2   Which **one** of the following substances would be oxidized to 2-methylpropanoic acid?

   A $(CH_3)_2CH.CH_2OH$

   B $CH_3.CH_2.CHOH.CH_3$

   C $(CH_3)_3COH$

   D $(CH_3)_2CH.CH_2.CH_2.CH_2OH$

   E $CH_3.CH_2.CH_2.CH_2OH$

3   Which **one** of the following is the best description for the reaction between C_3H_7COOH and C_3H_7OH?

   A addition

   B condensation

   C dehydration

   D elimination

   E substitution

4   Which **one** of the following carboxylic acids is the strongest?

   A $CHF_2.CH_2.COOH$

   B $CH_2F.CH_2.COOH$

   C $CH_3.CHF.COOH$

   D $CH_3.CHCl.COOH$

   E $CH_2F.CHF.COOH$

5   When phenylmethanoate is hydrolysed by excess $NaOH(aq)$, the products of the reaction are

   A phenol and methanoate ions

   B benzoate ions and methanol

   C phenate ions and methanol

   D phenoxide ions and methanoate ions

   E phenol and methanoic acid

6   To which **one** of the following classes of organic compounds does

$$CH_3-O-C{\overset{H}{\underset{O}{\diagdown}}}\ \text{belong?}$$

   A acid anhydrides

   B aldehydes

   C carboxylic acids

   D esters

   E ketones

Questions **7** to **13** refer to five carboxylic acids labelled **A, B, C, D** and **E** below.

A   methanoic (formic) acid
B   ethanoic (acetic) acid
C   propanoic (propionic) acid
D   ethanedioic (oxalic) acid
E   benzoic acid

Which **one** of these acids

7   has the lowest boiling point?

8   is the least soluble in cold water?

9   is the strongest electrolyte?

10   is the strongest reducing agent?

11   forms two different sodium salts?

12   decolorizes dilute acidified $KMnO_4(aq)$ only on warming?

13   readily dehydrates forming CO and CO_2?

234

For each of questions **14** to **18** one or more of the numbered alternatives (**1, 2** and **3**) listed below may be correct. Decide whether each of the alternatives is or is not correct and then choose

A if **1, 2** and **3** are all correct.
B if **1** and **2** only are correct.
C if **2** and **3** only are correct.
D if **1** only is correct.
E if **3** only is correct.

(No other combination is used as a correct answer.)

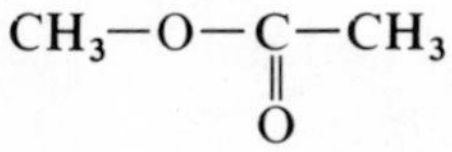

$$CH_3-O-\underset{\underset{O}{\|}}{C}-CH_3$$

1

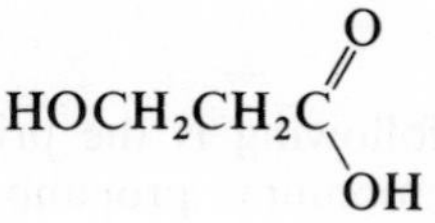

$$HOCH_2CH_2-\underset{\underset{OH}{}}{\overset{\overset{O}{\|}}{C}}$$

2

$$CH_3CH_2-\underset{\underset{O}{\|}}{C}-O-\underset{\underset{O}{\|}}{C}-CH_3$$

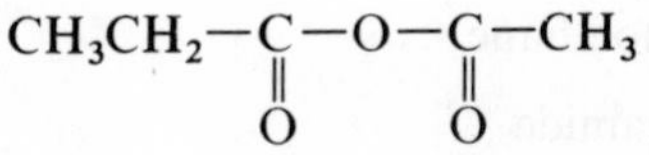

3

14 Which will form an acidic solution in water?

15 Which will form an immiscible layer with dilute NaOH?

16 Which can be obtained by reacting an acid with an alcohol?

17 Which will react with water forming two different organic products?

18 Which will react with ethanol forming an ester?

Organic nitrogen compounds

This test is composed of eighteen questions. For each question, five possible answers are suggested. These answers are labelled **A, B, C, D** and **E**. Select the most appropriate **one** of the answers and write its corresponding letter on a separate answer sheet.

1 Which **one** of the following reagents reacts in a similar fashion with both phenylamine ($C_6H_5NH_2$) and ethylamine ($CH_3CH_2NH_2$)?

   A $Br_2(aq)$

   B CH_3COCl

   C conc. H_2SO_4

   D cold $HNO_2(aq)$

   E conc. HNO_3

2 Phenylamine(aniline) can be prepared by reducing nitrobenzene with tin and concentrated hydrochloric acid followed by addition of alkali and finally steam distillation. The alkali is added to

   A prevent oxidation of phenylamine.

   B react with excess tin.

   C liberate free phenylamine from solution.

   D dissolve excess nitrobenzene.

   E dissolve the phenylamine.

3 Which **one** of the following statements best explains why trimethylamine $[(CH_3)_3N]$ has a lower boiling point than dimethylamine $[(CH_3)_2NH]$?

   A $(CH_3)_3N$ has a larger relative molecular mass.

   B $(CH_3)_3N$ molecules are symmetrical.

   C $(CH_3)_3N$ cannot hydrogen-bond with itself.

   D $(CH_3)_3N$ molecules are non-polar.

   E $(CH_3)_3N$ molecules have a larger volume.

4 The order of **increasing** strength as bases for dimethylamine, methylamine and phenylamine is

   A dimethylamine, methylamine, phenylamine.

   B methylamine, dimethylamine, phenylamine.

   C dimethylamine, phenylamine, methylamine.

   D phenylamine, dimethylamine, methylamine.

   E phenylamine, methylamine, dimethylamine.

5 When CH_3CN is refluxed with aqueous KOH the principal organic product is

   A CH_3OH

   B CH_3COO^-

   C CH_3CONH_2

   D CH_3CH_2OH

   E $CH_3CH_2NH_2$

6 Which **one** of the following is the principal organic product when ammonium propanoate is heated with P_4O_{10}?

   A ethanenitrile

   B isocyanoethane

   C propanamide

   D propanenitrile

   E propanoic acid

7 An organic nitrogen compound, X, gives ammonia on warming with dilute aqueous sodium hydroxide. X could be

   A ethanamide

   B ethylamine

   C aminoethanoic acid

   D phenylamine

   E dimethylamine

8 Which **one** of the following substances shows amphoteric character most prominently?

   A $CH_3.COONH_4$

   B H_2NCH_2COOH

   C $C_6H_5NH_2$

   D CH_3CONH_2

   E CH_3CN

9 Which **one** of the following terms best describes nylon?

   A regenerated natural fibre

   B synthetic protein

   C condensation polymer

   D semi-synthetic polypeptide

   E natural polyamide

10 Amino acids can be identified by paper chromatography and the measurement of R_f values. Which **one** of the following factors would have the greatest influence on R_f values?

A the kind of paper used for the chromatogram

B the temperature in the chromatography tank

C the solvent employed for chromatography

D the size of the original spot on the chromatogram

E the time taken for chromatography

Questions **11** to **18** relate to the five organic nitrogen compounds labelled **A, B, C, D** and **E** below.

A $CH_3CH_2NH_2$
B $C_6H_5NH_2$
C CH_3CONH_2
D $C_6H_5NO_2$
E $C_6H_5N_2Cl$

Which **one** of these compounds

11 is a strong electrolyte?

12 is the most volatile?

13 is the strongest base?

14 is insoluble in water, acid and alkali?

15 dissolves in dilute HCl, but not in water?

16 is explosive when pure?

17 gives nitrogen with HNO_2 at 5°C?

18 gives ammonia on warming with NaOH(aq)?

Appendix: Atomic numbers and relative atomic masses of the elements

Element	Symbol	Atomic number	Relative atomic mass
aluminium	Al	13	27
antimony	Sb	51	122
argon	Ar	18	40
arsenic	As	33	75
barium	Ba	56	137
beryllium	Be	4	9
bismuth	Bi	83	209
boron	B	5	11
bromine	Br	35	80
cadmium	Cd	48	112
caesium	Cs	55	133
calcium	Ca	20	40
carbon	C	6	12
cerium	Ce	58	140.1
chlorine	Cl	17	35.5
chromium	Cr	24	52
cobalt	Co	27	59
copper	Cu	29	63.5
dysprosium	Dy	66	162.5
erbium	Er	68	167.3
europium	Eu	63	152.0
fluorine	F	9	19
gadolinium	Gd	64	157.3
gallium	Ga	31	70
germanium	Ge	32	73
gold	Au	79	197
hafnium	Hf	72	178
hahnium	Ha	105	—
helium	He	2	4
holmium	Ho	67	164.9
hydrogen	H	1	1
indium	In	49	115
iodine	I	53	127
iridium	Ir	77	192
iron	Fe	26	56
krypton	Kr	36	84
kurchatovium	Ku	104	—
lanthanum	La	57	139
lead	Pb	82	207
lithium	Li	3	7
lutetium	Lu	71	175.0
magnesium	Mg	12	24
manganese	Mn	25	55
mercury	Hg	80	201
molybdenum	Mo	42	96
neodymium	Nd	60	144.2
neon	Ne	10	20
nickel	Ni	28	59
niobium	Nb	41	93
nitrogen	N	7	14
osmium	Os	76	190
oxygen	O	8	16
palladium	Pd	46	106
phosphorus	P	15	31
platinum	Pt	78	195
potassium	K	19	39
praseodymium	Pr	59	140.9
rhenium	Re	75	186
rhodium	Rh	45	103
rubidium	Rb	37	85
ruthenium	Ru	44	101
samarium	Sm	62	150.4
scandium	Sc	21	45
selenium	Se	34	79
silicon	Si	14	28
silver	Ag	47	108
sodium	Na	11	23
strontium	Sr	38	88
sulphur	S	16	32
tantalum	Ta	73	181
tellurium	Te	52	128
terbium	Tb	65	158.9
thallium	Tl	81	204
thorium	Th	90	232.0
thulium	Tm	69	169.9
tin	Sn	50	119
titanium	Ti	22	48
tungsten	W	74	184
uranium	U	92	238
vanadium	V	23	51
xenon	Xe	54	131
ytterbium	Yb	70	173
yttrium	Y	39	89
zinc	Zn	30	65
zirconium	Zr	40	91

Index